MODERN POWER PLANT ENGINEERING

JOEL WEISMAN
ROY ECKART

University of Cincinnati

Prentice-Hall, Inc./Englewood Cliffs, NJ 07632

Library of Congress Cataloging in Publication Data

Weisman, Joel, (date)
   Modern power plant engineering.

Includes bibliographies and index.
1. Electric power-plants.   I. Eckart, Roy
II. Title.
TK1191.W45   1985   621.31′2132   84-22840
ISBN 0-13-597252-3

Cover design: Judy Winthrop

© 1985 by Prentice-Hall, Inc., Englewood Cliffs, New Jersey 07632

All rights reserved. No part of this book may be
reproduced, in any form or by any means,
without permission in writing from the publisher.

Printed in the United States of America

10 9 8 7 6 5 4 3 2 1

ISBN 0-13-597252-3 01

Prentice-Hall International, Inc., *London*
Prentice-Hall of Australia Pty. Limited, *Sydney*
Editora Prentice-Hall do Brasil, Ltda., *Rio de Janeiro*
Prentice-Hall Canada Inc., *Toronto*
Prentice-Hall of India Private Limited, *New Delhi*
Prentice-Hall of Japan, Inc., *Tokyo*
Prentice-Hall of Southeast Asia Pte. Ltd., *Singapore*
Whitehall Books Limited, *Wellington, New Zealand*

Contents

Preface, v

1 Introduction to Power Generation Systems

1.1 History and present status of energy resources, 1 / **1.2** Electric power, 4 / **1.3** Energy sources for modern power plants, 11 / **1.4** Power systems and utility demand patterns, 12 / **1.5** Modern electric generating stations, 19 / **1.6** Economics of electric power production, 24

2 Power Plant Thermodynamics

2.1 The basic laws of thermodynamics, 37 / **2.2** Thermodynamic properties, 41 / **2.3** Thermodynamic power cycles, 45 / **2.4** Steam cycles for modern power plants, 53

3 Fossil Fuels

3.1 Petroleum and shale oil, 63 / **3.2** Properties of petroleum fuels, 69 / **3.3** Natural and petroleum gas, 73 / **3.4** Coal, 76 / **3.5** Coal cleaning and processing, 87 / **3.6** Synthetic fuels: gaseous fuels from coal, 90 / **3.7** Synthetic fuels: coal liquefaction and refining, 96

4 Combustion

4.1 Principles of combustion, 102 / **4.2** Burner design, 108 / **4.3** Mass and energy balances, 111 / **4.4** Chemical equilibrium, 126 / **4.5** Theoretical flame temperatures, 135 / **4.6** Calculation of actual flame temperatures, 141 / **4.7** Flame emissivity, 145 / **4.8** Fluidized-bed combustion systems, 148

5 Fossil-Fueled Steam Power Plants: Primary Systems

5.1 Introduction to power plant systems and components, 156 / **5.2** Fuel handling and preparation for burning, 158 / **5.3** Steam generator configuration, 166 / **5.4** Steam generator analysis, 178 / **5.5** Steam separation and purification, 187 / **5.6** Primary system heat exchangers, 190 / **5.7** Water and steam flow in the primary system: natural-convection systems, 202 / **5.8** Steam generator control, 205

6 Fossil-Fueled Steam Power Plants: Auxiliary Systems

6.1 Air circulating and heating system, 217 / **6.2** Water treatment systems, 229 / **6.3** Cooling towers, 234 / **6.4** Emission control systems, 246 / **6.5** Waste disposal, 255

7 Nuclear Power Stations

7.1 Principles of nuclear fission, 264 / **7.2** Nuclear core analysis, 274 / **7.3** Power reactor systems, 287 / **7.4** Fuel design and analysis, 305 / **7.5** Thermal analysis, 317 / **7.6** The nuclear fuel cycle, 320

8 Production of Mechanical Energy

8.1 The steam cycle, 345 / **8.2** Gas turbine power plants, 352 / **8.3** Combined cycles, 363 / **8.4** Steam and gas turbine configurations, 365 / **8.5** Turbine analysis, 371 / **8.6** Off-design turbine analysis, 384 / **8.7** Turbine control, 388 / **8.8** Hydraulic turbines and pumped storage, 393

9 Production of Electrical Energy

9.1 Principles of alternator design and operation, 410 / **9.2** Generator configurations, 420 / **9.3** Power plant electrical systems, 422 / **9.4** The electric power network, 426 / **9.5** Magnetohydrodynamics, 430 / **9.6** Fuel cells, 439 / **9.7** Thermionic generators, 443

10 Energy Alternatives

10.1 Energy storage, 451 / **10.2** Geothermal power, 459 / **10.3** Solar thermal power, 466 / **10.4** Electric power from wind energy, 476 / **10.5** Energy sources of the future, 486

Appendix

Table A.1 SI units and their conversion, 491 / **Table A.2** Thermodynamic properties of helium, 493 / **Table A.3** Mollier diagram for steam–water system, 494 / **Table A.4** Saturated steam properties, 495 / **Table A.5** Superheated steam properties, 497 / **Table A.6** Properties of air at low pressure, 502

Index, 503

Preface

Power engineering may be defined as the engineering and technology required for the production of central station electric power. This interdisciplinary area combines elements of chemical, electrical, and mechanical engineering. Because of the importance of electric power production to modern society, elective courses that help prepare and interest engineering students in careers in this field are highly desirable.

Development of basic courses in power engineering has been hampered in recent years by the lack of suitable texts. The only previously available modern text devoted fully to power engineering was deemed by some to be insufficiently quantitative in a number of subject areas. Related texts dealing with the entire energy area have been published but provide only limited information on power engineering. Excellent monographs on steam production are available from the Babcock and Wilcox Co. (specifically *Steam*) and from Combustion Eng., Inc. (*Combustion*), but these books provide far more detail than needed by the average student and deal with only part of the subject.

The present work aims to fill the need for a modern fully quantitative text devoted exclusively to the entire scope of power engineering. Sufficient descriptive material has been included here to assure general understanding of the nature of this technology, but the information needed to analyze the behavior of systems and components quantitatively has been stressed. We intend that this work will be useful to its readers both as a text and as a working reference for professionals in the power engineering field.

When used as a text, instructors will find slightly more material than can be covered easily in a single-semester course. This affords instructors the option of selecting the material they deem most pertinent in accordance with the background of their students. For example, if the students are predominantly electrical engineers, it may be desirable to omit Chapter 9 on electrical power generation, since it would probably be familiar to them.

SI units have been stressed here, and nearly all the physical properties in the Appendix are given in terms of SI units. Since SI units are not used exclusively by industry, however, other unit systems have been used where appropriate. The necessary conversion factors are provided in the Appendix.

J. WEISMAN
R. ECKART

1

Introduction to Power Generation Systems

1.1 HISTORY AND PRESENT STATUS OF ENERGY RESOURCES

Historic Energy Utilization

Abundant, cheap energy has been a decisive element in a creation of modern technological society. The primary energy sources providing the major fraction of society's requirements have changed radically during the last century. Figure 1.1 illustrates the changes that have occurred in the United States. Many experts believe that the pattern of energy use is about to change again and that the United States will shift from dependence on petroleum and natural gas to the use of a variety of other primary energy sources.

Historically, readily available and economical energy provided the basis for the development of the United States. Wood supplied the heat energy for the early settlers and was the only form of energy available during the initial westward migration. The burning of wood provided the heat for homes, factories, and for the boilers of early steam engines. By the mid-nineteenth century, coal appeared on the American scene as a new energy source. Coal was plentiful, easily found, and cheap. Coal was more energy intensive than wood, thus allowing for more efficient storage of energy supplies. In addition, the burning of coal could be better controlled over long time periods. Coal gradually replaced wood as a primary energy source and, as seen in Fig. 1.1, became the dominant energy source by the early twentieth century.

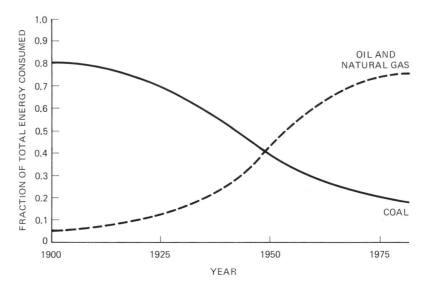

Figure 1.1 U.S. shift to different fuel use patterns. (Data from U.S. Bureau of Mines and Federal Energy Administration.)

Coal was the energy resource that allowed the industrialization of the United States and Western Europe.

Around the turn of the century, two new, more convenient energy sources appeared: petroleum and natural gas. The industrial revolution began with coal as the primary source of energy, but industrialization was completed with petroleum and natural gas. These fuels replaced coal in many existing uses, especially transportation, industrial applications, and space heating. The widespread use of automobiles and airplanes drastically expanded transportation usage. Oil and natural gas also began to compete with coal in the production of electric power. These two fossil fuels became the dominant sources of energy in the United States by the mid-twentieth century.

The Fuels Shortage

Today's so-called energy shortage is really a shortage of the two most popular fuels: petroleum and natural gas. Their popularity arises from the fact that these two fuels are clean and easily transportable. Until recently, these fuels were also considered abundant and cheap. Although petroleum and natural gas provide nearly 75% of the total U.S. energy consumption, the recoverable resource base of these two fuels is considerably smaller than that of coal. The relative abundance of recoverable (or usable) energy resources in the United States is examined in Table 1.1. Petroleum and natural gas constitute a mere 1% of the ultimately recoverable energy resources.

Sec. 1.1 History and Present Status of Energy Resources

TABLE 1.1 U.S. ENERGY CONSUMPTION AND RESOURCES IN 1977 (PERCENTAGES OF TOTAL)

Energy source	Current consumption	Relative abundance of energy resources	
		Without breeder	With breeder
Oil (including natural gas liquids)	49	0.5	0.3
Natural gas	26	0.5	0.3
Coal	19	83.9	44.7
Nuclear power (uranium)	3	1.1	47.3[a]
Oil shale	0	14.0	7.4
Nonmineral energy sources	3	—	—
Total	100.0	100.0	100.0

[a] This estimate is conservative, since it is based only on uranium resources economically recoverable for use in light-water reactors. Breeder reactors, however, would be able to use very low grade ores, which tends to magnify the potential resource base.

Source: U.S. Energy Supply Prospects to 2010, National Academy Press, Washington, D.C., 1979.

The domestic resource base is heavily weighted toward coal and uranium (with the development of breeder technology).

This situation is not unique to the United States. The fossil-fuel resources of the world are also reported to be heavily weighted toward coal, with much lesser amounts of petroleum and natural gas. Estimates of the energy content of the world's recoverable supply of fossil fuels are shown in Table 1.2.

Various models have been developed to estimate the probable length of time in which a depletable energy resource can be utilized. These models generally predict that production of a new energy resource will initially rise exponentially. This initial sharp rise is the result of new applications for the energy resource which increase production, thereby lowering cost. As production begins to outstrip discovery, the rate of expansion begins to slow down, and the production rate reaches a maximum

TABLE 1.2 RECOVERABLE ENERGY ESTIMATES OF THE WORLD'S FOSSIL-FUEL SUPPLY

Energy source	Percent of total energy
Coal and lignite	88
Petroleum liquids	5
Natural gas	4
Tar sand oil	1
Shale oil	2
Total	100

TABLE 1.3 ENERGY DEMAND FOR ELECTRICITY GENERATION (PERCENTAGE OF TOTAL ENERGY DEMAND)

Source of data	1977	1985	1990	1995	2000	2010
Edison Electric Institute (1976)	30				40–45	
Institute for Energy Analysis (1976)	30	38–39			47–51	47–52
Ebasco Services (1977)	30	37	42	49		
U.S. Bureau of Mines (1975)	30	38			48	

Source: U.S. Energy Supply Prospects to 2010, National Academy Press, Washington, D.C., 1979.

when about half the recoverable resource has been consumed. After this point, depletion of the resource results in decreasing production rates and higher costs. Resource utilization is further decreased as costs increase, resulting in still lower production rates. The resource continues to be depleted and its production falls steadily toward zero.

Data indicate that the peak in production for petroleum and gas may have occurred in the 1970s. This suggests that approximately one-half of the total recoverable resource base for these fuels has already been consumed. Coal presents a different picture, since only 5% of the recoverable coal resources had been consumed through the 1970s. With production estimated to peak at 4 billion tons per year (1980 production: 700 million tons), the resource lifetime is expected to extend well into the next century.

The shift in the primary sources of energy is coupled with a shift in the way that primary energy is used. An increasing fraction of our primary energy is being converted to electrical energy (see Table 1.3). Since electrical energy can be generated conveniently without the use of petroleum and natural gas, this trend will help reduce U.S. dependence on nondomestic primary energy sources.

1.2 ELECTRIC POWER

The Exponential Growth of Electric Power Production

The sale of electricity to the public began in 1879 with the California Electric Company in San Francisco. Full commercialization did not begin until three years later, when Thomas Edison's Pearl Street Station started operation in New York. For the first time, the incandescent lamp and the components for public supply of electrical power were brought together. Subsequently, 3500 electric utility systems were formed in the United States. The historic growth rate has been slightly over 6% per year, while the total energy growth rate has been about 3.2% per year. This

Sec. 1.2   Electric Power

growth is illustrated in Fig. 1.2, which shows the annual U.S. electric energy production from 1920 to 1976.

It is instructive to examine this constantly increasing demand for electrical energy. If the energy production, $\mathscr{E}$, increases at the same fractional rate, i, each year, the rate of change of energy production per year becomes

$$\frac{d\mathscr{E}}{dt} = \mathscr{E}i \tag{1.1}$$

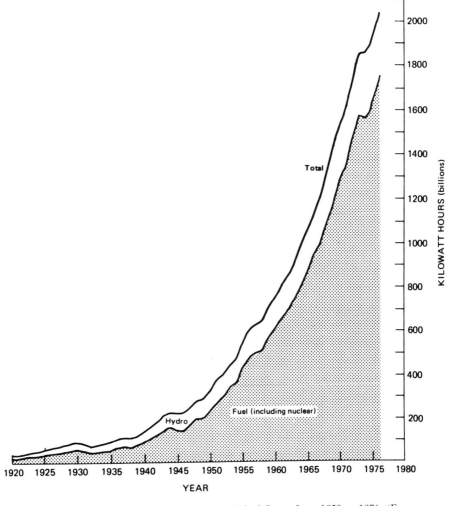

Figure 1.2 Electricity generation in the United States from 1920 to 1976. (From *U.S. Energy Prospects to 2010*, National Academy Press, Washington, D.C., 1979.)

After integration,

$$\ln \frac{\mathscr{E}}{\mathscr{E}_0} = it$$

or

$$\mathscr{E} = \mathscr{E}_0 e^{i(t-t_0)} \qquad (1.2)$$

where $\mathscr{E}_0$ is the energy production in base year t_0.

Electrical power has, in fact, exhibited the foregoing behavior. When the data of Fig. 1.2 are plotted on semilog paper, a close approximation of a straight line is obtained, as shown in Fig. 1.3. This exponential behavior gives rise to the term *doubling time*, the time required for electrical power output to double. Let $\mathscr{E}_1$ be the power generation at time t_1, and $\mathscr{E}_2$ be the power generation at time t_2. Then from

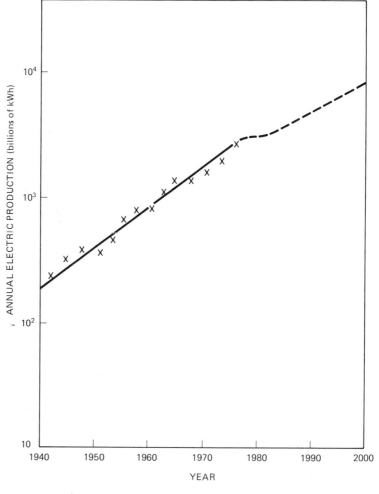

Figure 1.3 Exponential growth of electric power production.

Eq. (1.2), we obtain

$$\frac{\mathscr{E}_2}{\mathscr{E}_1} = \frac{e^{i(t_2-t_0)}}{e^{i(t_1-t_0)}} = e^{i(t_2-t_1)} \tag{1.3}$$

If we let t_d be the doubling time, then

$$t_d = t_2 - t_1 \quad \text{when} \quad \frac{\mathscr{E}_2}{\mathscr{E}_1} = 2 \tag{1.4}$$

Hence

$$\ln 2 = i t_d \quad \text{or} \quad t_d = \frac{0.693}{i} \tag{1.5}$$

Using this simple expression, the doubling time for various assumed or actual rates of increase can easily be found. Using the historic trend of an annual rate of increase in electrical power production of 6.2%, the doubling time is $(0.693/0.062) = 11.2$ years.

The implications of the exponential growth of electrical power on energy demand and resource consumption can be illustrated by one additional example. Consider the case shown by Fig. 1.4, where the power increases exponentially during

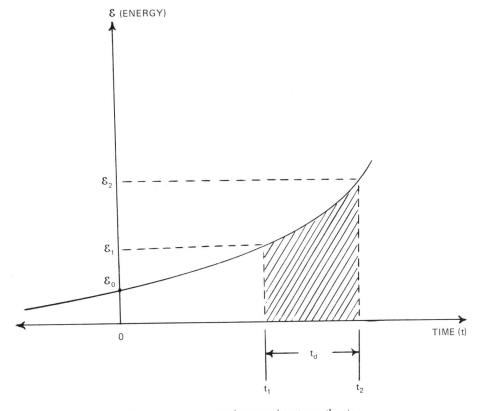

Figure 1.4 Energy use for a consistent growth rate.

the doubling-time interval $t_2 - t_1$. Since the area under the curve represents energy consumption, let $\mathscr{EC}_1$ represent the energy consumed during the entire period prior to t_1. Therefore,

$$\mathscr{EC}_1 = \int_{-\infty}^{t_1} \mathscr{E} \, dt = \int_{-\infty}^{t_1} \mathscr{E}_0 e^{it} \, dt = \frac{1}{i} \mathscr{E}_0 e^{it_1} \tag{1.6}$$

The energy consumed during the doubling time, t_d, is then given by

$$\mathscr{EC}_2 = \int_{t_1}^{t_2} \mathscr{E} \, dt = \int_{t_1}^{t_2} \mathscr{E}_0 e^{it} \, dt = \frac{\mathscr{E}_0}{i} \left(e^{it_2} - e^{it_1} \right) = \frac{1}{i} \mathscr{E}_0 e^{it_1} \left(e^{i(t_2 - t_1)} - 1 \right) \tag{1.7}$$

Since the quantity $t_2 - t_1$ represents doubling time,

$$e^{i(t_2 - t_1)} = e^{it_d} = 2 \tag{1.8}$$

After substitution into Eq. (1.8), we obtain

$$\mathscr{EC}_2 = \frac{1}{i} \mathscr{E}_0 e^{it_1} = \mathscr{EC}_1 \tag{1.9}$$

Since $\mathscr{EC}_2$ is equal to $\mathscr{EC}_1$, we conclude that the quantity of energy consumed in one doubling period is equal to the quantity of energy consumed in all of the time prior to that doubling period.

Future Projections

By extrapolation of the historic logarithmic growth curve, estimates of future electrical demand may be obtained; however, such estimates are highly uncertain. Although the average long-term growth rate has remained fairly constant over the

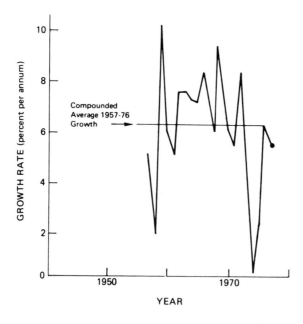

Figure 1.5 Variation in the annual growth rate of electric power. (From *U.S. Energy Prospects to 2010*, National Academy Press, Washington, D.C., 1979.)

Sec. 1.2 Electric Power

past decades, the growth rate in any given year has varied sharply from this average. This is illustrated in Fig. 1.5, which shows that between 1957 and 1977, annual growth rates ranged from nearly zero to more than 10%. Therefore, actual growth in any given year will be significantly different from predicted growth.

Most forecasters now believe that the demand for electricity is closely tied to the gross national product. This is illustrated in Fig. 1.6, where an essentially linear relationship is seen between electrical demand and gross national product (appropriately adjusted for monetary inflation). In time of economic stagnation, electrical demand may not increase at all, whereas in times of rapid economic growth, a rapid rise in electrical demand occurs. Projection of future demand is thus tied to estimates of economic growth.

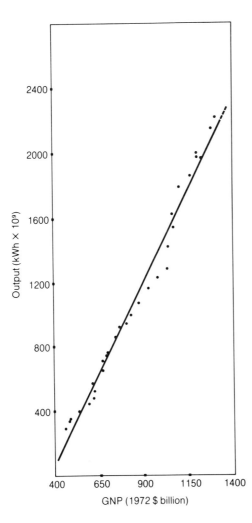

Figure 1.6 Electricity output versus GNP, 1947–1979. (From the *EPRI Journal*, Sept. 1980.)

10   Introduction to Power Generation Systems   Chap. 1

It will be observed in Fig. 1.6 that although the relationship to gross national product explains most of the electrical demand growth, there is an unexplained cyclical behavior. The data points tend to lie first on the side of the fitted line and then on the other. An understanding of this 14-year cyclic variation imposed on the basic curve remains to be obtained.

In addition to the uncertainties introduced by the effect of economic behavior

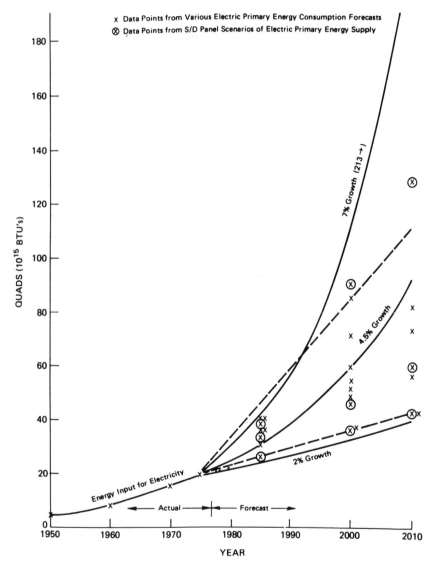

Figure 1.7 Future energy requirements. (From *U.S. Energy Prospects to 2010*, National Academy Press, Washington, D.C., 1979.)

on electrical demand, there are additional uncertainties generated by technological changes. Although the annual electrical energy growth rate was between 6 and 7% before 1973, it declined to near zero in 1973. The growth rate then showed some recovery in the late 1970s. Analysts believe that annual growth rate of 7% would be a maximum upper bound, while a growth rate of 2% would represent a lower limit. Most analysts expect a decrease in the historic growth rate to about 4.5% per year for the period between 1980 and 2010. Figure 1.7 shows the divergent projections for electrical energy usage in 2010 which result from these various possibilities. It may be observed that projected usage ranges from a low of under 40 quads to a high of 180 quads (1 quad = 10^{15} Btu = 1 quadrillion Btu). Even a modest growth rate of 4.5% corresponds to a doubling time of about 15 years. All these projections portend significant increases in electrical power demand in the future.

1.3 ENERGY SOURCES FOR MODERN POWER PLANTS

The means by which future electrical power demand is to be met is the topic of this book. Gas and petroleum will gradually be withdrawn as primary energy sources for electrical power. Thus, not only must new demands be met, but replacement must be found for the energy supplied previously by petroleum and gas.

Table 1.4 catalogs the potential sources for electric power generation. At present, the fossil fuels—coal, petroleum, and natural gas—together with nuclear fission and hydroelectric power, represent the only well-developed means of generating large quantities of electric power.

Base-load electric power generation for the near-term future will probably rely on coal and nuclear power, with slight increases in hydroelectric power output. Breeder reactors may be commercialized early in the next century. Research, development, and construction of experimental facilities are progressing in fusion, solar, wind, and ocean thermal power programs. All these new concepts appear to be a

TABLE 1.4 ENERGY SOURCES FOR ELECTRIC POWER GENERATION

Fossil fuels	Natural sources
Coal	Hydroelectric power
Oil	Geothermal
Natural gas	Wind
Substitute fuels	
Coal liquefaction	Solar
Coal gasification	Solar thermal
Oil from shale	Photovoltaic
Oil from tar sands	Ocean
Nuclear energy	Ocean thermal
Fission reactors	Ocean currents
Breeder reactors	Tidal
Fusion reactors	

number of years away from making possible the production of significant quantities of central station electric power.

1.4 POWER SYSTEMS AND UTILITY DEMAND PATTERNS

Power Systems

Power systems are established to supply electrical power to a specific geographic area. A power system must be capable of meeting all the load demands placed on it, either by its own generating capacity or through the purchase of power from a neighboring grid. All large companies endeavor to have sufficient reserve capacity to handle unusually high load demands or provide backup power in the event of equipment failure at a particular unit.

With the trend toward large power plants, smaller utility systems cannot afford sufficient backup power in the event that some of the large units are taken off line. In order to provide sufficient reserve electrical power, the country was divided into nine reliability regions, as shown in Fig. 1.8. Regional planning now supplements projections by individual systems to assure that each region will be able to meet the requirements placed on it.

Utility Demand Patterns

Because demand is continuously varying and power cannot be stored, the power output of a utility system must equal demand unless outside power is purchased. This variation in demand is most easily seen on a *load-demand curve*. The load-demand curve is a graph of power output (in kilowatts) over a given period of time. Figure 1.9 shows a typical load-demand curve over a 24-hour period in summer and winter. If such a plot were extended to cover an entire year, it would be apparent that there are a number of components to the demand variation:

1. *Diurnal variation*: daily variation caused by normal changes in activities during a day
2. *Seasonal variation*: effects of seasonal loads such as heating and air conditioning
3. *Random variation*: variation caused by factors such as short-term weather changes, sales of power to neighboring systems, and equipment failure
4. *Long-term growth*

As may be deduced from Fig. 1.9, a peak load is likely to occur during a hot midsummer afternoon. This peak is a combination of normal residential, commercial and industrial loads superimposed upon a high air-conditioning load. Air conditioning has caused a shift in the highest load demand from the winter to the summer season.

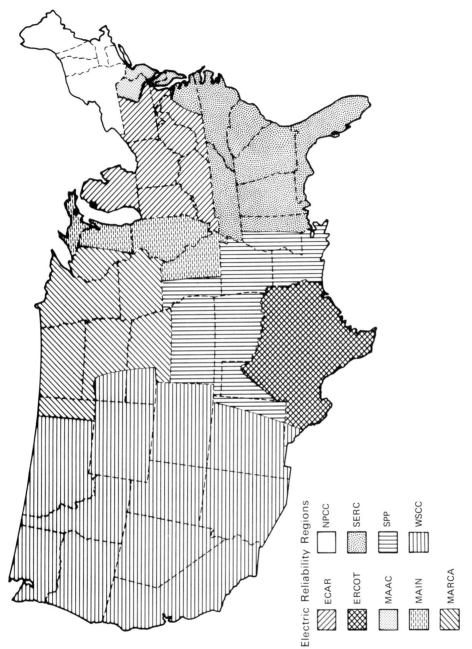

Figure 1.8 Electric reliability regions.

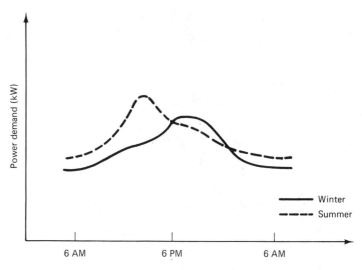

Figure 1.9 Daily variation in electric power demand.

To help quantify the observed demand variation, several factors are usually used. We define the *load factor* and the *capacity factor* as follows:

$$\text{load factor} \equiv \frac{\text{average load over given time interval}}{\text{peak load during the same time interval}}$$

$$\text{capacity factor} \equiv \frac{\text{energy actually produced by plant or system during a given time interval}}{\text{energy that could have been produced if the plant or system operated at net rated power during this time interval}}$$

For an individual plant, the availability factor is also useful:

$$\text{availability factor} \equiv \frac{\text{time period in which the power plant was in operational condition}}{\text{total time interval considered}}$$

A power plant may be ready to operate during one particular week, but may be held in reserve and thus not produce any power. The availability factor would be 1.0, but the capacity factor would be zero. Some nuclear power plants, for example, are required to operate at a reduced power level while problems are being investigated. Assume that a particular plant is required to operate at one-half its rated power for 1 month. During this time interval, the availability factor would be 1.0 and the capacity factor would be 0.5. The availability factor is always greater than or equal to the capacity factor.

Load-Duration Curve

The *load-duration curve* is plotted from the load data rearranged to show the number of hours during the period of interest in which the load is below the stated value. A load-duration curve for a single day may be constructed from the daily load curve as shown in Fig. 1.10. To construct the curve, the chronological load data are sorted in descending order without regard to the specific time of day.

A similar load-duration curve may be constructed using the load-demand curve for any specified period. An annual period is the most usual basis for such curves. For any given value of the ordinate, say P_i, the corresponding value of the abscissa obtained from the curve represents the number of hours per year in which the power demand exceeds P_i. The area under the annual load-duration curve of Fig. 1.11 represents the total energy supplied by the utility's generating system during the year. It is usually divided into three parts:

1. Base load
2. Intermediate load
3. Peaking load

The *base load* is the load below which the demand never falls and is supplied 100% of the time. The *peaking load* is generally considered the region to the left of the sharply sloping portion of the curve and usually represents loading seen less than 15% of the time. The *intermediate load* represents the remaining load region.

The load-duration curve may be interpreted in terms of a probability distribution. In Fig. 1.10 the ratio t_1/t_t may be considered to be the probability Pr that the load P will exceed P_0. That is,

$$\frac{t_1}{t_t} = \Pr(P > P_0) \tag{1.10}$$

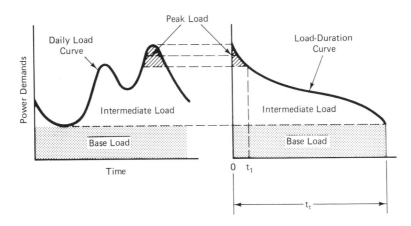

Figure 1.10 Construction of the load-duration curve.

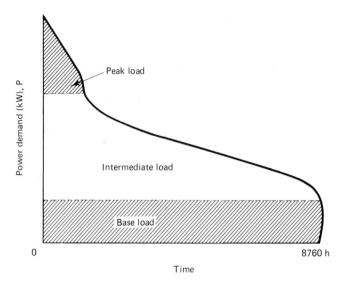

Figure 1.11 Power system load characteristics.

In statistical analysis, the cumulative distribution function, $F(P)$, is defined as the probability that P will be equal to or less than a given value, P_i. Since

$$\Pr(P > P_i) = 1 - \Pr(P \leq P_i) \tag{1.11}$$

we have for our load behavior

$$1 - F(P_0) = \Pr(P > P_0) \tag{1.12}$$

The cumulative distribution, $F(P)$, is obtained by integration of the frequency function, $f(P)$ between $-\infty$ and P_0. Hence

$$\Pr(P > P_0) = 1 - \int_{-\infty}^{P_0} f(P)\, dP \tag{1.13}$$

It has been found that the *log-normal* frequency function generally represents $f(P)$ very well. We then have

$$\Pr(P > P_i) = 1 - \frac{1}{\sqrt{2\pi}} \int_{-\infty}^{z_i} e^{-z^2/2}\, dz \tag{1.14}$$

where $z = (\ln P - \mu_l)/\sigma_l$. The quantities μ_l and σ_l represent the mean and standard deviation, respectively, of the logarithm of power demand. These quantities may be readily estimated by dividing the period for the load-duration curve into n equal time segments. If P_j represents the average power demand during time segment j, then

$$\mu_l \simeq \frac{\Sigma \ln P_j}{n} \tag{1.15}$$

and

$$\sigma_l = \left[\frac{\Sigma (\ln P_j)^2}{n} - \mu_l^2\right]^{1/2} \tag{1.16}$$

The Load-Duration Curve and Plant Selection

The pattern of utility load demand leads to the use of three types of power-generating plants. Since *peaking-load stations* are used only a small fraction of the time, fuel costs at these plants are not of major importance. Capital costs, on the other hand, must be charged whether the station is or is not used. Therefore, if a new plant is to be built for peaking service, it will be built in a way that will minimize capital costs. This usually means that the station will burn oil or natural gas. The resulting high fuel costs can be tolerated since the peaking stations are used only occasionally.

An exception to the case of low-capital-cost plants for peaking service is the use of hydroelectric plants for this purpose. Although the capital costs of such plants are high, in many cases a good fraction of these costs will have been charged to flood control or recreation. More important, the total quantity of water that can be impounded by a hydroelectric plant dam is limited. In many cases it is impossible to run the plant at rated load for most of the year, as the water required will exceed the total available. Such plants may then be used only during those portions of the year when the load is highest.

Base-loaded plants are plants that are always loaded as heavily as possible. Operating costs for such plants are very important, and a high capital cost is feasible if low operating costs can be maintained. Most new, large coal and nuclear power stations are base-loaded plants, due to the low operating costs of these units. Capital costs for nuclear plants are high due to the complex systems required to assure plant reliability and safety. Capital costs for large coal plants are also high due to the complicated systems required for high thermal efficiency, the coal-handling equipment needed, and the expenses imposed by the need to operate in an environmentally acceptable manner.

Intermediate-load plants tend to be the somewhat smaller and older units. For the most part, they are older coal plants where thermal efficiencies are significantly below those of the most modern units. They may also include older petroleum- and gas-fired stations in those areas of the country where petroleum and natural gas have been used for base-loaded plants. Operating costs for such plants tend to be higher than those of the base-loaded plants but more reasonable than those of peaking units.

Intermediate- and peaking-load plants are often termed *load-following plants*. This terminology is employed because the output of such plants responds to the diurnal variations of load on the system.

Figure 1.12 shows a load-duration curve for a typical utility. Superimposed on this curve (solid lines) are the relative total costs per kilowatt per year to produce power from different types of generation based on the output duration. The typical

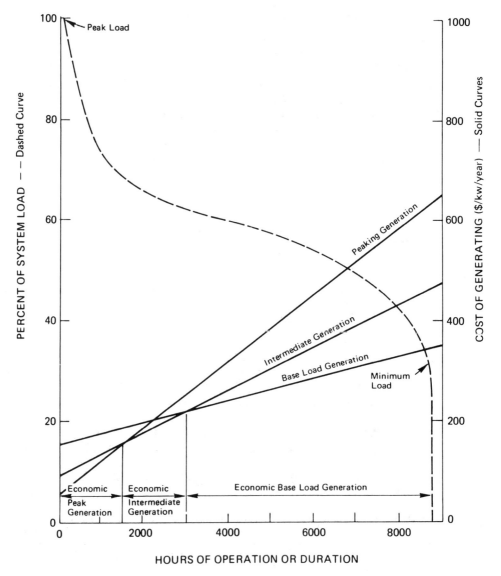

Figure 1.12 Effect of the load-duration curve on power cost. (From *U.S. Energy Prospects to 2010*, National Academy Press, Washington, D.C., 1979.)

Sec. 1.5   Modern Electric Generating Stations

costs shown for the three types of generation—peak, intermediate, and base—illustrate that maximum economy can be achieved by using the peaking units less than 1500 hours per year, using the intermediate-load units 1500 to 3000 hours per year, and using the base-loaded units for more than 3000 hours per year.

1.5 MODERN ELECTRIC GENERATING STATIONS

How Power Is Generated

Virtually all the electric power in the United States today is generated by one of the methods depicted schematically in Fig. 1.13. Approximately 90% of the electric power is generated by what is called the *thermal-mechanical generation concept*. In this concept, heat energy is converted into mechanical energy which powers an electric generator. Most of these power plants use the *Rankine cycle* for conversion of heat energy to mechanical power.

Fossil-fuel plants represent the most common application of the thermal-mechanical generation concept. Coal, gas, or a petroleum fraction is mixed with air and injected into a boiler where combustion takes place. In most plants, heat from combustion is used to convert water to steam. This steam, in turn, flows through large pipes to a multistage turbine. In some plants fired by petroleum or gas, the combustion products flow directly to the turbine. As the steam or hot gas passes through the turbine, energy is extracted from the vapor to turn the turbine. The turbine is mechanically connected to the generator, and it drives or rotates the generator, allowing the production of electrical energy.

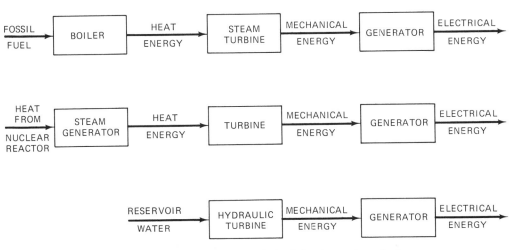

Figure 1.13 Modern methods of electric energy production.

Steam Power Plants

The most widely used fossil systems burn coal or a heavy petroleum fraction to supply the thermal energy, and employ a steam turbine to power the electric generator. A typical heat cycle for a modern coal plant utilizing the Rankine cycle is shown schematically in Fig. 1.14.

Coal has been, and for at least the next decade or so will continue to be, the largest source of fuel for electrical generating stations in the United States. As early as 1925, coal generated 57% of this country's electrical power. Coal generated an average of 52% of the electrical power during the period from 1925 to 1970. Utilities burn about 65% of all the coal mined in the United States. In 1985 it is projected that coal will generate approximately 47% of the electric power, and 875 million tons of coal will be consumed.

U.S. policy is to move away from the use of petroleum for electric power generation. This deemphasis on petroleum use for electric power generation will take years to have a substantial effect. Power generation by petroleum dropped slightly in 1974–1975 as a result of the original oil embargo, but in 1976 the use of petroleum rose to an all-time high of 556 million barrels for the generation of 320

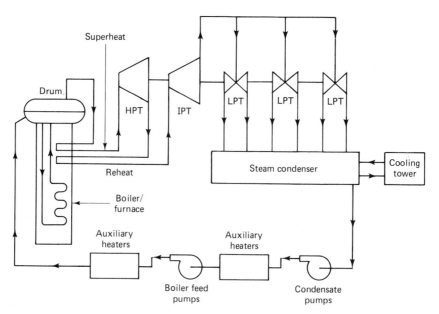

KEY:

HPT = high-pressure turbine
LPT = low-pressure turbine
IPT = intermediate-pressure turbine

Figure 1.14 Heat cycle for a coal-fired boiler.

Sec. 1.5 Modern Electric Generating Stations

billion kilowatthours. It has been estimated that completion of all the petroleum-fired generating units on order during the 1970s will result in a peak use of 900 million barrels of petroleum per year in the early 1980s. The use of petroleum for power generation should begin to decline thereafter.

As coal and nuclear plants assume a greater share of the base-loaded generation, oil plants will be increasingly used for peak or intermediate power production. However, it is expected that utilities will continue to use substantial amounts of petroleum well into the next century.

Gas Turbine Plants

Although natural gas or light petroleum fractions (e.g., kerosene) can be used to produce electric power by the Rankine cycle, the *Brayton cycle* is probably used more frequently. This thermodynamic cycle is a direct, once-through cycle where a gas turbine replaces the steam turbine (see the schematic diagram of Fig. 1.15). Air is compressed in the compressor and then mixed and burned with the natural gas in the combustor. The hot exhaust gases leaving the combustor drive the gas turbine, which supplies the power to drive the generator and compressor. The hot gases leaving the turbine are not recirculated.

The production of power from natural gas grew steadily from 57 billion kilowatthours in 1951 (21% of the electric power generated) to 376 billion kilowatthours in 1970 (29% of the electric power generated). The use of gas for electric power generation began to decline after 1971. By 1980, generation from natural gas had declined to about 10% of the market. Predictions are that natural gas will be used mainly for peak-load generation by the late 1980s and that its use in generating electricity will be negligible by the year 2000.

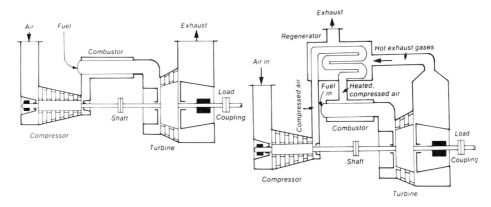

Figure 1.15 Basic gas turbine cycles. (Reprinted with permission from *Power*; copyright 1978 McGraw-Hill, Inc., New York.)

Nuclear Power Plants

A nuclear power plant produces electrical power in much the same manner as does a fossil-fueled power plant. The basic difference is that the heat for the boiler is supplied by the nuclear fission process rather than by the combustion of a fossil fuel. The energy released during the fission process is converted into heat in the reactor core, which then supplies heat to the boiler. This heat is removed by a coolant, which may be water, gas, or a liquid metal. These concepts are discussed more fully in Chapter 7. Most of the commercial power reactors use ordinary water, H_2O (called *light water* in the nuclear industry to distinguish it from "heavy" water, D_2O), as the coolant.

There are two basic nuclear power systems now in commercial use in the United States: the boiling-water reactor system (BWR) and the pressurized-water reactor system (PWR). In the *boiling-water reactor* system [Fig. 1.16(a)], the steam for the turbine is generated directly in the upper region of the reactor core area. In the *pressurized-water reactor* system the water coolant is highly pressurized and no bulk boiling is permitted in the exit line from the reactor [Fig. 1.16(b)]. The steam for the turbine is produced in the steam generator, which forms part of a secondary loop.

Chapter 7 describes other types of nuclear power systems that are presently used to produce electric power. In addition to these existing systems, consideration

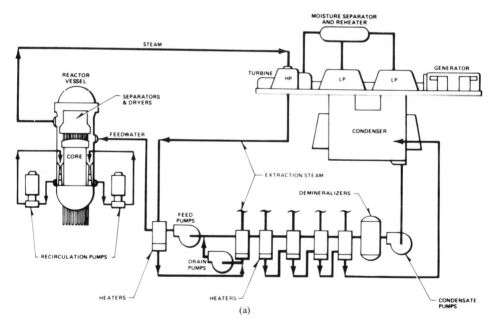

Figure 1.16 (a) Direct-cycle BWR system. (Reprinted with the permission of the General Electric Company, Nuclear Energy Operations, San Jose, Calif.)

Sec. 1.5   Modern Electric Generating Stations

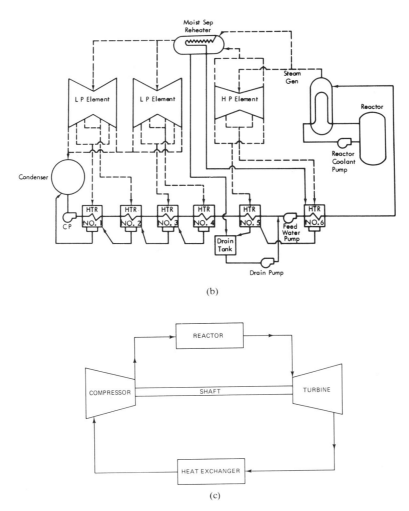

Figure 1.16 (b) Pressurized-water reactor system; (c) gas-cooled reactor system. [(b) from L. S. Tong and J. Weisman, *Thermal Analysis of Pressurized Water Reactors*, 1979; reprinted with the permission of the American Nuclear Society, LaGrange Park, Ill.]

is being given to advanced power cycles such as shown in Fig. 1.16(c), where a gas-cooled reactor, capable of achieving high-temperature performance, is used to drive a gas turbine which powers an electric generator. A system such as this could achieve potentially high thermodynamic efficiency.

In 1951, the Experimental Fast Breeder Reactor No. 1 generated the world's first electric power from nuclear energy. The status of commercial nuclear power 30 years later is shown in Table 1.5. It is expected that these totals will account for about 25% of the electrical power produced in the United States by 1990.

TABLE 1.5 COMMERCIAL POWER REACTORS IN THE UNITED STATES IN 1984

86 Reactors with operating licenses	69,847 MWe
52 Reactors with construction permits	57,638 MWe
2 Reactors on order	2,240 MWe
140 Total	129,725 MWe

Hydroelectric Power

In a hydroelectric power plant, the potential energy of water in a reservoir is converted to the mechanical energy of a rotating hydraulic turbine. The energy is transmitted to the turbine shaft, which is directly coupled to an electric generator. As noted previously, the power output of a hydroelectric station often may have to be curtailed due to limitations on the amount of water available.

Hydroelectric power grew from about 5 GW in 1920 to 59 GW in 1976. By 1945, hydroelectric power supplied 40% of the electricity in the United States. By 1960, despite modest growth, the market share had dropped to 14% of the total electrical power generation. Recent data indicate that hydroelectric power will generate about 300 billion kilowatthours in 1990, but this will represent an ever-declining share of the total electric power production.

1.6 ECONOMICS OF ELECTRIC POWER PRODUCTION

Components of Electric Power Production Cost

The principal components of electric power cost to the consumer are the generation or production cost, the transmission cost, the distribution cost, administrative costs, plus some allowed rate of return or profit on the investment. The production cost generally represents about 60% of the total cost of electric power to the consumer and is the main concern of power engineers responsible for economic comparisons. This section deals with current trends in generation costs and presents methods for evaluating the cost of power generated by steam power stations.

An economic comparison of the cost of producing electric power is made by comparing three major cost areas among competing alternatives. These three major areas are

1. Fixed charges
2. Operating and maintenance expenses
3. Fuel or energy costs

Sec. 1.6 Economics of Electric Power Production

If we let

CAP = capital cost of power station, dollars

$\mathscr{G}$ = net rated electrical capacity of power plant, kW

OM = annual operating and maintenance cost, dollars/yr

ϕ = levelized annual fixed-charge rate, %/(100 × yr)

e_c = fuel cost per unit of electrical energy, mills/kWh

C_p = expected plant capacity factor

then the total energy cost, e(mills/kWh), is generally given by

$$e(\text{mills/kWh}) = \frac{1000 \text{ mills}}{\text{dollars}} \times \frac{\text{years}}{8760 \text{ h}} \frac{1}{C_p \mathscr{G}} [\phi(\text{CAP}) + \text{OM}] \frac{\text{dollars}}{\text{yr}} + e_c$$

(1.17)

or

$$e = \frac{0.114}{C_p \mathscr{G}} [\phi(\text{CAP}) + \text{OM}] + e_c$$

The *fixed charges* are those costs that the utility must pay regardless of the quantity of electric power produced. As the name implies, these charges are constant over the lifetime of the plant. They are directly related to the capital cost of the power plant and all associated electrical generating equipment. Capital costs have become the dominant cost item in electric power production, especially for nuclear power plants.

Operating and maintenance (O & M) *costs* are those associated with the routine operation and maintenance of the power plant. These costs include the salaries of all plant operating and supervisory personnel, consumable supplies and equipment, outside support services, office supplies, training programs, the cost of requalifying the operators, annual operating fees, maintenance personnel salaries and the salaries of the engineering staff. Operating and maintenance costs are the smallest element of the production cost and vary with plant type and size.

The third element of the production cost is the *fuel cost*. In the case of nuclear power plants, this component includes the total fuel-cycle cost plus the interest charged on the working capital required to finance the fuel cycle (see Section 7.6 for a discussion of the nuclear fuel cycle). For fossil-fueled power plants, this includes the cost of gas, petroleum, or coal burned plus waste or sludge disposal. Because of the increasing cost of petroleum and coal, and high interest rates on money, an additional cost item is now included to cover the interest charges on the 60- to 90-day coal stockpile or petroleum reserve required at plant sites. A power plant engineer may be required to make a recommendation on the type of plant best suited for new base-load generation. In a study of this problem, the primary question is often whether or not the lower fuel-cycle costs for the nuclear power station can offset its higher initial capital costs.

Estimation and the Net Rated Power Output and the Plant Capacity Factor

The parameters $\mathscr{G}$ and C_p which appear in the denominator of Eq. (1.17) are common to both the fixed charges and the operating and maintenance expenses. The quantity $\mathscr{G}$ represents the net rated electrical plant capacity. This is the power plant's gross electrical output rating less power consumed at the site. The power consumed at the site may be estimated based on data from similar power plants.

Table 1.6 shows this comparison for typical power plants in the 800-MW size range. The higher site power for coal-burning plants is required to provide power for particle precipitators and sulfur dioxide removal equipment.

The capacity factor, C_p, was defined previously (see Section 1.4) as the ratio of the electrical energy produced by the plant to the rated plant capacity. Data on typical capacity factors for coal and nuclear units are readily available through industry sources such as the Electric Power Research Institute, the Edison Electric Institute, the Atomic Industrial Forum, and the Department of Energy. Industry data for normally operating base-loaded power plants show capacity factors that range from 0.50 to about 0.80. However, there is such a wide variation in individual unit capacity factors that a reasonable approach for preliminary economic comparisons would be to assume a constant capacity factor for all plants being evaluated. A value ranging from 0.6 to 0.7 is recommended for preliminary cost comparisons between coal and nuclear units.

Capital Cost of Electric Power Stations

Virtually all new power stations will be either coal or nuclear units. Capital costs for both types of units have been increasing at a rate higher than that for most industrial equipment. In 1965, the average estimated cost of a nuclear power station was $120/kWe. By 1974, the units ordered were estimated to cost $550/kWe, a compounded increase of 18% per year. The increase in capital costs has continued, with the most recent trends shown in Fig. 1.17.

The data in Fig. 1.17 show recent increases in the capital cost of coal-fired and nuclear power stations. These cost increases have been attributed to inflation, new

TABLE 1.6 COMPARISON OF SITE POWER CONSUMPTION

	Low-sulfur coal	High-sulfur coal	Light-water nuclear
Gross turbine output (MWe)	854	854	830
Site auxiliary equipt. power (MWe)	53	66	29
Net rated plant output (MWe)	801	788	801
Power consumed at site (%)	6.2	7.7	3.5

Source: U.S. Nuclear Regulatory Commission Report NUREG 0248, 1978.

Sec. 1.6 Economics of Electric Power Production

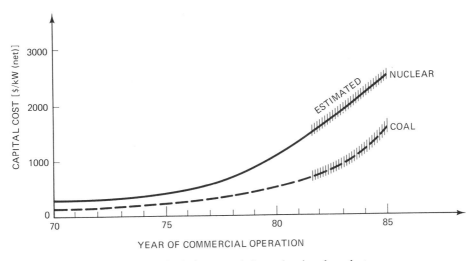

Figure 1.17 Capital cost trends for coal and nuclear plants.

environmental regulations, the high cost of borrowing money, enhanced safety regulations, and the increasing length of time required to construct a power plant. In 1967, studies indicated that the average time to construct a nuclear power station was 63 to 66 months. By 1973, these same studies showed the average construction time increasing to 90 to 99 months. Recent data indicate that the average construction time has increased further, to 100 to 120 months. This extremely long construction period, coupled with high interest rates, has resulted in interest during construction charges of $200 to 300 million. This is approximately equal to the total capital cost of first-generation commercial nuclear power stations such as Oyster Creek and Dresden I.

The estimation of plant capital costs has been greatly eased by the development of computer programs which base these estimations on historical data. The CONCEPT code is an example of a program of this type. CONCEPT utilizes the historic data from actual construction costs and projects these data trends into the future. All the costs associated with building fossil-fueled power stations (coal, oil, and natural gas) or nuclear power stations (BWR, PWR, and gas-cooled) are considered. Options are available to include SO_2-removal equipment. Regional variations in cost data, such as local labor rates and material prices, are considered.

Annual Charges on Capitalized Cost (Fixed Charges)

The levelized annual fixed-charge rate, ϕ (% per year/100), is unique to each utility. The fixed-charge rate includes depreciation charges on the capital investment, interest on borrowed money, return to stockholders, interim replacement of major power plant components, property insurance, and local, state, and federal taxes. Values of typical components of the levelized annual fixed-charge rate for a public

TABLE 1.7 LEVELIZED ANNUAL FIXED-CHARGE RATE

	Fixed-charge rate (%/yr)	
Component	Public utility	Private utility
Interest or return on bonds[a]	9.5	11.2[a]
Stock (equity)[a]	—	9.0[a]
Average cost of money	9.5	10.2
Depreciation[b]	0.7	0.6
Interim replacement of parts	0.4	0.4
Property insurance	0.3	0.3
Federal income tax	—	2.0
State and local taxes	1.0	3.0
Total fixed-charge rate	11.9	16.5

[a]Assumes that capital is acquired 55% through bonds and 45% through equity. Average cost of money would be $0.55(11.2) + 0.45(9.0) = 10.2\%$ per year.

[b]Assumes a 30-year sinking fund at average cost-of-money conditions.

and private utility are shown in Table 1.7. It may be seen from the table that the interest on return on investment is lower for publicly owned utilities. The government can borrow money at a lower cost than can investor-owned utilities, and, in addition, the public utilities pay no federal taxes and pay only minimal state and local taxes. Depreciation charges are based on a 30-year sinking fund.

Operating and Maintenance Expenses

Operating and maintenance (O & M) expenses are difficult to calculate a priori and are usually estimated based on past experience with specific power plant types. Fortunately, these expenses are the lowest-cost component of the generating cost. Table 1.8 illustrates typical O & M expenses based on 1979 data. In these estimates the O & M expenses were assumed to consist of operating labor (25%), supplies

TABLE 1.8 O & M EXPENSES, NATIONAL AVERAGE FOR 1979 (mills/kWh)

	Capacity factor (%)			
Type of plant	50	60	70	80
Nuclear	3.85	3.21	2.75	2.45
Coal				
Low sulfur	4.85	4.32	3.95	3.17
High sulfur	6.35	5.81	5.35	5.02

Source: U.S. Nuclear Regulatory Commission Report NUREG 0248, 1978.

Sec. 1.6 Economics of Electric Power Production

TABLE 1.9 ESTIMATED OPERATING AND MAINTENANCE EXPENSES FOR 1985 POWER PLANTS (mills/kWh)

Type of plant	Low O & M	High O & M
Nuclear	2.5	4.5
Coal		
Low sulfur	4.0	6.0
High sulfur	5.4	8.0

Source: U.S. Nuclear Regulatory Commission Report NUREG 0248, 1978.

(10%), and maintenance (65%). Future projections, shown in Table 1.9, reflect the anticipated cost increases due to additional government regulations for coal-burning power stations.

Fuel or Energy Costs

The final term in Eq. (1.17) represents the fuel costs for coal and nuclear power plants. To evaluate these costs, the overall power plant efficiency, η_o, must be estimated, and the fuel energy release per unit mass, Q'_{net}, or burnup, B, and total fuel cycle cost, F_c, must be evaluated. The fuel costs for coal and nuclear plants will be discussed separately, as there is a difference in the methods used to calculate e_c.

Fuel Costs for Fossil-Fueled Stations

The fuel or energy costs, e_c (mills/kWh), for a fossil-fueled power station are given by

$$e_c = \frac{F_c(\text{dollars/kg})}{\eta_o Q'_{net}(\text{kJ/kg})} \times \frac{1000 \text{ mills/dollar}}{\text{kWh/3600 kJ}} = \frac{3.6 \times 10^6 F_c}{\eta_o Q'_{net}} \quad (1.18)$$

where F_c = fuel cost, dollars/kg

Q'_{net} = net energy release per unit fuel mass, kJ/kg

η_o = overall thermal efficiency

The overall thermal efficiency, η_o, which for a modern coal-fired unit is in the range of 30 to 40%, must of course be estimated for the specific thermodynamic cycle used (see Chapter 2). It must also include the boiler efficiency. That is, it must allow for the fact that not all the heat released in combustion is captured in the boiler. Hence

$$\eta_o = \eta_b \eta_{th} \quad (1.19)$$

where

η_b = fraction of energy released from the fuel that is transferred to working fluid

η_{th} = actual thermodynamic efficiency of turbine cycle

The fuel cost, F_c, has three major components: the actual cost of the coal burned, the interest on the 60- to 90-day reserve supply kept at the plant, and the costs associated with waste and sludge disposal. When coal is used as a fuel, it should be recognized that it is a regional fuel and shipping costs constitute a large fraction of its price.

The fuel or energy cost of coal depends on both its price and its properties. The price charged for coal varies widely, the major factors being the type of mining, the distance from the mine to the power plant, the sulfur content of the coal, and the heating value of the coal. Western coal is characterized by low heating value ($\sim$ 18,000 to 23,000 kJ/kg) and low sulfur content ($\sim$ 0.5%), and is generally obtained by open-pit or surface mining. It may be cheaper per ton at the mine, but transportation costs may make it more expensive at the plant. Eastern coal is characterized by a higher heating value (23,000 to 30,000 kJ/kg), high sulfur content ($\sim$ 2.0 to 2.5%), and deep-mining techniques. Transportation costs to eastern power plants would obviously be lower.

The economics of the choice between a low-sulfur coal and a high-sulfur coal must be evaluated carefully. The use of a high-sulfur coal may result in lower fuel costs but higher capital costs for additional SO_2-removal equipment and higher O & M expenses. The power plant engineer must weigh all these factors carefully in calculating the fuel cost and must properly adjust capital cost and O & M expenses.

Typical fuel costs for coal-burning plants expected in the mid-1980s are shown in Table 1.10. These costs are very sensitive to the price of coal, and hence actual costs may vary considerably from those shown if the rate of inflation increases.

Example 1.1

A coal-fired power plant generates 854 MWe gross output with an overall plant efficiency of 40%. The plant burns coal, releasing 10,000 Btu/lb (lower heating value = 10,000 Btu/lb; see Chapter 3). The capacity factor for the plant is 0.60 for the year. Assume that money is worth 10% to the company. Calculate:

1. Cost of the coal burned for the year, dollars
2. Interest charges, I, on one 90-day reserve pile of coal, dollars
3. Fuel charges (mills/kWh) assuming that the price of the coal is $55/ton and that waste disposal costs are 7.5% of the coal price

TABLE 1.10 TYPICAL COAL FUEL COSTS

	Cost per ton	mills/kWh
Raw coal	$40.00	15.4
Transportation	$10.50	4.0
Indirect costs		0.6
		20.0

Sec. 1.6 Economics of Electric Power Production

Solution

(a) Cost of the coal burned:

$$\text{coal consumed} = 854 \text{ MWe} \times \frac{1}{0.40} \times \frac{\text{MW}_{th}}{\text{MWe}} \times 10^6 \frac{\text{W}}{\text{MW}_{th}} \times 3.4 \frac{\text{Btu}}{\text{W} \cdot \text{h}}$$

$$\times 8760 \frac{\text{h}}{\text{yr}} \times 0.6 \times \frac{1 \text{ lb}}{10,000 \text{ Btu}} \times \frac{1}{2000} \frac{\text{tons}}{\text{lb}}$$

$$= 1.9 \times 10^6 \text{ tons/year} \times 907.18 \text{ kg/ton}$$

or

$$\text{coal consumed} = 1.7 \times 10^9 \text{ kg/year}$$

$$\text{coal cost} = 1.9 \times 10^6 \frac{\text{tons}}{\text{year}} \times 55 \frac{\text{dollars}}{\text{ton}} = \$104 \times 10^6/\text{yr}$$

(b) Interest, I, on the 90-day reserve pile of coal: Since the 90-day stockpile reserve is approximately one-fourth of the annual cost, one estimate would be

$$I = \$104,500,000 \times \tfrac{1}{4} \times 0.1/\text{yr} = \$2.61 \times 10^6/\text{yr}$$

(c) Fuel costs:

F_c = total fuel cost (dollars/kg) = purchase price + waste disposal + interest

$$= \frac{\$104 \times 10^6/\text{yr} + 0.075(\$104 \times 10^6/\text{h}) + \$2.61 \times 10^6/\text{yr}}{1.9 \times 10^6 \text{ tons/yr} \times 907.18 \text{ kg/ton}}$$

$$= \$0.067/\text{kg}$$

$$Q'_{\text{net}} = 10^4 \frac{\text{Btu}}{\text{lb}} \times 2.2 \frac{\text{lb}}{\text{kg}} \times 1.054 \times 10^3 \frac{\text{J}}{\text{Btu}} \times \frac{\text{kJ}}{1000 \text{ J}} = 2.32 \times 10^4 \text{ kJ/kg}$$

$$e_c = \frac{3.6 \times 10^6 \, F_c}{\eta H} = \frac{(3.6 \times 10^6)(\$0.067/\text{kg})}{0.4 \times 2.32 \times 10^4 \text{ kJ/kg}} = 26.1 \text{ mills/kWh}$$

Fuel Costs for Nuclear Stations

Equation (1.18) could be used for estimating nuclear fuel costs, but it is customary to indicate the energy derived from nuclear fuel in terms of "MW days per 1000 kg of uranium (MWd/metric ton U)." This quantity is generally referred to as the *burnup*, B, and is of the order of 20,000 to 30,000 (MWd/metric ton) for current (1982) light-water reactors. With these revised units for fuel energy release, Eq. (1.18) becomes

$$e_c = \frac{41.67 F_c}{B} \qquad (1.20)$$

where F_c = total fuel cost, dollars/kg U

B = fuel burnup, MWd/metric ton U

and other quantities have their previous meanings.

TABLE 1.11 TYPICAL LWR FUEL COSTS COMPONENTS (1982 DOLLARS)

Purchase U_3O_8	\$21/lb U_3O_8
Conversion to UF_6	\$5/kg U
Fuel enrichment	\$141.1/SWU
Fuel fabrication	\$150/kg U
Spent-fuel transportation	\$20/kg U
Fuel reprocessing[a]	\$350/kg U
Bred fuel credit	\$30/g fissile plutonium

[a] If spent fuel is not reprocessed, interim storage and waste disposal charges are assessed at 1.0 mills/kWh under the Waste Policy Act of 1982.

The fuel cycle cost, F_c, refers to the total cost of the nuclear fuel. It includes purchase of the uranium in U_3O_8 form, various conversion steps necessary during the processing, enrichment of the uranium in the isotope uranium 235, fabrication of the nuclear fuel assemblies, all transportation costs, a provision for spent-fuel disposition after reactor operation, and indirect charges that account for interest charged on the working capital. Chapter 7 provides additional detailed information on the nuclear fuel cycle. Fuel cycle costs for light-water reactors are listed in Table 1.11. These costs are typical of all large, modern LWR systems. The table contains sufficient data to calculate fuel charges when used in conjunction with the material of Chapter 7.

Total Electric Power Production Cost

The cost of electric power produced by either fossil-fueled or nuclear power stations should be compared over the lifetime of the plant, usually assumed to be 30 to 40 years. The fixed costs are levelized over this period, as indicated previously. The average O & M and fuel costs must be evaluated, with proper allowance for inflation, and added to the levelized annual fixed charges.

TABLE 1.12 TECHNICAL AND ECONOMIC DATA FOR POWER PRODUCTION COST ANALYSIS[a] (854-MWe UNITS, 1982 DOLLARS)

	Coal	Nuclear
Capital cost (\$)	0.83×10^9	$\$1.25 \times 10^9$
Capacity factor	0.7	0.7
Levelized fixed charge (%/yr)	0.18	0.18
Net rated output (kWe)	800,000	810,000
Operating and maintenance cost (\$/yr)	7.5×10^6	5.0×10^6
Overall plant efficiency (%)	40.0	33.0
Fuel burnup (MWd/metric ton)	0.27	40,000
Additional data	Example 1.1	Table 1.11

[a] Plant startup in 1982.

Sec. 1.6   Economics of Electric Power Production

TABLE 1.13   POWER PRODUCTION COSTS, 1982 AND 1995[a]

	Coal (mills/kWh)		Nuclear (mills/kWh)	
	1982	1995	1982	1995
Fixed charges	30.4	30.4	45.2	45.2
O & M expenses	1.5	5.2	1.1	3.8
Energy costs	23.7	81.8	8.4	29.0
Total	55.6	117.4	54.7	78.0

[a] Escalation assumed at 10% per year.

This procedure requires a careful estimate of the fuel charges over this time period. Recent studies comparing coal and nuclear power costs over the lifetime of the power station show that the lifetime fuel costs become a major factor in the economic decision-making process. Nuclear power stations are characterized by high initial fixed charges and low fuel costs. Coal-fired power stations have lower fixed charges and higher fuel costs. Thus the cost of nuclear electric power is less sensitive to inflation than is coal-produced electric power. Tables 1.12 and 1.13 present data for the cost of electric power production from coal and nuclear stations. The 1995 production costs shown in Table 1.13 illustrate the sensitivity of both types of power stations to an annual cost escalation of 10% for both fuel charges and O & M expenses.

Incremental Costs

Although the total electric power production costs are considered in determining the type of new plant to be constructed, once the plant has been built the utility operator is generally most concerned about its incremental costs. After a power plant has been constructed, the fixed charges must be paid whether or not the plant is run at full capacity. Similarly, most of the operation and maintenance costs are for personnel and also must be paid regardless of the power produced. However, the fuel costs are proportional to power production.

The *incremental cost* of producing a unit of electric energy (mills/kWh) may be defined as the cost per unit energy which is incurred only when power is produced. The incremental cost primarily consists of the fuel cost, but does include that portion of the operation and maintenance cost which varies with power production.

We have noted previously that plants with the lowest operating costs are generally base loaded, while the plants with the highest operating costs are used for peak power. In making the decision as to the order in which plants will be loaded, the dispatcher will use the incremental costs of the various units. The next block of load is always taken by the unit not yet operating at peak capacity which has the lowest incremental cost.

It should be noted that incremental costs generally vary with the power level at which a plant is operated. The total thermal efficiency is generally a function of load, and this function varies from plant to plant. The relative order of the incremental costs of a group of coal-burning plants may thus vary with the power levels at which they are operating.

SYMBOLS

B	fuel burnup, MWd/metric ton
C_p	plant capacity factor
CAP	capital cost of power station, dollars
e	total energy cost, mills/kWh
e_c	fuel cost per unit of electric energy, mills/kWh
$\mathscr{E}$	electrical energy production, kWh
$\mathscr{E}\mathscr{C}_j$	total energy consumed prior to time j, kWh
$f(\cdot)$	frequency function
$F(\cdot)$	cumulative distribution function
F_c	Fuel cost (coal or uranium), \$/kg or \$/kg U
$\mathscr{G}$	net rated electrical capacity of power plant
i	fractional rate of increase per year
I	interest, dollars
OM	operating and maintenance expenses, \$/yr
P	electrical load, kW
Pr	probability
Q'_{net}	net energy released per unit fuel mass, kJ/kg
t	time
t_d	doubling time
z	$(\ln P - \mu_l)/\sigma_l$

Greek Symbols

η_o	overall thermal efficiency
η_b	fraction of the energy released from the fuel transferred to the working fluid
η_{th}	actual thermodynamic efficiency of turbine cycle
μ_l	mean of logarithm of power demand
σ_l	standard deviation of logarithm of power demand
ϕ	levelized annual fixed-charge rate, %/(yr $\times$ 100)

Chap. 1 Problems

Subscripts

0	initial condition
1	condition at time 1
2	condition at time 2
i, j	conditions at times i, j

PROBLEMS

1.1. A moderate-sized electric system has a generating capacity of 4000 MW and a peak load of 3100 MW. It is desired to maintain generating capacity at least 200 MW above the peak load. If the peak load has been growing at 3% per year, when should the next generating unit be added to the system?

1.2. The annual load variation of a particular utility has been fitted by a log-normal frequency function with P measured in megawatts and $\mu_l = 3.4$ and $\sigma_l = 0.4$. You may assume that peak power corresponds to the power that would be exceeded with a probability of 0.0015 or less.
(a) Sketch the load-duration curve.
(b) Determine the load factor.

1.3. The diurnal variation of load in a given system may be closely approximated by a sinusoidal variation. The peak load is approximately 5000 MW, while the minimum load is approximately 3200 MW. The peak load equals 80% of the system's generating capacity. Determine the load factor and the capacity factor.

1.4. An engineer is comparing an 800-MW nuclear power plant with a similar-sized coal-fired power plant. He has determined the following information:

	Nuclear plant	Coal-fired plant
Plant capital cost	$1100/kW	$700/kW
Operating and maintenance cost	3.2 mills/rated kWh	3.95 mills/rated kWh
Expected capacity factor	63%	70%
Fuel cost (per unit power generated)	13 mills/kWh	26 mills/kWh

What is the highest levelized annual fixed-charge rate at which the nuclear plant would be competitive with the coal-fired plant?

1.5. A coal-fired station burning low-sulfur coal can be constructed for $575/rated kW, while a station burning high-sulfur coal with stack gas scrubbers will cost $750/rated kW. Both plants may be assumed to have the same operating and maintenance costs and a thermal efficiency of 42%. Levelized annual fixed charges are 12%. The high-sulfur coal is estimated to release 11,000 Btu/lb and the low sulfur coal 10,500 Btu/lb when burned (both values are lower heating values). The station burning low-sulfur coal is expected to have a 70% capacity factor, while that burning high-sulfur coal is expected to have a 65%

capacity factor. Determine the premium in dollars/ton that can be paid for the low-sulfur coal.

1.6. A nuclear power plant is obtaining a fuel cost (exclusive of disposal charges) of 15 mills/kWh and a total power cost of 60 mills/kWh while using fuel to an average burnup of 24,000 MWd/metric ton. Fuel of a revised design which is capable of operating at 32,000 MWd/metric ton may be purchased at a premium of 30%. Use of the improved fuel is expected to increase plant capacity from 65% to 69% because of less frequent refueling shutdowns. Is purchase of the revised fuel desirable?

BIBLIOGRAPHY

CULP, A., JR., *Principles of Energy Conversion*. New York: McGraw-Hill, 1979.

FRAAS, A., *Engineering Evaluation of Energy Systems*. New York: McGraw-Hill, 1982.

HUDSON, C., and H. BOWERS, *CONCEPT 5*. Report ORNL 5470, Oak Ridge, Tenn.: Oak Ridge National Laboratory, 1980.

KRENZ, J.: *Energy, Conversion and Utilization*. Boston: Allyn and Bacon, 1976.

MCMULLAN, J., R. MORGAN, and R. B. MURRAY, *Energy Resources and Supply*. New York: Wiley, 1976.

MONETTE, J., "A Mathematical Model of an Electric Utility's Load Duration Curve," ORSA/TIMS Conference, Coloado Springs, Colo., 1980.

NATIONAL RESEARCH COUNCIL, COMMITTEE ON NUCLEAR AND ALTERNATIVE ENERGY SYSTEMS, *Energy in Transition*. San Francisco: W. H. Freeman, 1979.

The Need for Energy Facility Sites in the United States: 1975–1985 and 1985–2000, Environmental Policy Institute, June 1975.

NOVICK, S., "The Electric Power Industry," *Environment*, Vol. 17, No. 8 (Nov. 1975).

SHEAHAN, R., *Alternative Energy Sources*. Aspen System Corp., 1981.

Steam, Its Generation and Use. Lynchburg, Va.: The Babcock and Wilcox Company, 1978.

U.S. Energy Supply Prospects to 2010. Washington, D.C.: National Academy Press, 1979.

2

Power Plant Thermodynamics

2.1 THE BASIC LAWS OF THERMODYNAMICS

Energy utilization can be classified into two general areas: (1) energy used directly for heat and (2) energy used to do work. Fuels used for heating buildings or for industrial process heat fall into the first category. Fuel used to produce steam for the turbine of an electric power plant falls into the second category since the turbine converts heat energy into mechanical work. Although total energy is conserved, the conversion of heat into more useful forms of mechanical work requires an expenditure of heat energy in excess of the work output. This excess or *waste heat* generated by heat engines is an inevitable result of the principles of thermodynamics.

In 1840, James Prescott Joule demonstrated the now famous mechanical equivalency of work and heat through his paddlewheel experiment. Although Joule's early experimental results were inaccurate, recent measurements have precisely established Joule's constant as 4.184 J/cal (778.26 ft·lb/Btu).

The *first law of thermodynamics* is a statement that energy is conserved. For nuclear reactions, the first law is interpreted to include the equivalence of mass and energy. The overall energy balance established by the first law of thermodynamics applies to both reversible and irreversible processes. For example, Joule used a paddlewheel to heat water in an insulated container. The reverse is not possible. Simply heating the container will not cause the paddlewheel to rotate. Heating by mechanical work is an irreversible process. Although energy is conserved, it is impossible to recover the original mechanical energy.

A process is considered to be reversible only when (1) it is performed at an infinitesimal rate and (2) it is not accompanied by any dissipative effects such as friction. These are ideal conditions that cannot be met precisely by real systems.

The first law of thermodynamics places no restrictions on the direction or flow of heat and work. However, the first law is not sufficient to describe a thermodynamic system fully. Experimental evidence led to the formulation of the second law of thermodynamics. All thermodynamic cycles satisfy both the first and second laws of thermodynamics.

The *second law of thermodynamics* indicates that heat normally flows from a higher-temperature body to a lower-temperature body and that it is impossible to convert heat to work with no other effect. There will thus be some waste heat no matter what process is utilized for conversion of thermal energy. A heat engine will therefore require a low-temperature sink in addition to a heat source if it is to operate in a complete cycle. As will be shown subsequently, the second law is often used to determine the maximum efficiency possible of a device or system. The second law also implies that all natural processes are irreversible.

The *inequality of Clausius* is a corollary or consequence of the second law of thermodynamics. This inequality states that the integral of the quantity $\delta Q/T$ around a closed reversible thermodynamic path is zero. If we consider both reversible and irreversible system, we must write

$$\int \frac{\delta Q}{T} \leq 0 \qquad (2.1)$$

where Q = heat transferred across boundary of system
T = absolute temperature

The significance of this inequality will be apparent when the laws of thermodynamics are applied to analysis of power plant cycles and performance. The inequality of Clausius also leads to a property of the system called *entropy*.

Reviewing the First Law of Thermodynamics

The first law of thermodynamics is based on experimental evidence and cannot be derived mathematically. For engineering applications where only mechanical and thermal effects need to be considered, the first law for a closed system is often cast into the form[†]

$$_1Q_2 - {_1W_2} = E_2 - E_1 \qquad (2.2)$$

The quantity $_1Q_2$ is the heat transferred to the system during the process of changing from state 1 to state 2, E_1 and E_2 are the initial and final values of the total energy of the system, and $_1W_2$ represents the work done by the system during the process.

[†]Note the sign convention for Q and W. The quantity Q is positive when heat is transferred to the system and W is positive when work is done by the system on its surroundings.

Sec. 2.1   The Basic Laws of Thermodynamics

Note that Eq. (2.2) implies that heat and work are being expressed in the same units (e.g., joules or Btu or ft · lb). Where separate heat and mechanical energy units are used, the mechanical energy, W, must be divided by Joule's constant (the mechanical equivalent of heat).

The total energy, E, can be manifest in a variety of forms: kinetic energy, potential energy, and chemical energy. Often, the kinetic and potential energy are dealt with separately, and the remainder of the energy of the system is denoted by a single property called the *internal energy* of the system, U'. The total energy would then be written as the sum

$$E = \text{kinetic energy} + \text{potential energy} + \text{internal energy}$$

or

$$E = \text{KE} + \text{PE} + U' \tag{2.3}$$

The motivation for this separation is that the kinetic and potential energy can be evaluated by system macroscopic parameters such as mass, velocity, and elevation. The internal energy, U', which includes all other forms of energy of the system, is a thermodynamic property of the substance being considered.

In most engineering systems we are dealing with an open system (i.e., a system in which there is flow across its boundaries) rather than a closed system. We must then consider the pressure–volume (pV) work done by the system. Our first law energy balance then becomes

$$U'_1 + \text{KE}_1 + \text{PE}_1 + p_1V_1 + {}_1Q_2 = U'_2 + \text{KE}_2 + \text{PE}_2 + p_2V_2 + {}_1W_2 \tag{2.4}$$

The magnitudes of the kinetic and potential energy in most processes are much smaller than those of the other quantities. Hence we may usually write

$$U'_1 + p_1V_1 + {}_1Q_2 \simeq U'_2 + p_2V_2 + {}_1W_2 \tag{2.5}$$

We define the thermodynamic property *enthalpy* as

$$H = U' + pV \tag{2.6}$$

and therefore we have

$$H_1 + {}_1Q_2 \simeq H_2 + {}_1W_2 \tag{2.7}$$

or

$$\Delta H = \Delta Q + {}_1W_2$$

It is in this form that we generally use the first law.

Reviewing the Second Law of Thermodynamics

The second law of thermodynamics is also based on experimental evidence and cannot be derived. There are two statements of the second law:

1. *Kelvin–Planck*: It is not possible to construct a device that will operate in a cycle and produce no effect other than the raising of a weight and the exchange of heat with a single reservoir.

2. *Clausius*: It is not possible to construct a device that operates in a cycle and produces no other effect than the transfer of heat from a cooler body to a hotter body.

The Kelvin–Planck statement implies that it is impossible to build a heat engine that has a thermal efficiency of 100%. The Clausius statement leads to the conclusion that an input of energy is required to transfer heat from a cooler source to a hotter source, such as in refrigeration cycle.

A heat engine operates in a thermodynamic cycle and does net work as a result of heat transfer from a high-temperature source to a working fluid that discards some heat to a low-temperature sink. In a limited way, the simple steam power plant illustrated schematically in Fig. 2.1 is an example of a heat engine. In the figure the heat source is represented by Q_H. This heat can be supplied by the combustion of a fossil fuel or be provided by a nuclear reactor (via the fission process) or a boiler heated from an array of solar collectors. The heat from the energy source is transferred in the boiler to the working fluid. The turbine converts the internal energy of the working fluid into mechanical work. The condenser serves as a heat sink, Q_L, and completes the heat engine cycle. The pump is required to circulate the working fluid through the system. The net mechanical work performed by the system would be the turbine output work less the pump work.

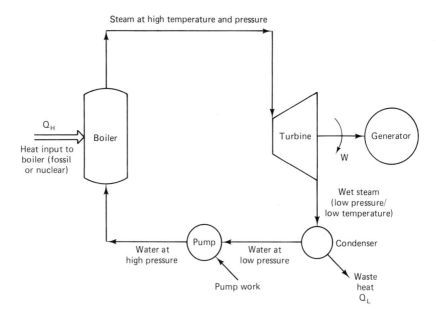

Figure 2.1 Simplified schematic diagram of a steam power station.

Thermodynamic Efficiency

Since a heat engine is incapable of converting all of the heat received into useful work, we define the thermodynamic efficiency of an engine as the fraction of input energy that can be converted to mechanical energy. Thermodynamic efficiency, η_{th} can be defined as

$$\eta_{th} = \frac{\text{work output}}{\text{heat input}} = \frac{W}{Q_2} \qquad (2.8a)$$

When the pumping work is negligible,

$$\eta_{th} = \frac{Q_2 - Q_1}{Q_2} = 1 - \frac{Q_1}{Q_2} \qquad (2.8b)$$

Although simple in concept, predicting the overall efficiency of a large modern plant is a formidable task. Techniques will be developed subsequently for this calculation.

2.2 THERMODYNAMIC PROPERTIES

Internal Energy, Enthalpy, and Entropy

In writing the first-law energy conservation equations, it was found useful to define the internal energy, U', and enthalpy, H, of a system. As defined previously, both H and U' are *extensive properties* of the system since they vary directly with the mass of fluid contained within the system. We may also define the *intensive properties* of specific internal energy, u', and enthalpy, h, as

u' = internal energy per unit mass (specific internal energy)

h = enthalpy per unit mass (specific enthalpy)

When outside influences, such as electrical and magnetic fields, are absent, both u' and h for a given single-phase fluid are defined when temperature and pressure are specified. Both u' and h are independent of the path taken to arrive at a given state and hence are thermodynamic properties.

A third thermodynamic property is based on the previously described notion that the integral of $\delta Q/T$ around a closed reversible thermodynamic path is zero. That is, for a reversible system

$$\oint \frac{dQ}{T}\bigg|_{rev} = 0 \qquad (2.9)$$

If the *entropy*, S, is defined by

$$dS = \frac{dQ}{T}\bigg|_{rev} \qquad (2.10)$$

then, in going from state 1 to state 2, we have

$$S_2 - S_1 = \int_1^2 \frac{dQ}{T}\bigg|_{rev} \qquad (2.11)$$

If we return to state 1 by any reversible path, Eq. (2.10) requires that

$$\int_1^2 \frac{dQ}{T}\bigg|_{rev} = -\int_2^1 \frac{dQ}{T}\bigg|_{rev}$$

and hence we will return to S_1. Thus S is also a thermodynamic property of the system.

As with enthalpy and internal energy, it is useful to define a specific entropy, s, where s is the entropy per unit mass (specific entropy). For a given single-phase fluid, s is specified when temperature and pressure are specified. Note that the units for s are usually (J/kg · °K) or sometimes (J/g · °K) or (Btu/lb_m · °R).

In any adiabatic, reversible process[†] the entropy of the working fluid remains constant since $\int dQ/T$ is zero. It will be seen subsequently that this leads to useful calculational techniques. All real processes are *irreversible* because of such factors as friction, heat transfer through a finite temperature difference, mixing of different substances, and hysteresis effects. However, in a number of adiabatic processes, these irreversibilities are small and the assumption of constant entropy is a good approximation.

Thermodynamic Properties of Permanent Gases

Thermodynamic cycle computations require values of enthalpy, h, entropy, s, and specific volume, v, as a function of temperature, T, and pressure, p. The *Gibbs phase rule* tells us that for a single-phase fluid consisting of a single component, there are 2 degrees of freedom. Hence both pressure and temperature must be specified to establish the state of the fluid. The thermodynamic properties are therefore tabulated as a function of pressure and temperature. Values of u' are not tabulated since these may be derived from h and pV if desired. A tabulation of helium properties is provided since helium is generally considered to be the preferred working fluid for closed-cycle gas turbine applications.

It is conventional to tabulate the thermodynamic properties of air and combustion gases in a manner somewhat different from that used for helium. The tabulation of these properties shown in the Appendix is explained in Chapter 8.

Thermodynamic Properties of Steam

Consider a piston–cylinder system that contains 1 kg of water, as shown in Fig. 2.2(a). Assume that a constant pressure, equal to atmospheric pressure, is maintained throughout the process of adding heat to this system.

[†]A *reversible process* is one such that at the conclusion of the process, both system and surroundings are restored to their initial state without any changes in the rest of the universe.

Sec. 2.2 Thermodynamic Properties

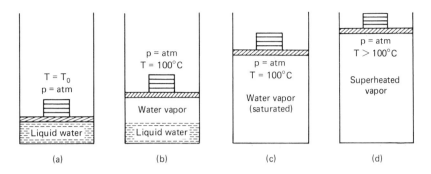

Constant-pressure change from liquid to vapor phase for a pure substance

T_0 = Initial temperature < 100°C

Figure 2.2 Constant-pressure change of a pure substance from the liquid phase to the vapor phase.

If heat is added to water in the liquid state, there will be a slight increase in the volume of the water. As the temperature of the liquid reaches the boiling point, additional transfer of heat to the system results in the formation of the vapor phase as shown in Fig. 2.2(b). Assume that the heat continues to be added while at constant pressure. The temperature will remain constant while the volume increases and until all the liquid phase has vaporized [Fig. 2.2(c)]. If heat continues to be added to the system as shown in Fig. 2.2(d), both the temperature and the volume will increase.

The term *saturation temperature* indicates the temperature at which vaporization takes place at a given pressure. Conversely, this pressure is called the *saturation pressure* for the given temperature. For water, the saturation temperature would be 100°C at standard atmospheric pressure. The temperature at which vaporization takes place is commonly called the *boiling point*. It is a function of pressure on the liquid, increasing as the pressure increases.

If the liquid is at exactly the saturation temperature and pressure, it is called a saturated liquid. Similarly, steam at the saturation conditions is called *saturated steam*. When the pressure is held constant and the temperature is lowered below the saturated liquid temperature, we obtain a *subcooled liquid*. If the temperature is held constant and the pressure is increased beyond the saturation pressure, we have a *compressed liquid*.

When we have a saturated liquid–vapor mixture as in Fig. 2.2(b), the thermodynamic properties of the mixture are derived from the properties of the saturated vapor and liquid. The quality x, of flowing systems is defined by

$$\text{quality} = x = \frac{\text{mass flow rate of vapor}}{\text{total mass flow rate}} \qquad (2.12)$$

The intensive thermodynamic properties of the mixture are obtained from

$$h_m = xh_g + (1-x)h_f$$
$$s_m = xs_g + (1-x)s_f \qquad (2.13)$$
$$v_m = xv_g + (1-x)v_f$$

where the subscripts designate

f = saturated liquid
g = saturated vapor
m = mixture

Table A.4 of the Appendix tabulates the properties of saturated vapor and liquid. Note that Gibbs phase rule indicates that for a two-phase mixture consisting of a single component, there is only 1 degree of freedom. Hence specifying either the saturation temperature or pressure is sufficient to specify the state of a given phase. Saturated phase properties are thus tabulated as a function of pressure only and the corresponding saturation temperature is read from the table. The table also provides values of h_{fg}, the latent heat of vaporization (change in h in going from liquid to vapor). Hence we can also calculate h_m from

$$h_m = h_f + xh_{fg} \qquad (2.14)$$

When we have only vapor at the saturation temperature, this is referred to as *dry and saturated steam*. When the vapor is heated to a temperature above the saturation temperature, we have *superheated steam*. Properties of superheated steam useful in analysis of power generation equipment are provided in the superheated steam table (Table A.5) in the Appendix.

A convenient graphical representation of the properties of superheated steam and steam vapor mixtures is the *Mollier diagram*. In such a diagram (see the simplified version in Table A.3 of the Appendix) enthalpy is the ordinate and entropy is the abscissa. Curves of constant pressure, enthalpy, and temperature are shown on the diagram. In addition, curves of constant moisture content are indicated.

The critical point for water occurs at a pressure of 221.19 bar (3209.2 psia) and 1096.4°C (2005.5°F). If the steam pressure is increased beyond this point, reference to saturated or superheated conditions are not meaningful. The overall behavior is usually considered to be that of a *fluid*, or *supercritical fluid*, rather than either vapor or liquid. The enthalpy of evaporation (h_{fg}) drops to zero and heat added to the fluid will produce a continuous increase in temperature. Changes do occur in the temperature-dependent properties, such as specific heat, viscosity, thermal conductivity, and specific volume, so that if the temperature is increased above the critical, these properties become more like that of the usual gas.

The operating pressure of conventional boilers has increased steadily over the years and some boilers are now designed to operate in the supercritical region. However, the majority of fossil-fuel-fired boilers and all nuclear power plant boilers

Sec. 2.3   Thermodynamic Power Cycles   45

operate in the subcritical region. In supercritical pressure boilers, there is really no boiling in the classical sense and no latent heat effect. The fluid temperature simply increases as heat is transferred to it from the combustion gases.

2.3 THERMODYNAMIC POWER CYCLES

The Carnot Cycle

The *Carnot cycle* is an ideal cycle used to compare heat engine performance and calculate the maximum theoretical efficiency for a heat engine operating between a high-temperature reservoir at temperature T_1, and a low-temperature reservoir at temperature T_2. The Carnot cycle serves as a useful comparison because the Carnot cycle is similar in many ways to a simple steam power cycle.

A simplified view of the Carnot cycle and the corresponding temperature entropy diagram is shown in Fig. 2.3. The Carnot cycle consists of four basic processes:

1. A reversible isothermal process in which heat is transferred from a high-temperature reservoir (1 → 2)
2. A reversible adiabatic process in which the temperature of the working fluid decreases from the high temperature to the low temperature (2 → 3)
3. A reversible isothermal process in which heat is transferred to or from the low-temperature reservoir (3 → 4)
4. A reversible adiabatic process in which the temperature of the working fluid increases from the low temperature to the high temperature (4 → 1)

Since each of the individual processes above is reversible, the complete cycle is reversible. The heat engine becomes a refrigerator by reversing each process.

We defined the thermodynamic efficiency, η_{th}, as the ratio of the work obtained to heat supplied. Hence, with use of the notation of Fig. 2.3(a), we have

$$\eta_{th} = \frac{W}{Q_H} \tag{2.15}$$

where Q_H is the heat transferred from the source. However, from the first law,

$$W = Q_H - Q_L \tag{2.16}$$

where Q_L is the heat transferred to the low-temperature reservoir (sink). Therefore,

$$\eta_{th} = \frac{Q_H - Q_L}{Q_H} = 1 - \frac{Q_L}{Q_H} \tag{2.17}$$

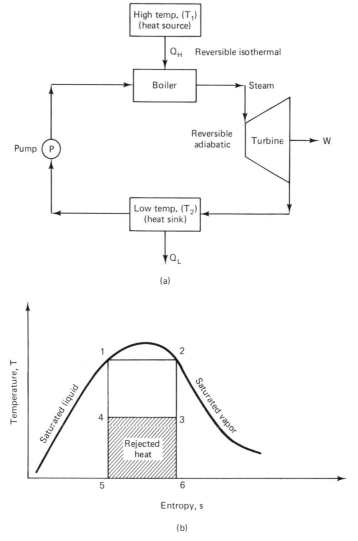

Figure 2.3 (a) Components of an ideal Carnot cycle; (b) temperature–entropy diagram for a Carnot cycle.

It may be shown that:

1. All such reversible engines operating between given temperature levels have the same efficiency.
2. This efficiency is a maximum.
3. The efficiency depends only on the source and sink temperatures.

Sec. 2.3   Thermodynamic Power Cycles

(This statement is called *Carnot's principle*.) Such a heat engine may therefore be used as a basis for an absolute scale of temperature since its efficiency depends only on the temperatures used and is independent of the working fluid. Lord Kelvin arbitrarily defined an absolute temperature scale such that

$$\frac{T_L}{T_H} = \frac{Q_L}{Q_H} \tag{2.18}$$

where T_L and T_H are the temperatures of heat sink and source, respectively. That is, the scale of temperature is such that the ratio of any two temperatures on it equals the ratio of heats absorbed and rejected by a reversible heat engine. Hence

$$\eta_{th} = 1 - \frac{T_L}{T_H} \tag{2.19}$$

If we operate a Carnot engine using an ideal gas ($PV = nRT^*$) as the working fluid, we find that

$$\eta_{th} = 1 - \frac{T_L^*}{T_H^*} \tag{2.20}$$

We therefore conclude that the absolute temperature scale defined by Eq. (2.18) is identical to the ideal gas scale ($T \equiv T^*$).

Equation (2.18) will hold for any arbitrary scaling factor. Hence a variety of absolute temperature scales may be used and R chosen to fit the selected scaling factor. The two most important absolute temperature scales are the Kelvin scale based on the Celsius scale ($T°K = °C + 273.16°$) and the Rankine scale based on the common Fahrenheit scale ($T°R = °F + 459.4°$).

Note that the temperature–entropy diagram of Fig. 2.3(b) provides a graphic representation of the Carnot cycle efficiency. The heat, Q_H, isothermally and reversibly absorbed from the high-temperature reservoir equals $T_H(S_2 - S_1)$ and hence is represented by the rectangle 1–2—6–5. Similarly, the heat rejected isothermally and reversably rejected at the low-temperature reservoir is given by $T_L(S_2 - S_1)$ and represented by the rectangle 3–4–5–6. The cycle efficiency is then the ratio of area 1–2–3–4 to area 1–2–6–5.

The Rankine Cycle

A steam engine operating on a Carnot cycle is impractical. After the expansion through the turbine, a precise quantity of heat, Q_L, would have to be removed from the liquid–vapor mixture at constant pressure and temperature, in order to just reach point 4 [Fig. 2.3(b)]. From point 4, an adiabatic compression of the liquid–vapor mixture would be required. This is impractical and modern closed-cycle steam plants condense the liquid–vapor mixture of point 3 at constant pressure. The pure liquid is then adiabatically compressed by the feedwater pump.

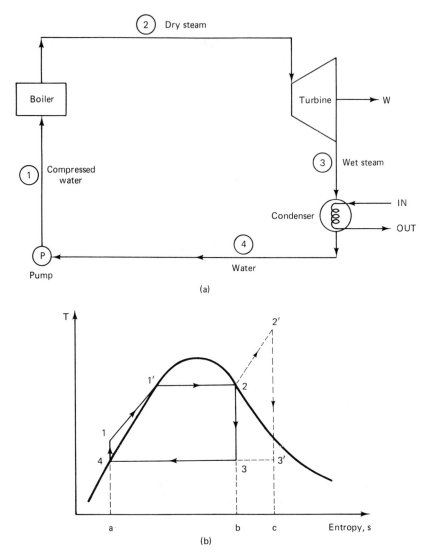

Figure 2.4 (a) Simple Rankine cycle; (b) temperature–entropy diagram for a simple Rankine cycle with superheat.

The Rankine cycle, shown in Fig. 2.4, is the ideal cycle that best represents a simple steam power plant. The processes that comprise the cycle include:

1. A reversible adiabatic pumping process, 4–1
2. Constant-pressure heat addition in the boiler, 1–1'–2
3. Reversible adiabatic expansion in the turbine, 2–3
4. Constant-pressure heat removal from the condenser, 3–4

Sec. 2.3 Thermodynamic Power Cycles

At point 2, the continued addition of heat would result in superheated vapor. The cycle would follow the dashed-line portion of the diagram to points 2′ and 3′.

In analyzing the Rankine cycle, it is helpful to think of the efficiency as depending on the average temperature at which the heat is supplied, T_1, and the average temperature at which the heat is rejected, T_2. Any changes that increase T_1 or decrease T_2 will result in increased system efficiency. The heat transfer and work can be related to the various areas of the T–S diagram. The heat transferred to the working fluid is represented by the area a–1–1′–2–b–a with no superheat or the area a–1–1′–2–2′–c–a in the case of superheat. The heat transferred from the working fluid is represented by respective areas a–4–3–b–a or a–4–3′–c–a. Since the first law defines the work done as the difference between the two corresponding areas of the T–S diagram, this would be 4–1–1′–2–3–4 for the case of no superheat. The thermodynamic efficiency for the Rankine cycle (no superheat) is therefore given by

$$\eta_{th} = \frac{\text{net work output}}{\text{total heat input}} = \frac{\text{area 4–1–1′–2–3–4}}{\text{area a–1–1′–2–b–a}} \qquad (2.21)$$

The Rankine cycle has a lower thermodynamic efficiency than a Carnot cycle operating between the same maximum and minimum temperatures. A major reason for this is the practical need to condense the mixture of water and steam into liquid before entering the pump. This reduces the average temperature of the cycle and hence lowers the thermodynamic efficiency. This is an important feature of the Rankine cycle since the compression is confined to the liquid phase only, thus avoiding the mechanical difficulties in building a device that will handle a mixture of liquid and vapor.

As noted above, the efficiency of a basic Rankine cycle can be surpassed by the use of superheat. This occurs because the ratio of the additional heat added (area 2–2′–3′–3) to the additional heat rejected (b–3′–c–b) is greater than the original Q_H/Q_L ratio. Further improvements in efficiency are possible by using the reheat and regenerative heating concepts described in Section 2.4.

For the basic Rankine cycle as shown in Fig. 2.4(a), the turbine work can be found by applying the form of the first law applicable to flow systems [Eq. (2.7)] to the turbine. Since the turbine is adiabatic, $\Delta Q = 0$, and

$$W_T = \dot{m}(h_2 - h_3) \qquad (2.22)$$

where $\dot{m}$ is the steam flow rate, kg/time. Similarly, the pump work is

$$W_P = \dot{m}(h_1 - h_4) \qquad (2.23)$$

The net work, W, of the cycle is the difference between the turbine work and the pump work, or

$$W = W_T - W_P = \dot{m}(h_2 - h_3) - \dot{m}(h_1 - h_4)$$
$$W = \dot{m}[(h_2 - h_3) - (h_1 - h_4)] \quad (2.24)$$

The heat is supplied to the boiler at approximately constant pressure and no work is being done. Hence

$$Q_H = \dot{m}(h_2 - h_1) \quad (2.25)$$

Since the cycle efficiency was defined previously by Eq. (2.15) as net work output divided by heat input, the efficiency, η_{th}, of the Rankine cycle is

$$\eta_{th} = \frac{W}{Q_H} = \frac{\dot{m}[(h_2 - h_3) - (h_1 - h_4)]}{\dot{m}(h_2 - h_1)} = \frac{(h_2 - h_3) - (h_1 - h_4)}{h_2 - h_1} \quad (2.26)$$

In cases where the pump work is small compared to the turbine work, Eq. (2.26) reduces to

$$\eta_{th} \simeq \frac{h_2 - h_3}{h_2 - h_1} \quad (\text{for } W_P \ll W_T) \quad (2.27)$$

The Brayton Cycle

The *Brayton cycle* is a simple, practical cycle used when permanent gases are the working fluids. The cycle may be used as an open or direct cycle [Fig. 2.5(a)] or as a closed system [Fig. 2.5(b)].

The Brayton cycle is most often used in conjunction with natural-gas-burning systems. A gas-cooled nuclear reactor could also serve as the heat source. The power transfer in this cycle is provided by a gas turbine rather than a steam turbine. Commercial gas turbines are presently designed for temperatures in the range of 1000°C. The temperature limit is expected to increase to 1600°C as advanced technology from jet engine development becomes available. Even with this high turbine inlet temperature, Brayton cycle efficiencies are only in the range 20 to 30%. Nevertheless, low capital cost of gas turbines, in terms of dollars per kilowatthour, make their use attractive for peaking power units.

The open Brayton cycle, shown in Fig. 2.5(a) operates as follows:

1. Air enters the compressor at atmospheric pressure, p_1, and temperature T_1. It is compressed adiabatically in the turbine to a higher pressure, p_2, and temperature, T_2.
2. Heat is added from the combuster or reactor at constant pressure, increasing the temperature to T_3.
3. The gas expands adiabatically through the turbine, reducing the temperature to T_4.
4. The hot gas is exhausted to the atmosphere.

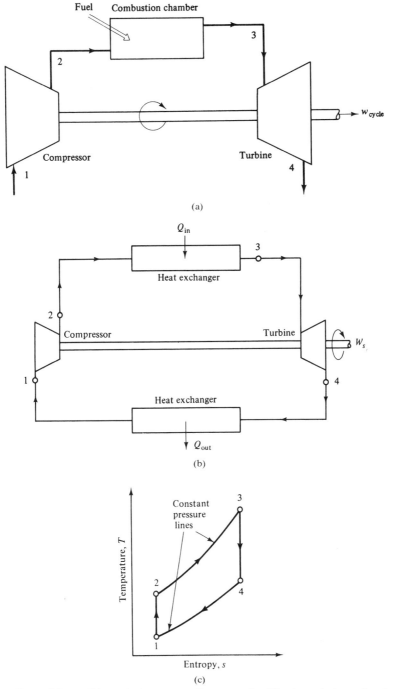

Figure 2.5 (a) Schematic for an open Brayton cycle; (b) schematic for a closed Brayton cycle; (c) T–s diagram for the Brayton cycle process. (From B. V. Karlekar, *Thermodynamics for Engineers*, © 1983, pp. 199, 324. Reprinted by permission of Prentice-Hall, Inc., Englewood Cliffs, N.J.)

A gas-cooled nuclear reactor would operate on a closed Brayton cycle [Fig. 2.5(b)] so as to prevent release of trace quantities of radionuclides to the environment. Note that the simple cycles shown in Fig. 2.5(a) and (b) would have identical efficiencies if the temperature of the sources and sinks were the same and the additional mass added in the combustor was negligible.

The net work output of the Brayton cycle is the turbine work output minus the compressor work input [see Fig. 2.5(c)].

$$W = W_T - W_C = \dot{m}(h_3 - h_4) - \dot{m}(h_2 - h_1) \qquad (2.28)$$

Since $dh = c_p\, dt$, we have for the case of constant c_p,

$$W_T = \dot{m} c_p (T_3 - T_4) \qquad (2.29)$$

where c_p is specific heat at constant pressure. For an ideal gas, we make use of the fact that in an adiabatic compression the ratio of the compressed gas temperature to the original temperature is

$$\frac{T_c}{T_0} = r_p^{(\gamma-1)/\gamma} \qquad (2.30)$$

where r_p = turbine pressure ratio, p_3/p_4
$\gamma = c_p/c_v$ (c_v = specific heat at constant volume)

Hence

$$W_T = \dot{m} c_p T_3 \left[1 - \frac{1}{r_p^{(\gamma-1)/\gamma}} \right] \qquad (2.31)$$

Similarly, the compressor work is found by

$$W_C = \dot{m} c_p (T_2 - T_1) \qquad (2.32)$$

By again using ideal gas assumptions, and noting that $T_c/T_0 = T_2/T_1$, we have

$$W_C = \dot{m} c_p T_2 \left[1 - \frac{1}{r_p^{(\gamma-1)/\gamma}} \right] \qquad (2.33)$$

The net work or the Brayton cycle becomes

$$W = W_T - W_C = [\dot{m} c_p (T_3 - T_2)] \left[1 - \frac{1}{r_p^{(\gamma-1)/\gamma}} \right] \qquad (2.34)$$

In order to compute efficiency, the net heat input to the gas must be obtained. This is simply

$$Q_H = \dot{m} c_p (T_3 - T_2) \qquad (2.35)$$

and the cycle efficiency, calculated by the ratio of W/Q_H, is

$$\eta_{th} = 1 - \frac{1}{r_p^{(\gamma-1)/\gamma}} \qquad (2.36)$$

2.4 STEAM CYCLES FOR MODERN POWER PLANTS

Reheat and Superheat

Production of power by the basic Rankine cycle is practical but relatively inefficient. In a modern steam plant, a number of modifications are made to this basic cycle to improve the efficiency. We have previously indicated that use of superheated steam provides significant improvements in efficiency. Adding heat to the cycle at the highest possible average temperature is the optimum way to raise cycle efficiency. In usual fossil-fueled plant practice, superheat is combined with *reheat*.

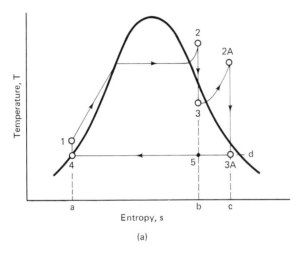

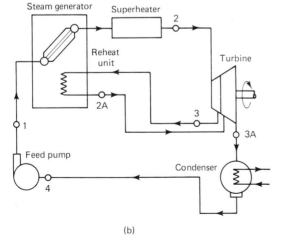

Figure 2.6 (a) *T–s* diagram for a Rankine cycle with superheat and reheat; (b) equipment schematic for a Rankine cycle with superheat and reheat. (Adapted from B. V. Karlekar, *Thermodynamics for Engineers*, © 1983, pp. 324–325. By permission of Prentice-Hall, Inc., Englewood Cliffs, N.J.)

When steam reheat is used, superheated vapor is partly expanded through the turbine. It is then returned to the boiler, where it is "reheated" to superheated vapor. The superheated steam is then returned to the turbine.

Figure 2.6 shows a simplified schematic diagram of a steam cycle that includes both superheat and reheat. A major reason for use of the reheat procedure is that it reduces the amount of moisture in the last stages of the turbine. This reduces erosion of the turbine blades and thus prolongs turbine lifetime.

The effect of reheat on cycle efficiency is best illustrated in the temperature–entropy diagram of Fig. 2.6(a). If we assume an ideal reversible cycle, then following our previous reasoning, the actual work obtained due to the reheat is represented by area $A_1 = 3-2A-3A-5-3$, while the heat added during reheat is represented by area $A_2 = 3-2A-c-b-3$. Since the ratio of the A_1/A_2 is greater than the ratio for the cycle without reheat, the cycle efficiency is significantly improved.

The same diagram may be used to quantify the effect of reheat on end-point moisture. Since the entropy of a steam–water mixture is $xs_g + (1-x)s_f$, the exit quality may be determined by the position of the exit point relative to the saturated liquid and vapor lines. Thus, without reheat, the fraction of moisture would be given by the distance ratio $(d-5)/(d-4)$, while with reheat it is reduced to $(d-3A)/(d-4)$.

Regenerative Feedwater Heating

A significant fraction of boiler heat input is required to heat the feedwater if it enters directly from the condenser. Further improvement in Rankine cycle efficiency can be obtained if the turbine steam is used to preheat the entering feedwater. Consider the hypothetical regenerative cycle in which the liquid leaving the condenser is compressed and then heated in an annular space around the turbine while flowing in the direction opposite to the steam flow. If the heat-transfer area is large enough, the liquid would always be facing vapor at the same temperature on the other side of the casing. It would then receive heat reversibly from the steam all along the path. The cycle, illustrated in Fig. 2.7(b), would have the temperature–entropy diagram of Fig. 2.7(a). The curves 1–2–3 and 5–4 are parallel; hence the distance 1–1' and 5–5' are equal. Thus the work obtained (area 1–2–3–4–5–1) is identical to that from an ideal Carnot cycle operating between the same temperatures (Carnot work = area 1'–3–4–5'–1'). Since the heat added (area a–2–3–4–5–c–a) is identical in both cycles, the hypothetical regenerative Rankine cycle has the same efficiency as an ideal Carnot cycle.

In actuality, it is not practical to build a regenerative cycle of the type described. The turbine casing would become unacceptably complicated. More important, the moisture content of the exit steam would be drastically increased. Although the hypothetical cycle cannot be built, we could closely simulate it by drawing off at each stage an amount of steam just equal to that which could be condensed by a water annulus in the hypothetical cycle. These steam flows could then be used in a series of external counterflow heat exchangers which reversibly

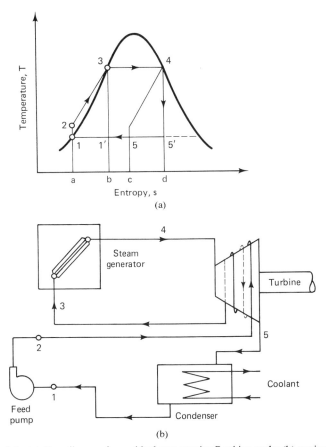

Figure 2.7 (a) $T-s$ diagram for an ideal regenerative Rankine cycle; (b) equipment schematic for an ideal regenerative Rankine cycle. (Adapted from B. V. Karlekar, *Thermodynamics for Engineers*, © 1983, p. 370. By permission of Prentice-Hall, Inc., Englewood Cliffs, N.J.)

heated the feedwater. This design would have essentially the same thermal efficiency as the cycle illustrated in Fig. 2.7. Of course, we cannot afford to have steam draw-offs at each stage nor to build heat exchangers with very large areas so that the $\Delta t \to 0$. We must then actually obtain our regenerative heating with a limited number of draw-offs and heat exchangers of reasonable size. Although we then cannot achieve Carnot cycle efficiencies, a significant improvement in efficiency is obtained.

The actual efficiency is obtained by performing a *heat balance* for the actual cycle. Consider the simplified cycle of Fig. 2.8 in which only two regenerative heaters are used. Our thermal efficiency is still obtained from

$$\eta_{\text{th}} = \frac{\text{net work output } (W)}{\text{heat added to the system } (Q_H)} = \frac{W_T - W_P}{Q_H}$$

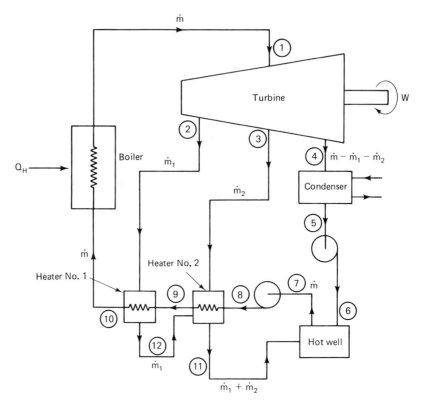

Figure 2.8 Regenerative power cycle.

For this simple regenerative cycle of Fig. 2.8, the net turbine work W_T and the heat added, Q_H, are

$$W_T = \dot{m}(h_1 - h_2) + (\dot{m} - \dot{m}_1)(h_2 - h_3)$$
$$+ (\dot{m} - \dot{m}_1 - \dot{m}_2)(h_3 - h_4) \qquad (2.37)$$
$$Q_H = \dot{m}(h_1 - h_{10})$$

Modern fossil power plants utilize several feedwater heaters. The optimum number of feedwater heaters can be found by balancing the increase in efficiency against the increase in capital cost of the system.

Condensation at Low Pressures

If we examine the T–s diagram for the Rankine cycle shown in Fig. 2.6, it is clear that we will increase the system efficiency by reducing the temperature at which the steam is condensed. Modern steam plants therefore operate their condensers at as low a temperature as possible. The allowable condensation temperature is set by the

Sec. 2.4 Steam Cycles for Modern Power Plants

temperature of the cooling water plus a reasonable temperature drop across the coolant films and condenser tubing. Such low condensing temperatues mean that the condenser must operate at pressure well below atmospheric. In practice, a vacuum system must be provided to maintain this pressure by removing air which unavoidably leaks into the condenser.

Types of steam cycles

Cycles for fossil plants. A simplified schematic diagram for a modern coal-fired steam plant is shown in Fig. 2.9. As in this typical design, regeneration, superheat, and reheat are used in almost all large steam plants to improve overall cycle efficiency. Typical steam conditions for an advanced design (1980) are 24.2 MN/m^2 (3500 lb/in.^{2}) and 810°K (1000°F) with reheat to 810°K (1000°F) and final feedwater temperature of 540°K (510°F). Thermal efficiencies for such plants are in the range of 40%. Turbine temperature limitations, currently about 810°K (1000°F), prevent achievement of higher efficiencies.

Cycles for nuclear plants. The basic differences between the steam cycles for nuclear power plants and fossil-fired power plants are in the source of heat and the steam conditions. The nuclear fission process provides heat instead of the combustion of a fossil fuel. Steam is supplied to the turbines at considerably lower temperatures than in fossil-fueled plants. As will be shown later, steam conditions

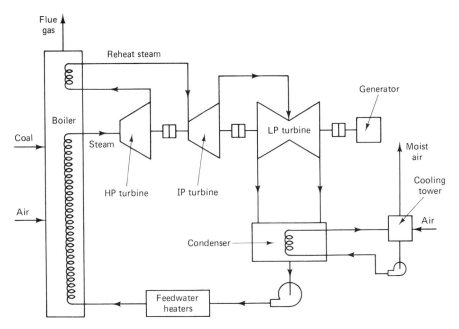

Figure 2.9 Basic components of a coal-fired power station.

vary between different types of nuclear steam supply systems. However, in the most common commercial nuclear steam supply systems, the steam delivered to the high-pressure turbine has at most 35°C of superheat.

The expansion of the nearly saturated supply steam through the turbine results in a wet steam flow through most of the high-pressure and low-pressure sections. These turbines must therefore be designed to handle large quantities of condensed moisture. Since high moisture content both increases turbine blade erosion and reduces expansion efficiency, action is required to reduce the moisture present. Most systems incorporate some type of moisture separator or separator plus reheater. These aspects of system design are included in Chapter 8.

Nuclear power stations have a system thermal efficiency of up to 34%. The lower efficiency is due to the fact that steam is supplied to the system at pressures of only 900 to 1100 psi. A combination of nuclear fuel limitations and economic considerations lead to the current choice of operating conditions.

Future cycles. Some consideration is being given to the use of multipurpose steam plants. In one such scheme, steam would be extracted from the turbine and used for industrial processes. In other proposals, waste heat would be used for district heating. Other schemes call for the addition of a magnetohydrodynamic topping cycle. Some designers believe that with careful utilization of all the heat, overall thermal utilization (not efficiency) could reach as high as 60%.

Simple Steam Cycle Calculations

The operation of an *ideal* turbine is adiabatic and hence the entropy, s, of the working fluid may be taken as constant. If the temperature and pressure of the superheated steam entering a given segment of an ideal turbine is known, the enthalpy, h, and the entropy, s, of the steam may be determined from the superheated steam tables (Tables A.4 and A.5). For a given exit pressure from the turbine section, we search the superheated steam tables at the given exit pressure and by interpolation determine the exit steam temperature (and associated enthalpy, h_2) which corresponds to the inlet enthalpy. More extensive tables of steam properties in SI units are given in the tables edited by Irvine and Hartnet (see the Bibliography at the end of the chapter).

Example 2.1

Assume that the steam entering a given turbine section is at 750°C and a pressure of 100 bar. If the pressure of the exit of the turbine is 60 bar, what is the corresponding steam enthalpy and temperature?

Solution From the superheated steam table we find that at 100 bar and 750°C that $s = 7.289$ kJ/kg · °K, and $h = 3993$ kJ/kg. We then search the superheated steam table listings at 60 bar and find

$T = 600°C$ $h = 3658$ kJ/kg $s = 7.165$ kJ/kg · °K
$T = 700°C$ $h = 3893$ kJ/kg $s = 7.422$ kJ/kg · °K

Sec. 2.4 Steam Cycles for Modern Power Plants

By linear interpolation, we find that the steam entropy will be 7.289 kJ/kg at $T = 648°C$. Again by interpolation, we find that the enthalpy corresponding to this temperature is 3772 kJ/kg.

We may find that there is no steam temperature above saturation corresponding to the entering entropy for the given exit pressure, p_2. This indicates that what we actually have is wet steam containing both liquid and vapor at the saturation temperature.

Example 2.2

Assume that the steam entering a turbine section is at the conditions of Example 2.1 but that the exit pressure is now 1.0 bar. Determine the exit steam temperature and enthalpy.

Solution When we examine the superheated steam tables at 1.0 bar, we see that the entropy $s_2 = s_3 = 7.289$ kJ/kg·°K is below the entropy of saturated vapor (7.361 kJ/kg·°K). Our steam temperature is at saturation (100°C). Therefore, we now write

$$s_2 = x_3 s_g + (1 - x_3) s_f$$

where x_3 = steam quality at turbine section exit = weight of steam vapor/total weight of vapor and liquid

s_f, s_g = entropy of saturated liquid and vapor, respectively, determined from saturated steam tables for exit pressure p_2

After obtaining s_f and s_g, from the saturated steam properties ($s_f = 1.3026$ and $s_g = 7.36$ kJ/kg·°K) we solve for x_3 ($x_3 = 0.987$) and then write

$$h_3 = x_3 h_g + (1 - x_3) h_f$$

where h_f and h_g are the enthalpy of saturated liquid and vapor, respectively, determined from saturated steam tables for exit pressure p_2. With the value of x_3 just determined, we determine h_3, the enthalpy of the exit steam.

$$h_3 = 0.987(2675) + 0.013(417.4) = 2645 \text{ kJ/kg}$$

Example 2.3

Determine the ideal thermodynamic efficiency of a simple Rankine cycle, as illustrated by Fig. 2.4, if turbine inlet and outlet conditions are as specified in Example 2.2.

Solution

$$\eta_{th} \simeq \frac{h_2 - h_3}{h_2 - h_1}$$

We take h_1 as being approximately equal to the enthalpy of saturated liquid leaving the condenser. (The enthalpy will be only very slightly changed by the pump since the liquid is almost incompressible.) We then have

$$\eta_{th} = \frac{3993 - 2645 \text{ (kJ/kg)}}{3993 - 417.4 \text{ (kJ/kg)}} = 0.379$$

SYMBOLS

A	area, m^2
c_p	specific heat at constant pressure, kJ/kg · °C
c_v	specific heat at constant volume, kJ/kg · °C
E	total energy of the system, J
h	specific enthalpy, J/kg
H	enthalpy of system, J
KE	kinetic energy, J
$\dot{m}$	flow rate, kg/s
n	number of moles
p	pressure, Pa (N/m^2) or bar
PE	potential energy, J
Q	heat transferred across the boundary of a system, J
Q_H	heat transferred from the source
Q_L	heat transferred to the sink or low-temperature reservoir, J
r_p	pressure ratio across turbine, p_3/p_4
R	gas constant
s	specific entropy, J/(kg · °K)
S	entropy of a system, J/°K
T	temperature, °K (unless specified otherwise)
T_c	temperature of the compressed gas, °K
T_H	temperature of the heat source, °K
T_L	temperature of the heat sink, °K
u'	specific internal energy, J/kg
U'	internal energy of a system, J
v	specific volume, m^3/kg
V	volume of system, m^3
W	work, J
x	quality = mass flow rate vapor/total mass flow rate

Greek Symbols

γ	heat capacity ratio, c_p/c_v
η_{th}	thermodynamic efficiency of the system

Subscripts

0	initial or original condition
1, 2, ...	state or condition 1, 2, ... of the system

Chap. 2   Problems

	C	compressor
	f	saturated liquid
	fg	change in going from saturated liquid to saturated vapor
	g	saturated vapor
	m	mixture
	P	pump
	T	turbine

PROBLEMS

2.1. Use the steam tables (Tables A.4 and A.5 in the Appendix) to determine the change in internal energy when 1 kg of water is vaporized at 10 bar.

2.2. Determine the rate of heat input required to superheat 10,000 kg/h of saturated steam at 100 bar by 320°C.

2.3. An inventor claims to have devised a highly efficient heat pump. He claims that his heat pump, which derives its energy from outside air, is capable of providing 1000 W of heat to a 20°C room while requiring only 100 W of work (electrical energy) when the ambient air is at −15°C. Show quantitatively whether or not this claim is reasonable.

2.4. (a) Calculate the maximum possible efficiency of a simple Rankine-cycle high-pressure steam turbine which operates under the following conditions: inlet, 110-bar saturated steam and outlet, 35-bar wet steam.

(b) What is the maximum efficiency to be expected if the steam at the inlet were superheated by 200°C?

2.5. (a) A closed Brayton cycle is being designed for an advanced nuclear reactor system. Materials limitations set the maximum reactor coolant exit temperature at 550°C. The gas leaving the turbine is cooled in an air cooler to 150°C. If the working fluid may be considered to be an ideal gas with $\gamma = 1.66$ (value of γ for monatomic gas), what is the maximum cycle efficiency possible?

(b) For the same temperatures as indicated in part (a), what would the maximum thermodynamic efficiency be if the working fluid had a $\gamma = 1.4$ (value for diatomic gas)?

2.6. A fossil-fueled steam plant provides 30,000 kg/h of steam at 160 bar, which is superheated to 175°C above saturation. The steam flows to the high-pressure stage of a turbine which exhausts to a reheater.

(a) At what pressure should the exhaust be set if the moisture content of the steam, upstream of the heater, is not to exceed 5%?

(b) If the turbine may be assumed to be ideal, how many megawatts of power can the turbine provide?

2.7. (a) A boiling-water reactor produces saturated steam at 85 bar. The steam flows to a turbine exhausting at 0.3 bar. To prevent excessive end-point moisture in the turbine, moisture separators are incorporated at 40 bar and 15 bar. Estimate the maximum possible thermodynamic efficiency of such a system.

(b) Estimate the increase in efficiency of the foregoing system if a single feedwater heater, using steam at 25 bar, were incorporated. The temperature of the feedwater leaving the heater is to be 15°C below the temperature of the heating steam.

BIBLIOGRAPHY

Combustion, 3rd ed. Windsor, Conn.: Combustion Engineering, Inc., 1981.

EL-WAKIL, M., *Nuclear Power Engineering*. New York: McGraw-Hill, 1962.

FRAAS, A., *Engineering Evaluation of Energy Systems*. New York: McGraw-Hill, 1982.

IRVINE, T. F., and J. P. HARTNETT (eds.), *Steam and Air Tables in SI Units*. New York: Hemisphere, 1976.

Steam, Its Generation and Use. Lynchburg, Va.: The Babcock & Wilcox Company, 1978.

VAN WYLEN, G., and R. SONNTAG, *Fundamentals of Classical Thermodynamics*. New York: Wiley, 1978.

WEISMAN, J. (ed.), *Elements of Nuclear Reactor Design*, 2nd ed. Melbourne, Fla.: R. E. Krieger, 1983.

ZEMANSKY, M., and H. VAN NESS, *Basic Engineering Thermodynamics*. New York: McGraw-Hill, 1976.

3

Fossil Fuels

3.1 PETROLEUM AND SHALE OIL

Formation

Petroleum is believed to have been formed many millions of years ago from marine plankton deposited on undersea beds in still water. The plankton were subsequently covered by a fine-grained mud that produced an anaerobic condition preventing rapid decay. Slow decomposition by anaerobic bacteria converted the plankton into sapropel, an amorphous organic material. The slow decay process is estimated to have occurred over millions of years as the seabeds and decaying plankton were covered by many layers of deposits. The resulting environmental conditions, high pressure and moderate temperature, produced petroleum. The composition of the petroleum produced depends somewhat on the age of the deposits. Older deposits of oil, such as some of the Middle Eastern deposits, tend to have a greater proportion of lighter hydrocarbons.

In nearly all cases, the initial deposit of fine-grained mud, which later became sedimentary rock, was eventually covered with coarser-grained sand and silt. The sedimentary rock in which the oil formation process occurred is called the *source rock*. For some reason not entirely clear to geologists, the petroleum generally migrated from the tightly packed sedimentary rock to porous and permeable rock formed from the coarse-grained sediments. This rock formation, from which the oil can flow at appreciable rates, is referred to as the *reservoir rock*.

The migration of oil through the reservoir rock continues until some barrier prevents further migration. Because of its porous structure, the reservoir rock

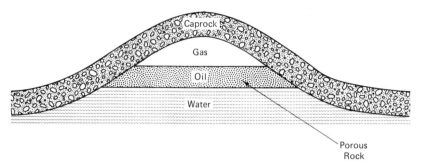

Figure 3.1 Simplified view of an oil trap.

cannot act as a barrier. Where usable petroleum concentrates are found, a *barrier base*, usually in the form of a fine-grained layer of silt, eventually covered the coarse sand. The fine-grained silt solidified and prevented further migration. The barrier is known as a *cap rock*. Unless a barrier such as a cap rock is present, accumulation is limited as the oil will seep to the surface. In the very early days of oil discovery, ground seepage was one of the principal methods of locating oil.

The combination of the reservoir rock and cap rock form an *oil trap*. The oil trap contains natural gas (the more volatile products of organic decomposition), petroleum, and water, usually under some pressure. Figure 3.1 is a very simplified view of such an oil trap. Although the figure appears to show the natural gas, oil, and water in discrete *pools*, this is not the condition that normally exists in the wells from which petroleum is extracted. The natural gas, oil, and water are trapped in sedimentary or porous rock in a relative concentration, as indicated in Fig. 3.1. There is rarely a "pool" of liquid petroleum.

When a drill bit penetrates the cap rock, the high pressure exerted by the trapped oil and gas often results in the rapid flow of oil or gas from the well, known popularly as a "gusher." This pressure can be as high as 700 bar (10,000 psi). As the oil and gas are extracted from the well, this pressure gradually diminishes and older wells must be pumped to extract the petroleum. Only about one-third of the petroleum trapped underground is recoverable by normal drilling and pumping operations. Many companies are now investigating methods to increase the yield from oil wells. One method of enhanced recovery is to increase the water pressure in the well artificially, thus encouraging more oil flow. Other methods include the use of steam or a chemical solvent to release additional trapped oil from the reservoir rock.

Petroleum Chemistry

Petroleum is composed primarily of hydrocarbon compounds with the general formula $H_x(H_2C)_y$. The carbon and hydrogen atoms are arranged in many different ways to form a variety of compounds. The chemistry is further complicated by the presence of hydrocarbon derivatives, which may contain nitrogen, oxygen, or sulfur in addition to hydrogen and carbon.

The bulk of the hydrocarbons in petroleum belong to the paraffin series. These compounds, sometimes called *alkane hydrocarbons*, are a saturated group with the general formula C_nH_{2n+2}. The lighter, gaseous[†] members of the family are

Methane, CH_4
Ethane, C_2H_6
Propane, C_3H_8
Butane, C_4H_{10}

The heavier members of the family are liquids at standard temperature and pressure (STP) conditions and include such fuels as

Pentane, C_5H_{12}
Hexane, C_6H_{14}
Octane, C_8H_{18}

Paraffins can be classified according to the arrangement of the chain of carbon atoms. When the prefix *n* is used in front of the hydrocarbon name, all of the carbon atoms are connected in one long chain. An *n*-pentane molecule would be

$$\begin{array}{c} H\ \ H\ \ H\ \ H\ \ H \\ |\ \ \ |\ \ \ |\ \ \ |\ \ \ | \\ H-C-C-C-C-C-H \\ |\ \ \ |\ \ \ |\ \ \ |\ \ \ | \\ H\ \ H\ \ H\ \ H\ \ H \end{array}$$

or $CH_3-CH_2-CH_2-CH_2-CH_3$.

In addition to the normal-chain paraffin compounds, there are (1) the branched chains (which are often designated as isoparaffins) and (2) cyclic chains (designated as cycloparaffins). Figure 3.2 illustrates these chains plus examples of the aromatic (benzene-like compounds) and the aromatic cycloparaffins which are also found.

Crude oil is classified in one of three categories, depending on the type of residue that remains after the lighter fractions are distilled. The three classifications are paraffin-based crude, asphalt-based crude, or mixed-base crude. Table 3.1 illustrates the composition of a typical paraffin-based crude oil.

The proportion of the various hydrocarbon groups and derivatives found varies widely with geographic location, and substantial variations are seen in composition between U.S. crudes as well as between U.S. and foreign crudes. The elemental analysis, however, is fairly constant. The carbon mass fraction ranges

[†] At STP conditions.

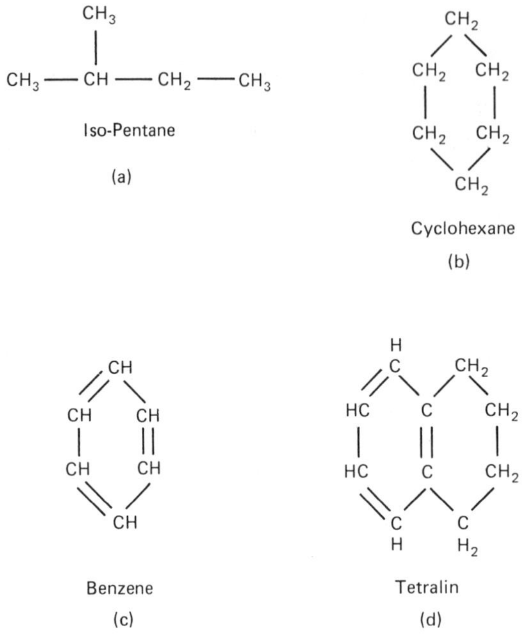

Figure 3.2 Components found in petroleum in addition to straight-chain paraffins: (a) iso- or branched-chain paraffin; (b) cycloparaffin; (c) aromatic; (d) aromatic cycloparaffin.

TABLE 3.1 TYPICAL COMPOSITION (%) OF A PARAFFIN-BASED CRUDE OIL

Normal paraffins	25
Isoparaffins	13
Cycloparaffins	40
Aromatics	7
Aromatic cycloparaffins	8
Balance (resins, asphaltenes)	4
	100

from 83 to 88% and the hydrogen mass fraction from 10 to 16%. It should be noted, however, that the presence of sulfur, nitrogen, and metal compounds usually require processing of crude oil before it is sold as a fuel. Desulfurization, hydrogenation, cracking, and other refining processes may be carried out on selected fractions before they are blended for fuel use.

Petroleum Resources

Resource estimates for conventional petroleum vary widely between reporting groups. Table 3.2 lists typical estimates of world oil reserves prepared by the Oil and Gas Institute. It is clear that although the demand of the world's industrial nations—the

TABLE 3.2 WORLD PETROLEUM RESOURCES (SELECTED COUNTRIES, BILLIONS OF BARRELS OF OIL)

Area	Proven reserves	Additional speculative reserves	Annual production
Western Hemisphere			
Argentina	5.0	10	0.5
Canada	10.0	15	1.5
Columbia	1.5	5	0.3
Ecuador	6.0	10	0.2
Mexico	3.0	20	0.5
United States	40.0	75	8.5
Venezuela	15.0	25	3.5
Europe			
Norway	2.0	5	0.1
United Kingdom	7.5	5	1.5
Poland	8.0	7	1.0
USSR	80.0	100	7.0 (est.)
Middle East			
Abu Dhabi	20.0	15	2.0
Iran	70.0	25	5.0
Iraq	30.0	20	2.0
Kuwait	65.0	40	3.0
Saudi Arabia	150.0	50	5.0–10.0
Africa			
Algeria	47.0	60	1.5
Egypt	6.0	20	0.3
Libya	35.0	40	3.0
Nigeria	15.0	25	2.0

Source: Data from *OCS Oil and Gas*, U.S. Council on Environmental Quality, Washington, D.C., 1979.

United States, Europe, and Japan—is high, these countries have far less resources than the Middle Eastern countries.

Shale Oil

Because of limited petroleum resources, attention is now being given to the processing of *oil shale*. This name is somewhat of a misnomer since the rocks are not shale and do not contain oil. Oil shale is a marlstone deposit interspersed with layers of organic material called *kerogen*. Shale oil was formed by living organisms at the bottom of stillwater lakes about 50 million years ago during a period of about 7 million years. The so-called shale oil is now found in sedimentary rock (a fine-grained structured rock) containing the solid organic material (kerogen).

Kerogen is an organic polymer with a large molecular weight having the approximate composition ($C_{200}H_{300}SN_5O_{11}$). Kerogen decomposes when heated into gases, an oil resembling petroleum, and a carbonaceous residue. About 65% is converted to oil, 10% to gas, and 25% to carbonaceous residue. The ultimate composition of the kerogen is

	w/o
Carbon	80.5
Hydrogen	10.3
Nitrogen	2.4
Sulfur	1.0
Oxygen	5.8

The quality of the shale is usually measured by the number of gallons of oil that can be obtained from a ton of shale. A very high quality shale would yield up to 50 gal/ton. The average yield of "good"-quality shale is 25 gal/ton.

The United States is reported to have vast resources of shale. Resource estimates vary widely because very little detailed exploration has been conducted for shale oil. Most of the resource base is located in the Green River formations of Colorado, Utah, and Wyoming. High-grade shale is located in 17,000 square miles of four major basin areas: Piceance Creek, Uinta, Washakie, and Green River. There is believed to be low-grade shale (less than 25 gal/ton) located in the central United States from Texas to Pennsylvania.

There are two general approaches to the recovery of the oil from shale: (1) mining, crushing, and aboveground retorting and (2) in situ recovery.

In the conventional or *aboveground retorting process*, the shale is crushed and fed to a refractory-lined retort vessel. The crushed shale is heated to 540°C (1000°F) by hot retorting gases as it falls vertically through the vessel. This heating or retorting process yields the oil, gas, and residue fractions previously described. The liquid hydrocarbon products must still be processed in special refineries to be converted to usable liquid fuels.

In the *in situ process*, a void is first created in the shale deposit by some conventional mining method. Explosives are used to fracture the shale just above the void. Combustion is started in the shale rubble and supported by air injected into the well. Liquid and gaseous products are removed from the bottom of the chamber.

Estimates place the in-ground oil shale resources between 600 and 2000×10^9 bbl of oil in deposits of shale at least 10 ft thick. Although a clear incentive exists for development of this resource, relatively high processing costs, as well as environmental problems associated with the mining and disposal of the shale, have prevented large-scale commercial exploitation.

3.2 PROPERTIES OF PETROLEUM FUELS

Crude oil can be used directly as a power plant fuel. In the 1970s, a number of seacoast power plants were modified for use of crude oil. Under the conditions of the 1980s, such use is less common. The high demand for the lighter fractions for transportation uses provides an incentive for processing of the crude oil. In addition, since no processing is performed on the crude oil, no sulfur has been removed and environmental criteria for stack releases may be difficult to meet.

Crude oil is usually distilled into a number of fractions and the lightest (having the lowest boiling point) liquid fractions are used primarily for transportation fuels. The heavier fractions are used for fuels and chemical production. In the United States, commercial refined fuels are divided into five grades. The lightest and least viscous grade, No. 1, is a petroleum distillate, boiling between 215 and 250°C, used for vaporizing pot-type burners. Grade No. 2 is a petroleum distillate with a maximum boiling point of 340°C used for general-purpose domestic heating. The highest boiling and most viscous fuel, No. 6, is largely obtained from the still bottoms. The still bottoms, called *residual fuels*, are diluted with 5 to 20% distillate to reduce the fuel viscosity. Nevertheless, this fuel, which is the petroleum fuel most widely used in power plants, must be heated to 30 to 50°C to reduce the viscosity sufficiently for handling. Further heating to 75 to 95°C is required before atomization.

It should be noted that the marine designation "Bunker C Fuel Oil" is being abandoned. Originally, Bunker C fuel corresponded to the viscous end of No. 6 fuel oil, but now Bunker C is considered an alternative name for No. 6 fuel oil.

By diluting the residual fuel with 20 to 50% distillate, the less viscous No. 4 and No. 5 fuels (there no longer is a No. 3) are produced. These fuels are widely used commercially in schools, apartment houses, and office buildings. Residual fuel is sometimes used directly, without dilution with distillate, as a power plant fuel.

Both No. 6 and residual fuel oil may contain 10 to 500 ppm nickel and vanadium in complex molecules. Small amounts of sand, rust, dirt, and water are also present. Typical ash contents of these fuels are between 0.01 and 0.5% by weight.

Physical Properties

The graded commercial fuel sold in the United States is produced in accordance with the American Society for Testing and Materials (ASTM) specifications which has been adopted by the U.S. Bureau of Standards. The required physical properties are shown in Table 3.3. Note that in addition to the distillation range, specific gravity, and viscosity, pour point and flash point are shown. The pour point is the lowest temperature at which the oil will flow under standard pressure conditions. The flash point is the minimum temperature at which the oil may be ignited.

The specific gravity of fuel oils is usually determined at ambient conditions with specialized hydrometers which read in "degrees API" (API stands for American

TABLE 3.3 SOME PROPERTIES OF COMMERCIAL FUEL OILS AS SPECIFIED BY THE AMERICAN SOCIETY FOR TESTING AND MATERIALS

Fuel oil grade	Description	Flash point (°C) Min.	Pour point (°C) Max.	Kinematic viscosity (cS)				Specific gravity[a] (°API) Min.
				At 38°C		At 50°C		
				Min.	Max.	Min.	Max.	
No. 1	Distillate oil for vaporizing pot-type burners	38	−17	1.4	2.2			35
No. 2	Distillate oil for general-purpose domestic heating	38	−7	2.0	3.8			30
No. 4	Preheating not usually needed for handling or burning	55	−7	5.8	26.4			
No. 5 (light)	Preheating may be needed	55		32	65			
No. 5 (heavy)	Preheating may be needed for burning and required for handling in cold climates	55				42	81	
No. 6	Preheating needed for burning and handling	65				92	633	

[a]Unofficial standard based on trade practice.

Source: Adapted from *ASTM Specification D396*, 1980; copyright ASTM, 1916 Race Street, Philadelphia, PA 19103. Reprinted with permission.

Petroleum Institute). "Degrees API" are inversely related to the specific gravity, s.g., at 60°F by

$$\text{degrees API} = \frac{141.5}{\text{s.g.}} - 131.5 \qquad (3.1)$$

A variety of fuel properties (hydrogen content, heat of combustion, specific heat, etc.) may be related to the specific gravity of the fuel. Hydrogen content may be estimated by

$$H' = a' - 15 \text{ s.g.} \qquad (3.2)$$

where H' = weight percent hydrogen
a' = constant dependent on specific gravity (°API)

a'	°API
24.5	0– 9
25.0	10–20
25.2	21–30
24.45	31–45

The specific heat, c_p, of petroleum liquids between 32 and 400°F can be closely estimated from

$$c_p = \frac{0.388 - 0.0045T}{\text{s.g.}}$$

where T = temperature (°F), s.g. has its previous meaning, and c_p is in (Btu/lb · °F) or (cal/g · °C).

Heating value (heat of combustion). The heat of combustion of a given material is the theoretical amount of heat given off when a unit mass of material is burned at standard conditions (usually 20°C and 1 atm pressure). In most cases the actual combustion cannot occur at a temperature as low as 25°C. Hence, in an experimental determination the fuel is generally ignited by a spark, the reaction allowed to proceed, and then the products of combustion are cooled down to room temperature. Since heat is given off in this process, by the nomenclature of thermodynamics the heat of combustion is negative.

The heating value of a fuel is numerically equal to the standard heat of combustion but with an opposite sign (heating values are positive). The heating value obtained directly from an experiment in which the products of combustion are cooled to room temperature, and hence the water vapor condensed, is called the *higher heating value* (HHV). In actuality not all of this heat will really be available, since in a real combination process the final state of the water will be a vapor. To take this into account, the *lower heating value* (LHV) is defined as the higher heating value less the latent heat of water vaporization (at 25°C) for all the water formed.

TABLE 3.4 HEAT OF COMBUSTION OF FUEL OIL COMPONENTS[a]

Compound	Formula	Heat of combustion	
		kcal/kg mol	kJ/kg mol
Benzene	C_6H_6	−783,400	−3,280,000
Cyclohexane	C_6H_{12}	−939,000	−3,931,000
n-Hexane	C_6H_{14}	−990,600	−4,147,000
2-Methylpentane	C_6H_{14}	−993,700	−4,160,000
Cycloheptane	C_7H_{14}	−1,087,300	−4,552,000
n-Heptane	C_7H_{16}	−1,149,000	−4,810,000
3-Ethylpentane	C_7H_{16}	−1,151,100	−4,819,000
n-Octane	C_8H_{18}	−1,305,000	−5,463,000
2,2-Dimethylhexane	C_8H_{18}	−1,304,600	−5,462,000
2,2,3-Trimethylpentane	C_8H_{18}	−1,305,800	−5,467,000
n-Nonane	C_9H_{20}	−1,464,000	−6,128,000
n-Decane	$C_{10}H_{22}$	−1,620,000	−6,782,000
n-Undecane	$C_{11}H_{24}$	−1,776,000	−7,437,000
n-Dodecane	$C_{12}H_{26}$	−1,933,000	−8,091,000
n-Tridecane	$C_{13}H_{28}$	−2,089,000	−8,745,000
n-Hexadecane	$C_{16}H_{34}$	−2,558,000	−10,708,000
n-Heptadecane	$C_{17}H_{36}$	−2,714,000	−11,362,000
n-Octadecane	$C_{18}H_{38}$	−2,870,000	−12,016,000
n-Nonadecane	$C_{19}H_{40}$	−3,026,000	−12,670,000
n-Eicosane	$C_{20}H_{42}$	−3,183,000	−13,324,000

[a] Initial state—liquid; final products: $CO_2(g)$ and $H_2O(l)$.

If the exact chemical composition of a fuel oil is known, the heat of combustion can be determined from the heat of combustion of its components. Table 3.4 gives the heats of combustion of a number of the chemical compounds found in fuel oils.

When the chemical composition of a fuel oil is not known, a reasonable estimate of its larger higher value (HHV) may be obtained from its specific gravity:

$$\text{HHV (kJ/kg)} = 2.326[17{,}450 + 170\,(°\text{API})] \tag{3.3}$$

When no experimental data are available, an approximate estimate of the HHV may be obtained simply from knowledge of the fuel oil grade. Approximate values for various diesel and fuel oils are given in Table 3.5.

Kerosene

Kerosene is a medium-boiling (175 to 325°) nonviscous, petroleum fraction. Appropriately processed, this fraction is widely used for jet fuel. It is also widely used as the fuel for small and medium-size stationary gas turbine units. The units, which are based heavily on aerospace technology, have found use as peaking units on a

Sec. 3.3 Natural and Petroleum Gas

TABLE 3.5 HIGHER HEATING VALUES OF FUEL OILS

Fuel oil grade	Specific gravity (°API)	Higher heating value	
		kJ/kg	kcal/kg
No. 1	52.0	46,000	10,950
No. 2	34.0	45,200	10,800
No. 4	22.5	43,800	10,400
No. 5	18.0	43,100	10,300
No. 6	14.5	42,300	10,100

large number of utility systems. Hydrogen content, heat of combustion, and specific heat can be estimated from the specific gravity in the same manner as for fuel oils.

3.3 NATURAL AND PETROLEUM GAS

Formation and Availability

Current geological theories indicate that natural gas was formed from decaying vegetable matter in the same manner, and generally along with petroleum. However, in some regions, wells drilled into the geologic formation containing the gas and trapped oil release gas predominately. In other regions, both oil and gas are released together. Even today, in some fields, the natural gas is flared or burned at the wellhead because of lack of facilities to transport the gas.

Natural gas is a popular energy source that is in relatively short supply. As in the case with oil, resource estimates vary widely. The United States has been the leading producer and consumer of natural gas. Table 3.6 shows the total production of natural gas up to 1975 as well as the proven reserves available.

Most natural gas is used as domestic or commercial heating fuel. However, in regions close to the source of natural gas, it has also been used as a power plant fuel.

Transportation of natural gas within the United States is almost entirely by pipeline. Pipelines of up to 48 in. in diameter and over 2000 miles in length have been constructed.

Composition and Properties of Natural Gas Fuels

Natural gas is the cleanest burning of all the fossil fuels. It is free of ash and mixes well with air to undergo complete combustion and producing very little smoke. Natural gas has a high hydrogen content and produces a considerable amount of water vapor when burned. The heat of combustion of natural gas varies from 33.5 to 40 MJ/m^3 (950 to 1150 Btu/ft^3). The specific gravity is about 0.63, relative to air.

TABLE 3.6 WORLD NATURAL GAS RESOURCES (TRILLIONS OF CUBIC FEET)

Area	Proven reserves	Additional speculative reserves	Annual production
Western Hemisphere			
Argentina	9.0	15.	0.5
Canada	55.0	25.	3.5
Columbia	3.0	7.	0.5
Ecuador	0.5	10.	0.2
Mexico	12.0	35.	1.0
United States	250.0	220.	22.0
Venezuela	35.0	50.	2.0
Europe			
Norway	15.0	17.	1.0
United Kingdom	45.0	30.	0.7
Poland	5.0	13.	0.3
USSR	700.0	800.	15.0 (est.)
Middle East			
Abu Dhabi	12.0	7.	0.2
Iran	200.0	100.	2.0
Iraq	25.0	15.	1.0
Kuwait	35.0	10.	0.8
Saudi Arabia	50.0	30.	2.5
Africa			
Algeria	110.0	75.	0.5
Egypt	8.0	10.	0.3
Libya	30.0	20.	1.0
Nigeria	40.0	25.	0.4

Source: Data from *OCS Oil and Gas*, U.S. Council on Environmental Quality, Washington, D.C., 1979.

TABLE 3.7 COMPOSITION OF NATURAL GAS (v/o)

	Source of gas			
Constituent	U.S.[a]	Kuwait	Algeria	North Sea
Methane, CH_4	86	77	84	95
Ethane, C_2H_6	7	13	7	3
Propane, C_3H_8	3	5	2	0.5
Butanes, C_4H_{10}	1	2	1	0.5
Pentanes and higher	0.5	1	0.5	0.5
Other[b]	2.5	2	5.5	0.5

[a] Actual composition varies appreciably depending on source within given country.
[b] Miscellaneous gases, such as N_2, CO_2, H_2S, and He.

Sec. 3.3 Natural and Petroleum Gas

TABLE 3.8 HEAT OF COMBUSTION OF NATURAL GAS COMPONENTS[a]

Compound	Formula	State[b]	Heat of combustion	
			kcal/kg mol	kJ/kg mol
Methane	CH_4	g	−212,800	−890,800
Acetylene	C_2H_2	g	−312,000	−1,306,000
Ethylene	C_2H_4	g	−332,000	−1,390,000
Ethane	C_2H_6	g	−368,400	−1,542,000
Allylene	C_3H_4	g	−464,600	−1,945,000
Dropylene	C_3H_6	g	−490,200	−2,052,000
Propane	C_3H_8	g	−526,300	−2,203,000
Isobutylene	C_4H_8	g	−647,200	−3,710,000
Isobutane	C_4H_{10}	g	−683,400	−2,861,000
Isopentane	C_5H_{12}	g	−843,500	−3,531,000
Isopentane	C_5H_{12}	l	−838,300	−3,510,000
n-Pentane	C_5H_{12}	g	−838,300	−3,510,000
n-Pentane	C_5H_{12}	l	−833,400	−3,489,000

[a] Final products: $CO_2(g)$ and $H_2O(l)$.
[b] g, gas; l, liquid.

Natural gas consists primarily of a mixture of the most volatile paraffins—methane through pentane. Table 3.7 indicates typical compositions from a number of sources.

Gas is termed *dry* if it contains less than 0.1 gal of gasoline vapor per 1000 ft^3 and *wet* if it contains more than this amount. The terms *sweet* and *sour* are used to designate the absence or presence of H_2S.

The formula weight and normal state of the light hydrocarbons that make up nearly all of natural gas are given in Table 3.8. The heating value of natural gas is determined from the heat of combustion of its components. These values (from which the HHV is derived) are also shown on Table 3.8.

Liquefied Natural Gas and Petroleum Gas

Since the major constituent of all natural gas is methane (critical temperature −83°C), cryogenic temperatures are required to maintain the gas as a liquid at moderate pressures (e.g., −100°C at 36 bar). Storage and shipment of the liquid has a large advantage in that 1 m^3 of liquid is equivalent to more than 600 m^3 at standard atmospheric conditions.

Transportation of *liquid natural gas* (LNG) by special tankers from overseas locations to U.S. ports is now commonplace. Storage of the LNG in spherical pressure vessels allows the gas to be used as needed, particularly for meeting peak load requirements.

Liquefied petroleum gas (LPG) refers to hydrocarbons (such as propane, propylene, butane, butylene, and isobutane) which are liquefied under moderate pressures and at normal temperatures but which are vapors at the usual atmospheric conditions. LPG can be obtained from either (1) the separation of the heavier components from natural gas or (2) from refinery gases. LPG derived from both sources is mainly paraffinic, but the LPG from refinery gases usually contains some olefinic (unsaturated) hydrocarbons.

Liquefied petroleum gas is widely used to supplement natural gas flow during peak usage periods. In a few cases, natural gas flow may be supplemented by "oil gases." These gases are made by thermal decomposition of petroleum fractions ranging from naphtha to residual fuel.

3.4 COAL

Evolution and Formation

Coal and peat are the partially decomposed remains of vegetation that grew millions of years ago. Coal formation began when this partially decayed vegetation was covered by inorganic silt, which prevented further decay. As inorganic sediment layers built up, pressure and heat converted the vegetation to coal.

Coal is a generic term that describes a rather wide variety of solid hydrocarbon products formed from prehistoric terrestrial vegetation. The type of vegetation from which the coal was produced significantly affects the characteristics of the coal ultimately formed. The composition of a coal, including impurities such as sulfur, is generally characteristic of the region in which the coal was found.

Peat is formed from the bacterial decomposition of vegetation that is rich in celluose. In the presence of bacterial action, the chemical decomposition proceeds by a process called *humification*:

$$2C_6H_{10}O_5 \xrightarrow[\text{action and heat}]{\text{bacterial}} C_8H_{10}O_5 + 2CH_4 + H_2O + 2CO_2$$
$$\text{vegetation} \qquad\qquad\qquad \text{peat}$$

The accumulation of compounds that resist further decomposition (peat, resins, waxes, and dead bacteria) eventually forms the principal part of the deposit.

In locations where coal was formed, the peat became covered with layers of sand and silt. As the layers of sand and silt built up, the pressure on the peat increased, forcing out the water and volatile products, such as methane. The remaining material became depleted in oxygen and rich in hydrogen, thus promoting additional hydrocarbon formation. The longer this process continues, the more hydrocarbons are formed and the less water is present. Ultimately, the soft, spongy peat is converted into hard, brittle coal, with high carbon content and low impurity and water content. It is this "aging" process that gives each type of coal its basic characteristics and properties. As will be discussed subsequently, these characteristics and properties are cataloged by "class" or "rank."

Sec. 3.4   Coal

Principal Locations and Reserve Estimates

Coal resources are widely distributed throughout the world, but major deposits are limited to a few regions of the world. The United States, the USSR, and China hold almost 90% of the world's coal resources. Table 3.9 list the major coal resources of the world. The estimated total resources indicated by this table include those resources whose presence has not been verified by detailed exploration. Data for assessing these total resources are obtained from geologic information, visual inspection, seismic surveys, and very limited exploratory drilling. The exploration drilling samples are obtained from widely scattered geographic locations and are subject to considerable uncertainty. Measured or proven reserves have been verified by detailed exploration and drilling. These resources are reasonably well assured. The economically recoverable resources are estimated from the proven reserves on the basis of practical experience. A common assumption is that between 85 and 90% of all surface mined coal can be recovered and between 45 and 50% of deep-mined coal can be recovered.

As seen from Table 3.9, the United States has large reserves of coal. Locations of these reserves are shown in Fig. 3.3 and their distribution among the states is shown in Table 3.10. The eastern bituminous coal fields are the leading producers of coal, but new environmental regulations have resulted in decreased production of eastern coal. Western coal, usually lower in sulfur content, is now in great demand for many applications. Most eastern coal is deep mined, while much of the western coal is surface mined.

TABLE 3.9   MAJOR WORLD COAL RESOURCES (MILLIONS OF METRIC TONS)

Area	Total known and estimated recoverable resources
USSR	6,000,000
Unites States	3,200,000
China	1,000,000
Canada	500,000
West Germany	300,000
England	200,000
Australia	200,000
India	80,000
Poland	60,000
South Africa	50,000
Other nations	500,000

Source: Data from *Mineral Resources and the Environment*, National Academy Press, Washington, D.C., 1974; and *Alternate Energy Demand Futures to 2010*, National Research Council, Washington, D.C., 1979.

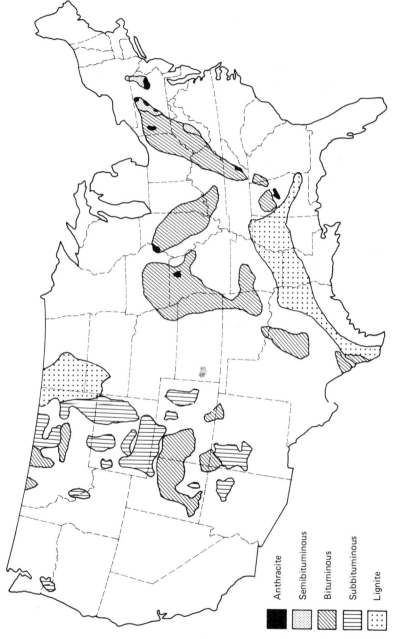

Figure 3.3 Coal resource locations in the United States.

TABLE 3.10 ESTIMATED COAL RESOURCES OF THE UNITED STATES (MILLIONS OF METRIC TONS)

State	Known resources				Additional estimated resources below ground
	Bituminous coal	Subbituminous coal	Lignite or (anthracite)	Total	
Alabama	13,500	0	20	13,520	26,500
Alaska	20,000	111,000	0	131,000	135,000
Arkansas	1,650	0	350(450)		4,000
Colorado	62,000	18,000	(78)	80,078	300,000
Illinois	140,000	0	0	140,000	100,000
Indiana	35,000	0	0	35,000	22,000
Iowa	6,500	0	0	6,500	14,000
Kansas	18,700	0	0	18,700	5,000
Kentucky	66,000	0	0	66,000	52,000
Missouri	23,000	0	0	23,000	10,000
Montana	2,000	132,000	88,000	222,000	160,000
New Mexico	11,000	51,000	0	62,000	47,500
North Dakota	0	0	351,000	351,000	180,000
Ohio	42,000	0	0	42,000	10,000
Oklahoma	4,000	0	0	4,000	20,000
Pennsylvania	60,000	0	(12,000)	72,000	10,000
South Dakota	0	0	2,000	2,000	1,000
Tennessee	2,700	0	0	2,700	5,000
Texas	6,000	0	7,000	13,000	15,000
Utah	32,000	0	0	32,000	80,000
Virginia	9,700	0	(1,000)	10,007	5,000
Washington	2,000	4,000	0	6,000	45,000
West Virginia	102,000	0	0	102,000	0
Wyoming	13,000	110,000	0	123,000	425,000
Other states	8,000	75,000	900(100)	84,000	75,000

Source: Data from *U.S. Energy Supply Prospects to 2010*, National Academy Press, Washington, D.C., 1979; and *Mineral Resources and the Environment*, National Academy Press, Washington, D.C., 1974.

Classification

The physical and chemical nature of coal varies widely, depending on the nature of the original vegetation, the temperature and pressure during formation, and the time of formation (age). The most important factor in determining coal characteristics appears to be the age of the coal. The carbon content increases with time, while the water and oxygen contents decrease. As previously indicated, the deposits are gradually converted from peat, a soft, high-water-content form, to a hard, brittle, dry coal.

Four major types of coal are easily recognizable. In order of increasing age, or carbon content, these are:

Peat
Lignite or brown coal
Bituminous coal
Anthracite

Peat is characterized by well-preserved remains of the vegetation from which it was formed. It is soft and spongy, and in its natural state has an exceptionally high water content, often 92 to 94%. Even after drying the water content is very high and it is often burned while still containing as much as 50% water. Because peat represents an early stage of decomposition, the carbon content is low (about 50% of the dry weight is a typical figure), and there is a high proportion of volatile matter, including water and hydrocarbons. Consequently, it burns with a long flame. It is used as a fuel only in relatively few parts of the world, where its principal uses are as a domestic heating fuel and occasionally for electricity generation.

Lignite or brown coal is a partially transformed peat. It still contains recognizable woody material from the lignin of the original plant tissues. Chemically, it is not much different from peat and it contains a large percentage of water and volatiles. Hence its burning characteristics are similar to those of peat. Mechanically, it is soft but not noticeably spongy. Lignite is used extensively in Germany and central Europe as an industrial fuel, for electricity generation, and for domestic fuel to a lesser extent. It has also been an important source of chemical feedstock.

Bituminous coal varies from soft to hard, is black in color, and is often banded. It contains recognizable vegetable material but only in fossilized form. About 30 to 40% of the coal is volatile matter, and it is low in water content. It burns with a smoky yellow flame, and has a higher heat content than lignite or peat. Bituminous coal is an important source of fuel for electrical power generation. Its volatile content is mainly hydrocarbons, and for this reason it can be used for the manufacture of low-Btu gas. By controlled distillation of the volatile matter of certain grades of bituminous coal, a very hard, strong residue, called *coke*, is produced. This is used as a reducing agent by the steel industry and sometimes as a smokeless commercial or industrial fuel.

Anthracite coal is hard, brittle, and stony in appearance. Because of the absence of volatiles, it burns with a short blue smokeless flame. The calorific value is high, but that of some bituminous coals can be higher because they have a greater hydrogen content. It has little or no tendency to form a coke. Anthracite is used only as a fuel, and because of its high cost it was used mainly as home heating fuel, where it was regarded as a "premium-grade" fuel with natural smokeless properties. The carbon content can be over 90%.

A fifth coal category, *cannel*, is sometimes used. Cannel is characterized by a very high percentage of hydrocarbon volatiles. It can be regarded as a special type of bituminous coal and is usually marketed in that category rather than on its own. It is particularly notable for the ease with which it can be distilled to yield oils of a petroleum type.

A more useful classification of coal is that of *rank*. Rank carries the meaning of degree of maturation and is a measure of carbon content. Lignite would be considered to be low rank and the anthracite to be high rank. Table 3.11 shows the ASTM rank classification system. The classifications are based on fixed carbon content (for high-rank coals) and heating or calorific value (for low-rank coals).

It is also found that the hydrogen content decreases slightly with increasing rank. This is because of the reduction in the volatile fraction, which is largely composed of hydrocarbons and is thus rich in hydrogen. However, the decrease is very slight, and much less than might be expected in view of the great decrease in volatiles.

The sulfur content is variable and cannot be correlated with rank. A high sulfur content is undesirable because sulfur burns to sulfur dioxide during combustion. Excessive sulfur content can make the coal liable to spontaneous combustion in storage. Similarly, the ash content is highly variable. It appears that in some cases high ash contents were derived from muddy conditions in the original coal swamp. There is no correlation with rank.

Properties

There are several properties of coal that are important in power plant applications: sulfur content, combustion properties, weatherability of the coal, ash softening temperature, grindability of the coal, and heat content.

The sulfur content is now an important consideration in the selection of coal. Sulfur is one of the combustible constituents of coal, thus generating some energy. However, the product of combustion, sulfur dioxide, (SO_2), is considered to be a major source of atmospheric pollution. It is difficult and expensive to remove the sulfur dioxide from the combustion products. In selecting a coal source, detailed assessment must be made considering current environmental regulations covering plant SO_2 emissions, the sulfur content of the coal, the cost of the coal, and the capital cost and operating cost of SO_2 removal equipment. Emission control system selection is discussed in Section 4.1.

TABLE 3.11 ASTM COAL CLASSIFICATION BY RANK

Class	Rank	Group	Fixed carbon limits (%) (dry mineral-matter-free basis)		Volatile matter limits (%) (dry mineral-matter-free basis)		Calorific value limits [Btu/lb (kJ/kg)] (moist mineral-matter-free basis)	
			Equal or greater than	Less than	Greater than	Equal or less than	Equal or greater than	Less than
I. Anthracitic	1. Meta-anthracite		98	—	—	2	—	—
	2. Anthracite		92	98	2	8	—	—
	3. Semianthracite		86	92	2	8	—	—
II. Bituminous	1. Low-volatile bituminous coal		78	86	14	22	—	—
	2. Medium-volatile bituminous coal		69	78	22	31	—	—
	3. High-volatile A bituminous coal		—	69	31	—	14,000 (32,400)	—
	4. High-volatile B bituminous coal		—	—	—	—	13,000 (30,100)	14,000 (32,400)
	5. High-volatile C bituminous coal		—	—	—	—	11,500 (26,600)	13,000 (30,100)
III. Subbituminous	1. Subbituminous A coal		—	—	—	—	10,500 (24,200)	11,500 (26,600)
	2. Subbituminous B coal		—	—	—	—	9,500 (21,900)	10,500 (24,200)
	3. Subbituminous C coal		—	—	—	—	8,300 (19,200)	9,500 (21,900)
IV. Lignitic	1. Lignite A		—	—	—	—	6,300 (14,600)	8,300 (19,200)
	2. Lignite B		—	—	—	—	—	6,300 (14,600)

Source: *ASTM Specification D388*, 1980; copyright ASTM, 1916 Race Street, Philadelphia, PA 19103. Reprinted with permission.

Sec. 3.4 Coal

The combustion properties of coal are related to how well the coal burns. In a stationary bed, such as a chain grate stoker furnace, the coal must not cake as it burns. Coal that does not cake, called *free-burning coal*, breaks apart during combustion. This exposes the unburned center of the coal to the combustion air, thereby enhancing the combustion process. Conversely, a caking coal does not burn completely. Caking coals are used to produce coke. In power plant applications using a stationary bed, caking coal beds must be mechanically agitated to prevent large masses of partially burned coal from forming. A qualitative evaluation method, called the *free-swelling index*, has been devised to aid in the selection process. Information is contained in ASTM Standard D720. A free-burning coal is characterized by a high value of the free-swelling index. This property is of minor importance when modern pulverized coal burners are used.

The *weatherability* of a coal is a measure of how well it can be stockpiled for long periods of time without crumbling. Modern coal-fired power plants normally stockpile 60 to 90 days' supply of coal in a large pile near the power plant. A typical pile would contain up to 300,000 tons of coal. The coal is generally unloaded from a train or barge and spread and packed in a long trapezoidal-shaped pile. Excessive crumbling due to cyclic weather conditions would result in the formation of small particles of coal that can be dispersed by wind or rain. Environmental problems would result if these particles pollute the local rivers or groundwater table.

Grindability is another important consideration when selecting coal. This property is measured by the grindability index, ASTM Standard D409. Grindability of a standard coal (bituminous class II group 1) is defined as 100. The index is established to be inversely proportional to the power required to grind the coal to a specified particle size for burning. If the coal selected for use at a power plant has a grindability index of 50, it would require twice the grinding power of the standard coal to produce a specified particle size. The grindability index for commercial coals varies from about 25 to 105.

The *ash softening temperature* is the temperature at which the ash softens and becomes plastic. This is usually somewhat below the melting point of the ash. This coal characteristic is important to the ash discharge system. For furnaces that discharge firebox ash as a slag, coal with a low ash softening temperature would be desirable. For firebox systems that discharge the ash in solid form, a high ash softening temperature would be required. A stoker-type furnace, which discharges a solid-type ash, must utilize a coal with a high ash softening temperature or "clinkers" will form in the firebox. Clinkers, which are large masses of fused ash, can be very troublesome to discharge.

Composition

The composition of coal is generally described in terms of either the proximate or ultimate analyses. The *proximate analysis* is the simplest test and is performed by weighting, heating, and burning a small coal sample. The proximate analysis determines the moisture content [M'_0, in weight percent (w/o)] by driving off the free moisture at ~107°C. The volatile matter content (V', w/o) is determined by driving off volatile hydrocarbons CO, CO_2, and combined H_2O at ~950°C. The coal is then burned and the inorganic residue is the ash content (A'_s, w/o). The fixed carbon (C'_f, w/o) is then calculated by difference:

$$C'_f(\text{w/o}) = 100 - \left[M'_o(\text{w/o}) + V'(\text{w/o}) + A'_s(\text{w/o})\right] \tag{3.4}$$

Although not strictly part of the proximate analysis, most such analyses are accompanied by a chemical determination of the coal's sulfur content (S', w/o).

The proximate analysis gives a physical picture of the coal and is a relatively simple test. On the other hand, the ultimate or elemental analysis is a quantitative evaluation of total carbon (C'), hydrogen (H'), nitrogen (N'), sulfur (S') and oxygen (O') percentages after removal of the moisture and ash. The ultimate analysis is performed using classic oxidation, decomposition, and/or reduction techniques to determine C, H, N, and S. Oxygen, $O'(\text{w/o})$, is calculated by difference:

$$\begin{aligned} O'(\text{w/o}) = 100 - \big[&C'(\text{w/o}) + H'(\text{w/o}) + N'(\text{w/o}) \\ &+ S'(\text{w/o}) + M'_o(\text{w/o}) + A'_s(\text{w/o})\big] \end{aligned} \tag{3.5}$$

The data obtained from the ultimate analysis are reported in weight or mass percent.

Results of the *ultimate analysis* are often converted to a dry mineral-matter-free basis just as done for the proximate analysis. However, for combustion calculations and coal-handling calculations, these results are needed on an as-burned or as-received basis. The analysis on the *as-received basis* includes the total moisture content of the coal received at the plant. The *as-burned basis* includes the total moisture content of the coal as it enters the furnace. The relationship between the *dry ash-free mass fraction* (DAFMF) and the *as-burned mass fraction* (ABMF) is

$$\text{ABMF} = \text{DAFMF}\left(1 - \frac{M'_o + A'_s}{100}\right) \tag{3.6}$$

The test results of a proximate and an ultimate analysis on a typical eastern bituminous coal are shown in Table 3.12. Note that the carbon content of the coal, determined in the ultimate analysis, reflects the carbon content of the volatile matter in the coal in addition to the fixed carbon.

Once the results of the proximate analysis have been obtained, the rank of the coal may be determined if the coal is of medium to high rank. In order to do so, we must calculate the dry material-matter-free weight percentages of fixed carbon (C') and volatile matter (V'). This is required since moisture and ash content varies widely even for a given coal source. They are computed using the *Parr formulas*:

$$C'(\text{w/o}) = \frac{(C - 0.15S)100}{100 - \left(M'_o + 1.08A'_s + 0.55S\right)}$$

$$V' = 100 - C' \tag{3.7}$$

Sec. 3.4   Coal

TABLE 3.12 ANALYSIS OF EASTERN BITUMINOUS COAL

Proximate analysis		Ultimate analysis	
Component	w/o	Component	w/o
Moisture	2.7	Moisture	2.7
Volatile matter	35.3	Carbon	73.6
Ash	8.4	Hydrogen	4.6
Fixed carbon	53.6	Sulfur	2.5
Total	100.0	Nitrogen	1.3
		Ash	8.4
		Oxygen	6.9
		Total	100.0

Example 3.1

A proximate analysis of a coal indicates the following: moisture 3.0%, volatile matter 18.2%, ash 7.5%, sulfur 2.1%, and fixed carbon 69.2%. Determine the coal's rank.

Solution

$$C'(w/o) = \frac{[69.2 - 0.15(2.1)]100}{100 - [3.0 + 1.08(7.5) + 0.55(2.1)]} = \frac{68.9(100)}{100 - 12.25} = 78.5$$

$$V' = 100 - 78.5 = 21.5$$

From Table 3.11 we see that the values of C' and V' fall within the limits set for a low-volatile bituminous coal and the rank is then II-1.

Heating Value or Calorific Value

The calorific value of coal is a property of fundamental importance. The heating value is a complex function of the elemental composition of the coal. Test or experimental values are obtained by burning a 1- or 2-g sample in oxygen at 25 MPa pressure in a bomb calorimeter (ASTM Standard D2015-73). Two different heating values are cited for coal. The value obtained directly from the calorimeter is called the *gross heating value*, or *high heating value*, since all water formed remains a liquid. A second heating value, called either the *net heating value* or the *lower heating value* is found if the water is allowed to evaporate.

This lower, or net, heating value is the more practical of the two since it is the heating value of the fuel minus the latent heat of vaporization of the water vapor present in the exhaust gases. The lower heating value can be found by subtracting 2.395 kJ/(grams of water in the combustion products) from the calorific value given in kJ/g. Alternatively we may subtract 1030 Btu per pound of H_2O per pound of coal if the calorific value is given in Btu/lb. That is,

$$Q'_{low} = Q'_{high} - 2.395 w_m \quad kJ/g$$

or

$$Q'_{low} = Q'_{high}(\text{cal/g}) - 570 w_m \quad \text{cal/g} \tag{3.8}$$

where $w_m = $ (kg H_2O in combustion products)/(kg original coal). Note that the total amount of water, initial moisture in the sample plus water formed on combustion, must be considered in determining w_m. The procedure for obtaining Q'_{low} from Q'_{high} applies to any type of coal or fossil fuel. Unless indicated otherwise, it should be assumed that stated heating values are high heating values.

Several empirical relationships have been developed to calculate the heating value of coal. If the ultimate analysis results are available, the dry, ash-free heating value can be calculated to within 2% accuracy by the use of the *Dulong–Berthelot formula*:

$$Q'_d = 81.37 C' + 345\left(H' - \frac{O' + N' - 1}{8}\right) + 22.2 S \tag{3.9}$$

Q'_d has units of calories per gram and the elements are in units of weight percent. Both Q'_d and the compositions of Eq. (3.9) are on a dry, ash-free basis. The as-burned heating value, Q'_{AB}, may be calculated from Q'_d by

$$Q_{AB} = Q'_d\left(1 - \frac{M'_o - A'_s}{100}\right) \tag{3.10}$$

where M'_o and A'_s are the mass percentages of moisture and ash, respectively, at the point of combustion.

In practice, the determination of the ultimate analysis is relatively difficult, whereas the determination of the heating value in a bomb calorimeter is relatively simple. It is therefore quite frequently found that the heating value plus the proximate analysis (including sulfur) is available for a given coal but that the ultimate analysis is not readily obtained. In such a situation, the heating value and coal rank may be used to obtain an approximate composition that can be used in subsequent combustion calculations.

Table 3.13 indicates the approximate value of the hydrogen available for combustion, called H'_a, available hydrogen content, for the several classes of coal. We can use the estimated value of H'_a from Table 3.13 and a simplified equation for

TABLE 3.13 HYDROGEN CONTENT OF COAL

Coal class	H'_a = available hydrogen content[a] (kg H/kg C)
Anthracite	0.029
Semibituminous	0.049
Bituminous	0.054
Subbituminous	0.045
Lignite	0.037

[a] See Section 4.1 for a definition of available hydrogen.

Sec. 3.5 Coal Cleaning and Processing

Q'_d to obtain an approximate value of C'. That is, we solve Eq. (3.11) for C' using a bomb measurement of Q'_d.

$$Q'_d = 81.37C' + 345H_a(C') + 22.2S' \qquad (3.11)$$

where H'_a is in kg/kg · °C. The value of C' so obtained allows approximate combustion calculations to be conducted.

3.5 COAL CLEANING AND PROCESSING

Objectives and Benefits of Coal Cleaning

Coal cleaning removes some of the noncombustible mineral content, rock, sulfur, and miscellaneous impurities from the coal prior to burning. Cleaning can improve power plant performance and reduce particulate and SO_2 emission cleanup costs. Although coal-cleaning technology is highly developed, less than half of the coal burned in power plants is cleaned. However, the use of cleaned coal is increasing due to stringent environmental regulations and the relatively low cost of cleaning. In addition, cleaning may also lead to reduced maintenance for the boiler and associated equipment because of the reduced corrosion and fouling. This increases boiler availability and capacity.

The benefit derived from coal cleaning depends on the type of coal, especially the sulfur and ash content. Coal contains sulfur principally in two forms: organic sulfur and mineral sulfur. Organic sulfur is an integral part of the coal matrix, while mineral sulfur is found in particle form. Mineral sulfur is the only kind that can be even partially removed by cleaning. Ash, the noncombustible mineral matter in coal, can also be partially removed by cleaning. Some moisture which collects on the exposed surfaces of coal is also removable.

Coal cleaning can reduce ash content by 10 to 70%. Sulfur content can be reduced by up to 35%, while heating-value improvements of 2 to 25% can be achieved. In terms of levelized power costs, cleaned coal costs about 1.0 to 2.0 mills/kWh more than raw coal.

Coal-Cleaning Operations

Coal may be cleaned at the mine site or at the power plant. The cleaning procedure depends on the type of coal being processed. The cleaning ranges from a simple crushing operation to a complete and thorough crushing, gravity separation, dewatering, and drying. Figure 3.4 schematically shows the complete range of cleaning operations that may be required by a high-sulfur eastern bituminous coal. A low-sulfur western subbituminous coal may only require crushing.

Five cleaning levels are recognized:

1. This level is the simple crushing operation to remove large rock and liberate some of the mineral impurities in the coal.

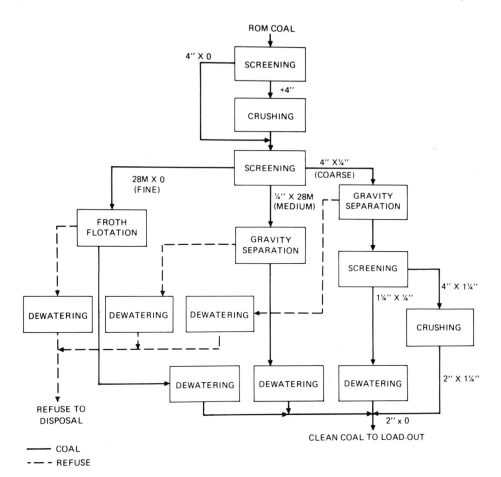

Figure 3.4 Major steps in coal cleaning. (Reproduced with permission from *Proceedings of the American Power Conference*, Vol. 41, 1979.)

2. A sizing, cleaning, and dewatering operation is performed on the coarse coal. Finer coal is recovered as crushed.
3. A second sizing operation is performed to separate out medium-size coal. Medium-size coal is cleaned. Only very fine coal is recovered as crushed.
4. All three standard sizes are screened and cleaned as indicated in Fig. 3.4.
5. Additional sizing to ultrafine coal, including screening and cleaning operations beyond those indicated in Fig. 3.4.

The tabulated levels are cumulative in the cleaning cycle. For example, level 3 cleaning would include the cleaning of levels 1 and 2 as well as level 3. When level 4 or 5 cleaning is performed, the crushed coal is separated into three size ranges. The

Sec. 3.5 Coal Cleaning and Processing

TABLE 3.14 SPECIFIC GRAVITY COMPARISON

Coal	1.1–1.3
Clay	1.8–2.2
Shale	2.0–2.6
Iron pyrites	4.8–5.2

coarse and medium sizes are cleaned by a gravity separation technique, while the fine size is cleaned by froth flotation.

In the *froth flotation method*, the small coal particles are buoyed up on top of a controlled surface froth. The heavier impurities, such as iron pyrites slate and shale, sink to the bottom of the mixture. The segregated layers can easily be separated from the mixture.

The *gravity separation process* achieves separation due to the difference in specific gravity between coal and its impurities. Coal has a specific gravity between 1.1 and 1.3. As Table 3.14 shows, the specific gravity of the major impurities in coal is considerably greater. Water is generally utilized as the separating medium. Particles with a specific gravity of 1.5 or less are removed as clean coal, while particles with a specific gravity greater than 1.5 go to refuse.

Dewatering and drying comprise a major part of the cleaning process whenever flotation or gravity separation is used. Modern dewatering and drying techniques utilize pressure filters and centrifuges with high-g forces. Mechanical dewatering is preferred over thermal drying. Thermal drying is used only where the mechanically dewatered product cannot meet moisture specifications.

Coke Production

Production of coke by carbonization of bituminous and semibituminous coal has been of importance ever since the widespread replacement of charcoal by coke in steelmaking in the early eighteenth century. Coke production for blast furnace and foundry use takes place in modern high-temperature ($\sim 900°C$) slot-type coke ovens. These ovens are chambers 50 to 55 ft long, 18 to 22 ft high, with slot widths of 12 to 22 in. A number of such chambers, alternating with similar cells that contain heating flues with hot combustion gases, are built over a common firing system. At the end of the cycle a large ram pushes the hot carbonized residue (coke) out of the chambers onto a quenching platform.

The volatile material driven off consists of coal gas and coal tars. The coal gas, formerly important for cooking and illuminating gas, is now used largely as industrial fuel. The coal tar, consisting largely of aromatic hydrocarbons, forms the base for a variety of organic chemicals.

Prior to the end of World War II, the demand for coal gas and coal tars was considerably higher than at present. Production of these materials was maximized by using low-temperature carbonization (at 600 to 700°C) for their production. The coke so produced is more porous and is more suitable than high-temperature coke for domestic fuel. There is little current demand for this fuel.

3.6 SYNTHETIC FUELS: GASEOUS FUELS FROM COAL

The gasification of coal for use as a power plant fuel is being considered as the supply of natural gas diminishes. There are a number of technically feasible processes being studied. Some of the most promising processes for making low, medium, and high heat content gas from coal are examined in this section.

We have previously noted the widespread use of coal gas, produced by destructive distillation of coal, for illuminating and cooking purposes in the nineteenth century and the first part of the twentieth century. In most municipal gas distribution systems, the coke produced in the distillation operation was then used to produce low-heat-content gas which substantially augmented the coal gas flow. The coke was placed in large beds and burned for a period with less than the stoichiometric quantity of air to give a *producer gas*:

$$\underset{\text{coke}}{2C} + \underset{\text{air}}{(O_2 + 3.76N_2)} \rightarrow \underset{\text{producer gas}}{2CO + 3.76N_2} \quad (3.12)$$

When the bed was air heated to a high temperature, the flow of air was replaced by a flow of steam and *water gas* was produced:

$$\underset{\text{coke}}{C} + \underset{\text{steam}}{H_2O} \rightarrow \underset{\text{water gas}}{CO + H_2} \quad (3.13)$$

After the bed cooled because of the endothermic water gas reaction, the steam flow was replaced by air and the exothermic partial combustion reheated the bed. The alternate production of water and producer gas continued until the coke bed was exhausted.

Modern gasification techniques are based on the producer and water gas reactions. However, more efficient, continuous production processes have been devised. Further, modern processes have also addressed the problem that natural gas has a heating value substantially higher than that of water and producer gas. Without further treatment such gases cannot be used as a direct replacement for natural gas. However, they may be used in power plant equipment especially designed for "low-Btu" (low heating value per unit volume) gas.

Modern Coal Gasification Chemistry

The coal is first ground into a sand-like powder and then preheated and dried to reduce caking during conversion. A high-temperature pretreater is often used to give the coal particles a thin coating of oxygen to prevent sticking. Figure 3.5 illustrates the basic steps in the gasification process. The process illustrated by the lower flow sheet yields a low-heat-content gas (which is a mixture of water and producer gas)

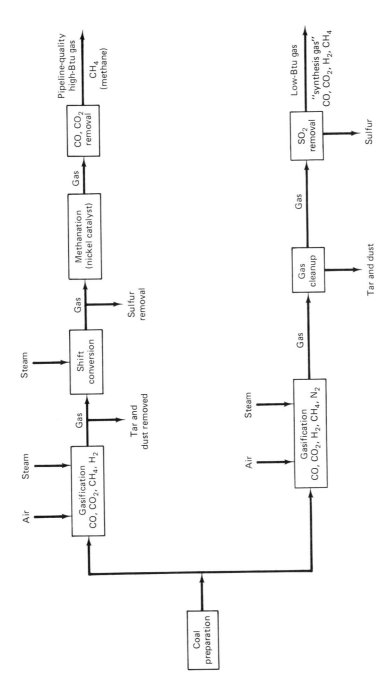

Figure 3.5 Coal gasification processes.

called *synthesis gas*. The process illustrated by the upper flow sheet yields a higher-Btu gas called *pipeline gas* with properties close to that of natural gas.

The primary reaction sequence of a well-known procedure, illustrated in Fig. 3.5, is:

1. An initial endothermic reaction of coal and steam in the presence of oxygen at 1800°F:

$$\left. \begin{array}{c} 2C + (O_2 + 3.76N_2) \rightarrow 2CO + 3.76N_2 \\ C + H_2O \rightarrow CO + H_2 \end{array} \right\} \text{synthesis gas or syngas} \quad (3.14)$$

2. If the product desired is a medium or high-Btu gas, a synthesis gas "shift reaction" or conversion is used to produce additional hydrogen (upper flow sheet of Fig. 3.5) by reacting some of the CO with steam and removing CO_2 from the products:

$$CO + H_2O \rightarrow CO_2 + H_2 \quad (3.15)$$

The medium-Btu gas can be used at this point if desired.

3. The final step in producing a pipeline quality gas is called *catalytic methanation*. The products of the second reation are reacted over a nickel catalyst at a temperature of ~1100°C (2000°F) and pressure 6.8bar (100 psi).

$$CO + 3H_2 \rightarrow CH_4 + H_2O \quad (3.16)$$

The product gas is a higher-quality gas containing 37.7 MJ/m³ (1030 Btu/ft³) and is directly substitutable for natural gas.

Another process under development which differs substantially from the processes listed above is called *hydrogasification*. In this process, fluidized coal is gasified directly with hydrogen-rich steam to a methane-rich gas that requires very little additional shifting. The overall reaction is of the form

$$CH_{0.8} + 0.55H_2O + 1.15H_2 \rightarrow 0.575CH_4 + 0.425CO_2 \quad (3.17)$$

The hydrogasification process is an attempt to improve the overall efficiency of the conversion process.

Commercial Coal Gasification Systems

One of the first successful commercial coal gasification systems was the *Lurgi pressure gasifier*. Figure 3.6 shows the basic features of the Lurgi process. The Lurgi process is based on steam–oxygen gasification in a mechanically stirred fuel bed. The product from this process is a synthesis gas (syngas). The Lurgi process can produce a low- to medium-heat-content (low- to medium-Btu gas), depending on the oxidant (air or oxygen).

Sec. 3.6 Synthetic Fuels: Gaseous Fuels from Coal

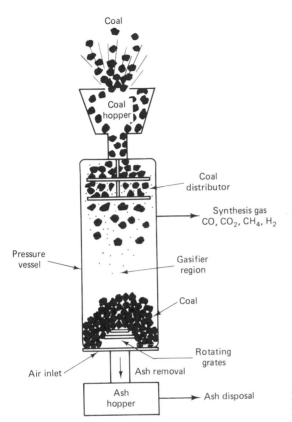

Figure 3.6 Lurgi coal gasification process.

In the Lurgi process, coal is fed into the unit in batches from the top. Steam and air (or oxygen) are injected through a rotary bottom grate and blown upward. Ash is removed from the bottom. The process operates at a pressure of 3 MPa (430 psi) and 1000°C (1832°F) temperature. The product gas is cleaned in a scrubbing column where the CO_2 is removed by absorption. Water is then removed by a drying process. The product is a synthesis gas with a heating value ranging from 200 to 500 Btu/ft^3 (7.3 to 18 MJ/m^3). The product gas composition is typically 50% H_2, 35% CO, and 15% CH_4.

Individual Lurgi units are 12 to 16 ft in diameter, produce up to 10 to 15×10^6 SCF of gas per day from 600 to 1000 tons of coal per day. A plant consists of the number of units required to achieve desired output.

One of the main drawbacks of Lurgi gasifiers is their limited throughput per unit and the consequent need to operate multiple generators in parallel. This carries the penalties of higher operating labor and more frequent mechanical breakdown. In addition, Lurgi gasifiers are, like all surface plants, surrounded by the usual paraphernalia of coal storage yards, cranes, conveyors, scales, sampling devices, and unloading facilities for coal and ash disposal. The gas cleaning section of the plant is similarly complex and consists, apart from the scrubbers and wash columns already mentioned, of solvent recovery equipment and the processing of by-products such as benzene and other aromatics, hydrogen sulfide and sulfur, carbon dioxide, and condensate. It should be noted, however, that the need to deal with coal gasification by-products is not a function of the gasification process, and practically all coal gasifiers will require equipment of the same type and size to produce a similar volume of syngas.

There are many other gasification systems being developed for commercial use. In addition to the Lurgi process, other well-developed processes are Wellman–Galusha gasifier, Winkler gasifier, and the Koppers–Totzek process. The gas produced by all of these processes contains H_2S which must be removed. This removal is generally accomplished by scrubbing the gas with an absorbant. This operation is much less expensive than removing the sulfur-bearing compounds from flue gas after combustion, since the gas volume to be handled is much lower.

Developmental Coal Gasification Processes

Now coal gasification systems under development strive to attain better efficiency and greater throughputs via a continuous gasification process. There are three major classes of gasification systems under development in the United States today.

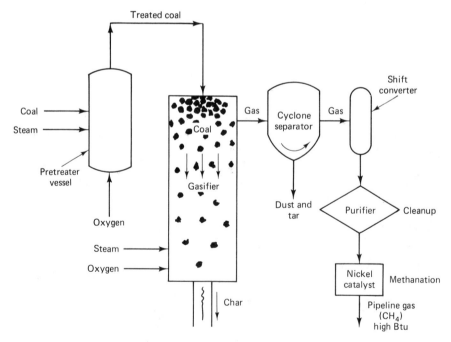

Figure 3.7 Synthane coal gasification process.

1. Fluidized-bed or gas-entrained processes that produce a synthesis gas product. Shift reactions and methanation are required to produce high-quality gas.
2. Hydrogasification systems.
3. Systems that utilize liquid derivatives of coal to produce a synthesis gas.

One typical process that is being developed is the U.S. Bureau of Mines *Synthane process*, shown schematically in Fig. 3.7. Dry crushed coal is fed from a pressurized hopper into a pretreater. The coal is reacted with oxygen and steam at 400°C and 7 MPa to devolatilize it and prevent caking. The coal mixture flows to the gasifier, where it is partially gasified in a dense phase at 800°C and 7 MPa. The coal and gaseous products continue to flow downward to the dilute section of the gasifier, where they are reacted with additional oxygen and steam at 1000°C and 7 MPa. Unreacted char and ash are removed from the bottom of the vessel. The synthesis gas leaves the upper section of the vessel and passes through a cyclone separator to remove dust and char. After further cleaning and removal of sulfur-bearing compounds, the gas can be used as a low-Btu gas. Alternatively, it can be further reacted by passage through a catalytic methanation process to produce a pipeline-quality synthetic natural gas.

Underground (In Situ) Coal Gasification

In addition to the coal gasification projects, described previously called *remote-site gasification*, studies and tests are under way to evaluate *underground coal gasification*. Figure 3.8 illustrates the principle involved. Two deep drilled holes are spaced close enough to permit air and gas to flow between them. The combustion takes place at the bottom of the air hole and the combustion zone proceeds toward the product gas hole. In the combustion zone, carbon dioxide is formed in the reaction:

$$C + O_2 \rightarrow CO_2 \qquad (3.18)$$

Slightly ahead of the combustion zone is a reduction zone where carbon monoxide is formed:

$$C + CO_2 \rightarrow 2CO \qquad (3.19)$$

Some hydrogen can also be formed from moisture in the coal:

$$C + H_2O \rightarrow CO + H_2 \qquad (3.20)$$

Volatile matter is also released from the coal as the process proceeds. The product gases are very similar to the coal gas or water gas previously described. Thus the heating value is low, about 5.5 MJ/m^3 (150 Btu/ft^3). Considerable development work is required before this method is commercialized.

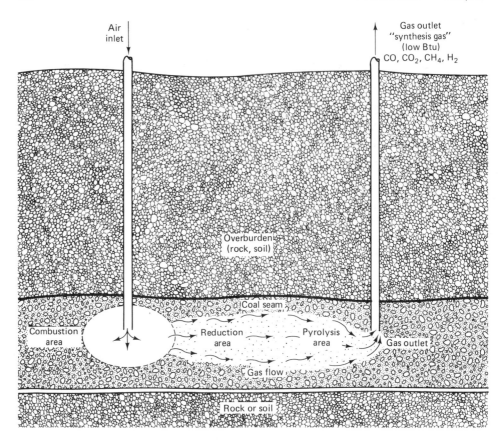

Figure 3.8 In situ coal gasification.

3.7 SYNTHETIC FUELS: COAL LIQUEFACTION AND REFINING

The growing shortage of world petroleum supplies and the rapidly increasing cost of the oil has revived interest in producing a liquid fuel from coal. Coal liquefaction has been technically feasible for many years. Germany produced large quantities of gasoline from coal during World War II. A few countries today, with limited petroleum resources but abundant coal supplies, depend on coal liquefaction for their liquid fuel and gasoline. In the United States, coal liquefaction is viewed as a positive step toward reducing dependence on foreign petroleum. The liquefied product can be used as a heating oil or steam power plant fuel or refined to produce transportation fuel.

The conversion of coal into a liquid fuel requires the addition of hydrogen to the coal. Coal has a ratio of hydrogen atoms to carbon atoms of only 0.8 to 1, while in petroleum this ratio is 1.75 to 1. In addition to being a major technical goal, the addition of hydrogen to coal is also the major cost item in the conversion process.

There are three basic methods that have been used to liquefy coal. These are:

1. Hydrogenation or direct liquefaction
2. Catalytic conversion (also called indirect liquefaction)
3. Hydropyrolysis

In the *hydrogenation process*, coal and catalyst are suspended as a slurry using a process-derived oil. The mixture is then reacted with hydrogen at high pressure and moderate temperature to form liquid hydrocarbons. The hydrogenation process is the major thrust of the coal liquefaction program.

In the *catalytic conversion process*, a synthesis gas is produced from the coal by one of the methods previously described. The hydrogen and carbon monoxide in the gas are then combined in the presence of a catalyst to form a liquid hydrocarbon fuel.

When coal is heated to above 450°C for a brief period, its component molecules fragment extensively. If the products are promptly stabilized, the fraction of coal volatilized can greatly exceed the analytically determined volatile matter. In *hydropyrolysis*, hydrogen-entrained pulverized coal is flash pyrolyzed. Under proper conditions up to 50% of the coal can be liquefied.

Most of the liquefaction processes under development are based on the use of the liquid-phase hydrogenation process with low-rank coals. A typical process using low-rank coal operates at 450°C and 25 MPa. The hydrogenation of bituminous coal is performed at even higher pressures, up to 70 MPa (10,000 psi). The coal–oil slurry feedstock is cycled through a series of reactors, and gases and light liquid hydrocarbons are removed when the mixture passes between reactors. A catalyst of iron oxide or ferrous sulfate is often combined with the coal–oil slurry to increase the rate of the reaction process.

Synthetic Solid Fuels

The widespread availability of very high organic sulfur coal in the eastern United States has stimulated research into ways the sulfur content can be reduced prior to combustion. The *solvent refining of coal* (SRC) is one way in which this can be accomplished.

The SRC process is essentially a prematurely arrested liquefaction process. Pulverized coal is slurried with recycled solvent and pressurized with gaseous hydrogen at ~450°C. After hydrogenation, unreacted hydrogen and H_2S are removed by degassing and filtration removes solids (ash). Solvent is then separated by vacuum distillation and the SRC product (mp 150 to 200°C) is solidified by cooling. The product is almost entirely sulfur and ash free.

The process has been proven to be technologically sound, but the costs imposed by treating coal in this manner are considerably higher than those imposed by SO_2 gas scrubbers. Current development efforts center on conversion of the SRC product to transportation fuels.

Utility Applications for Synthetic Fuels

Low- and medium-Btu gas offers considerable potential as a fuel for both new and existing power plants. In future power plant construction, gasifiers can be integrated with steam turbines or gas turbines or combined cycle operation. However, high-Btu synthetic gas is expensive and probably not practical as a power plant fuel. Integration of a suitable gasifier with a high-temperature gas turbine offers the potential for increased overall system efficiency.

Since the sulfur has been largely removed from the gas prior to combustion, there is also the potential for savings in SO_2 removal and disposal systems. These savings would be reflected in both capital and operating costs for the power plant.

There are significant differences between the combustion properties of these synthetic fuels and natural gas. The heat content of the substitute gas is considerably lower than that of natural gas, 100 to 500 Btu/standard ft^3 versus 1030 Btu/standard ft^3 for natural gas. This requires a proportionally increased volume of gas flow to the burner system and sharply increased pressure drops throughout the gas system.

Because of its composition, low- and medium-Btu gas requires less air for combustion. In a typical low-Btu gas the stoichiometric air/fuel mixture may increase by only 30% and the flue gas volume may increase by only 20%.

The flame temperature is lower for these substitute gases than with natural gas. A synthetic gas flame temperature is approximately 1760°C (3200°F) versus a flame temperature of 1960°C (3560°F) for natural gas. The ignition of low- and medium-Btu gas poses no additional problems. There is actually a wider range of flammability limits than natural gas due to the higher hydrogen content.

Existing furnaces can be modified to utilize low- and medium-Btu gas. In general, boilers designed for pulverized coal firing would be least affected by a change to low- or medium-Btu gas. Typical modifications that would be required are:

> Installation of a new fuel-gas piping system
>
> Combustion equipment modification
>
> Enlargement of burners
>
> Uprating of forced- and induced-draft fans
>
> Some modifications of heat-transfer surfaces

Oil- or gas-fired boilers that have been sized for conventional oil and natural gas utilization cannot be easily modified to burn low-Btu gas. The furnace is much too small to complete combustion of the fuel without derating the unit by as much as 50%. Burner enlargement is also a problem. Much less severe problems are encountered in the conversion of oil and natural gas systems to medium-Btu gas.

The prospects for utilizing synthetic liquid fuels do not appear as bright as for synthetic gaseous fuels. Primary use would appear to be as a fuel for gas turbine peaking units that utilize liquid fuel. Some applications are possible in combined cycle plants which are expected to share base-load duty with conventional coal and nuclear plants in the future.

The coal-derived liquids are more aromatic and contain more nitrogen and oxygen and less hydrogen than conventional petroleum products. The low hydrogen-to-carbon ratio means that the combustion will be smoky. Improvements will be required in burner design and combustion control if these fuels are to be utilized. These fuels can be combined with conventional oil to minimize these problems. Combustion characteristics are similar to those of No. 2 fuel oil except that nitrogen oxide emissions tend to be slightly higher, while SO_2 emissions are generally lower. System and modifications are less than those required to utilize low- and medium-Btu gases.

SYMBOLS

a'	constant dependent on specific gravity of oil
A'_s	ash content of coal, w/o
c_p	specific heat capacity at constant pressure, J/kg·°K
C'	total carbon, w/o
C'_d	fixed carbon–dry, mineral matter free, w/o
C'_f	fixed carbon, w/o
H'	hydrogen content, w/o
H'_a	available hydrogen content, kg H/kg coal
M'_o	moisture content of coal, w/o
N'	nitrogen content of coal, w/o
O'	oxygen content of coal, w/o
Q'_{AB}	as-burned heating value of coal, cal/g or kcal/kg
Q'_d	dry mineral-matter-free heating value, cal/g or kcal/kg
Q'_{high}	high heating value of coal, cal/g or kcal/kg
Q'_{low}	low heating value of coal, cal/g or kcal/kg
S'	sulfur content of coal, w/o
s.g.	specific gravity, dimensionless
T	temperature, °K (unless specified otherwise)

V' volatile matter content of coal, w/o
V'_d volatile matter content of coal—dry, mineral matter free, w/o
w_m mass H_2O in combustion products/original mass coal, kg H_2O/kg coal

PROBLEMS

3.1. The density of a fuel oil is found to be 45°API and its viscosity at 50°C is 55 centistokes (cS). Estimate
(a) w/o hydrogen,
(b) w/o carbon, and
(c) higher heating value.
(d) Determine the fuel oil grade.

3.2. A petroleum is found to have the following composition:

	w/o
n-Decane	10
n-Dodecane	30
n-Tridecane	40
n-Nonane	10
2,2,3-Trimethylpentane	5
2,2-Dimethylhexane	5

Determine the higher heating value of this petroleum fraction.

3.3. (a) A natural gas fuel contains 80 v/o methane, 10 v/o ethane, and 10 v/o propane. Determine the heat released by combustion of 1 m³ of the gas. Assume that products of combustion are CO_2 and liquid H_2O.
(b) How would the answer to part (a) change if the water produced were vaporized?

3.4. Hydrogen has a heat combustion of 286,000 kJ/kg mol and CO has a heat combustion of 283,000 kJ/kg mol. Determine the heat released by combustion of 1 m³ of a mixture which is 50 v/o H_2 and 50 v/o CO. Determine both higher and lower heating values per cubic meter of gas. How does your answer compare with those of Problem 3.3?

3.5. For coal having the proximate and ultimate analyses given in Table 3.12:
(a) Determine dry mineral-matter-free percentages of fixed carbon and volatile matter.
(b) Estimate the higher and lower heating values.

3.6. A coal sample has the following proximate analysis:

	w/o
Moisture	5.4
Volatile matter	22.1
Ash	9.9
Fixed carbon	62.6

In addition, a bomb measurement of its heat of combustion indicates that 32,600 kJ is released per kilogram burned (higher heating value). Estimate the total w/o carbon.

3.7. Show that the heat of combustion in Btu/lb may be obtained from the standard molal heat of combustion in cal/g mol by

$$\text{heat of combustion (Btu/lb)} = \frac{\text{molal heat of combustion (cal/g mol)}}{1.8 \times \text{mol wt}}$$

BIBLIOGRAPHY

BERKOWITZ, N., *An Introduction to Coal Technology*. New York: Academic Press, 1979.

BUDER, M., ET AL., "The Effects of Coal Cleaning on Power Generation Economics," *Proc. American Power Conference*, Vol. 41, 1979, p. 623.

Combustion, 3rd ed. Windsor, Conn.: Combustion Engineering, Inc., 1981.

CONN, A., and J. CORNS, *Evaluation of Project H-Coal*, Report PB-177068. Washington, D.C.: U.S. Dept. of Interior, Office of Coal Research, 1967.

FURLONG, L., ET AL., "Coal Liquefaction," *Chemical Engineering Progress*, Vol. 72, No. 8 (1976).

HUETTENHAIN, H., "Coal Preparation and Cleaning for the Power Industry," *Proc. American Power Conference*, Vol. 40, 1978, p. 515.

Steam, Its Generation and Use. Lynchburg, Va.: The Babcock & Wilcox Company, 1978.

4

Combustion

4.1 PRINCIPLES OF COMBUSTION

Combustion Reactions

Combustion is defined as the rapid chemical reaction of oxygen with the combustible elements of a fuel. There are three significant combustible chemical elements in coal and petroleum: carbon, hydrogen, and sulfur. The basic chemical equations for complete combustion are

$$C + O_2 \rightarrow CO_2$$
$$2H_2 + O_2 \rightarrow 2H_2O \tag{4.1a}$$
$$S + O_2 \rightarrow SO_2$$

When insufficient oxygen is present, the carbon will be burned incompletely with the formation of carbon monoxide.

$$2C + O_2 \rightarrow 2CO \tag{4.1b}$$

In order to burn a fuel completely, four basic criteria must be satisfied:

1. An adequate quantity of air (oxygen) must be supplied to the fuel.
2. The oxygen and fuel must be throughly mixed.
3. The fuel/air mixture must be maintained at or above the ignition temperature.
4. The furnace volume must be large enough to give the mixture time for complete combustion.

Sec. 4.1 Principles of Combustion

The furnace burner supplies the air and provides the mixing action for the combustion process. Since complete mixing of the air and fuel is virtually impossible, excess air must be supplied to the combustion process to ensure complete combustion. The mixing process and the quantity of excess air supplied will determine whether the exhaust gases will contain both the products of complete and incomplete combustion. The products of incomplete combustion include unburned fuel, carbon monoxide, and partially oxidized fuel. Most of the products of incomplete combustion are atmospheric pollutents.

Heating Value of a Fuel (Heat of Combustion)

In Chapter 3 it was pointed out that the heating value of a fuel is numerically equal to its standard heat of combustion but with an opposite sign. It was also indicated that fuel oil heating values may be most accurately obtained from the heats of combustion of the components (see Table 3.4) when the chemical composition is known. Alternatively, ways of estimating the heating value from knowledge of the fuel oil type or its specific gravity were indicated. The heating value of natural gas is almost always determined from the heating value of its component chemicals.

Section 3.4 indicates how approximate heating values for coal may be obtained on the basis of the coal's rank. When the ultimate analysis is known, the heating value for complete combustion may be accurately obtained from the Dulong–Berthelot equation [Eq. (3.9)]. Alternatively, particularly in problems involving incomplete combustion of coal, it may be desirable to obtain the heating value of coal directly from the heats of combustion of its constituents. Heats of combustion for the major constitutents of coal are tabulated in Table 4.1. Although the heat released when carbon is burned to CO is not given in Table 4.1, it is readily determined from the difference between the heats of combustion of carbon and CO tabulated in the table.

In computing the fuel available for combustion from the ultimate analysis of a coal, it is generally assumed that all the carbon and sulfur are present in elemental form and that they are available for combustion. It is, however, assumed that all oxygen and nitrogen reported in the ultimate analysis are combined with hydrogen. The *available hydrogen* is that hydrogen available for combustion and is the total

TABLE 4.1 HEATS OF COMBUSTION OF COAL CONSTITUENTS

	Formula and state	Product of combustion and state	Heat of combustion	
			cal/g mol	kJ/kg mol
Carbon (coke)	C(s)	CO_2(g)	−97,000	−407,000
Carbon	C(s)	CO_2(g)	−94,400	−397,000
Carbon monoxide	CO(g)	CO_2(g)	−67,620	−283,000
Hydrogen	H_2(g)	H_2O(l)	−68,310	−286,000
Sulfur	S(s)	SO_2(g)	−69,400	−291,000

hydrogen reported less than that required to combine with the oxygen and the nitrogen in the coal as $H_2O^\dagger$ and H_2N, respectively. If these assumptions are made, and all carbon is assumed to be burned to CO_2, the heats of combustion of Table 4.1 can be used to derive an equation that is very close to the Dulong–Berthelot formula.

For 1 g of coal containing C' w/o carbon, the heat released by the carbon combustion (taking carbon to have heat of combustion of coke) at standard conditions is

$$\text{heat released by carbon combustion} = \left(\frac{C'}{100}\right) g \times \frac{97{,}000 \text{ cal/g mol}}{12 \text{ g/g mol}} = 81.0 C \text{ cal/g}$$

Similarly, we have for the sulfur (when S' is w/o sulfur),

$$\text{heat released by sulfur} = \left(\frac{S'}{100}\right) g \times \frac{69{,}400 \text{ cal/g mol}}{32 \text{ g/g mol}} = 21.8 S \text{ cal/g}$$

However, if we have H' w/o hydrogen and O' w/o oxygen, the available hydrogen is $H' - O'/8$; hence

$$\begin{matrix}\text{heat released by} \\ \text{Hydrogen combustion} \\ \text{per 1 g of coal}\end{matrix} = \left(H' - \frac{O'}{8}\right)\left(\frac{1}{100}\right) g \times \frac{68{,}310 \text{ cal/g mol}}{2 \text{ g/g mol}}$$

$$Q = \text{total heat released} = 81.0 C' + 341.5\left(H' - \frac{O'}{8}\right) + 21.8 S' \text{ cal/g}$$

The quantities C', H', O', and S' are percent by weight of carbon, hydrogen, oxygen, and sulfur, respectively.

Theoretical Air-to-Fuel Ratio

The oxygen for the combustion process comes from the air supplied to the burner. In the design of the boiler furnace, it is customary practice to supply sufficient air for complete combustion plus excess air to allow for incomplete mixing. For any fuel, the moles of dry air theoretically required for complete combustion are determined from the moles of oxygen required. For fuel containing carbon, hydrogen and sulfur we may write a balanced chemical equation of the form

$$C_{Z_i} H_{Z_{i+1}} S_{Z_{i+2}} + \left(Z_i + \frac{Z_{i+1}}{2} + Z_{i+2}\right) O_2 \rightarrow Z_i(CO_2) + \frac{Z_{i+1}}{2} H_2O + Z_{i+2} SO_2 \tag{4.2}$$

Since air consists of 79% N_2 and 21% O_2, the ratio of N_2 moles to O_2 moles is

---

† The quantity of water containing the equivalent of the oxygen in the fuel is called the *combined water*.

79/21 = 3.76. Hence if our fuel is burned in air, we have

$$C_{Z_i}H_{Z_{i+1}}S_{Z_{i+2}} + \left(Z_i + \frac{Z_{i+1}}{2} + Z_{i+2}\right)(O_2 + 3.76N_2) \rightarrow$$

$$Z_i(CO)_2 + \frac{Z_{i+1}}{2}H_2O + Z_{i+2}(SO_2) + \left(Z_i + \frac{Z_{i+1}}{2} + Z_{i+2}\right) \times 3.76N_2 \quad (4.3)$$

It is obvious that for each mole of carbon and sulfur in the fuel, 4.76 mol of air is required. For each kilogram atom of available hydrogen in the fuel, 4.76/2 = 2.38 mol of air is needed. Since the molecular weight of air is 28.9 g/g mol, the mass of air required per gram of carbon, $m'_{a/C}$, is

$$m'_{a/C} = \frac{1 \text{ g}}{12 \text{ g C/g mol C}} \times \frac{4.76 \text{ g mol air}}{\text{g mol C}} \times 28.9 \text{ g/g mol air} = 11.47 \frac{\text{g air}}{\text{g carbon}} \quad (4.4)$$

In a similar fashion, we find that 4.3 g of air is required per gram of sulfur and 34.4 g of air per gram of available hydrogen. Hence

$$m'_a = \left[11.47C' + 34.4\left(H' - \frac{O'}{8}\right) + 4.3S'\right]\left(\frac{1}{100}\right) \quad (4.5)$$

where m'_a represents grams of air required for complete combustion of 1 g of fuel (mass air/mass fuel).

The theoretical air-to-fuel ratios for coal range from 8.5 to 12.5. The theoretical air-to-fuel ratio for liquid or gaseous fuels can be calculated using Eq. (4.4) or an alternative form utilizing the number of atoms of each combustible element in a mole of liquid or gas. That is,

$$M'_{a/F} = 4.76Z_C + 2.38Z_H + 4.76Z_S \quad (4.6)$$

where $M'_{a/F}$ = moles of air per mole of fuel
Z_C, Z_H, Z_S = atoms C, H, and S per mole of fuel

The theoretical weight (or volume) of air required per mole of fuel burned must always be determined in the analysis of any combustion processes. Excess air, above this theoretical minimum, must be supplied to ensure complete combustion. Excess air is expressed as a percentage or through the use of a dilution coefficient. The precent excess air supplied during operation is

$$\% \text{ excess air} = \frac{(m'_a)_{op} - (m'_a)_{theo}}{(m'_a)_{theo}} \times 100 = \frac{\text{mass of air used} - \text{mass of air theoretically required}}{\text{mass of air theoretically required}} \times 100$$

while d, the dilution coefficient, is given by

$$d = \frac{(m'_a)_{op}}{(m'_a)_{theo}} = \frac{\text{mass of air to be used}}{\text{mass of air theoretically required}}$$

The percentage of excess air varies between 15 and 30% for most large utility boiler/furnace systems.

Example 4.1

A liquid fuel having the chemical composition $C_{10}H_{22}$ is burned with 30% excess air. Determine the air-to-fuel ratio used.

Solution Basis for solution: 1 kg mol of fuel. We note that $Z_C = 10$ and $Z_H = 22$ and hence from Eq. (4.6),

$$M'_{a/F} = 4.76(10) + 2.38(22) = 99.96 \text{ kg mol air/kg mol fuel}$$

theoretical weight of air required = 99.96 kg mol × 28.9 kg/kg mol = 2889 kg

actual air used = 1.3 × kg mol theoretical air

weight of fuel used = $Z_C(12) + Z_H(1) = 120 + 22 = 142$ kg

$$\text{air/fuel ratio} = \frac{\text{weight air}}{\text{weight fuel}} = \frac{2889 \text{ kg} \times 1.3}{142 \text{ kg}} = 20.3$$

Actual Air-to-Fuel Ratio

Example 4.1 illustrates the calculational procedure that is used by the designer to set the air-to-fuel ratio for a given combustion process. The actual ratio achieved after the burner is built must be estimated from experimental measurements of the gaseous components of the flue gas. There are several ways to perform these measurements, but one of the most frequently used is the simple *Orsat portable gas analyzer*. The Orsat gas analyzer is used to determine the volumetric or molar fractions of carbon monoxide, carbon dioxide, and oxygen in the dry exhaust gas.

In a typical measurement, a 100-cm³ flue gas sample is collected over water at ambient conditions, and passed through a series of chemical solutions. It is usually assumed that the sample is free of water vapor and SO_2, since any water vapor would have condensed during the collection process and the SO_2 in the flue gas will react with the water in the collecting container. The flue gas sample is analyzed for carbon dioxide, carbon monoxide, oxygen, and nitrogen. The reagents normally used are a KOH solution to remove the CO_2, pyrogallol solution to remove the O_2, and a cuprous chloride mixture ($CuCl_2$) to remove the CO. The final remaining, unabsorbed gas is assumed to be nitrogen.

For liquid fuels, the ultimate and the Orsat analyses are sufficient to evaluate the actual air-to-fuel ratio. A refuse analysis is also required to determine the actual air-to-fuel ratio when burning coal. The refuse analysis is an experimental evaluation of the higher heating value of the refuse. The results of this analysis are reported in units of energy per unit of mass of refuse (kJ/kg). The refuse analysis assumes that all of the ash and unburned carbon is collected in the ash pit below the furnace. The refuse analysis can also be reported as the mass fraction of carbon in the refuse (kilograms of carbon in refuse/kilograms of refuse) or percent combustible carbon in the refuse. If 32,778 kJ/kg is taken as the HHV (high heating value) for pure carbon, then β, the mass fraction of carbon in the refuse (kg C/kg refuse) is

$$\beta = \frac{\text{HHV} \cdot \text{refuse}}{\text{HHV} \cdot \text{pure carbon}} = 3.05 \times 10^{-5} \, (\text{HHV} \cdot \text{refuse}) \qquad (4.7)$$

provided that (HHV · refuse) is in units of kJ/kg.

Sec. 4.1 Principles of Combustion

The actual air-to-fuel ratio, $(m'_a)_{act}$, in the combustion process can be calculated from the refuse data and the Orsat and ultimate analysis data. From the refuse analysis we can obtain m'_b, the mass of carbon burned per unit mass of coal. From the Orsat analysis and ultimate analysis, we obtain the air used per unit mass of carbon burned. We can then write

$$(m'_a)_{act} = \frac{\text{kg air used}}{\text{kg coal burned}} = \frac{\text{mol air} \times \dfrac{\text{kg air}}{\text{mol air}}}{\text{mol C} \times \dfrac{\text{kg C}}{\text{mol C}}} \times m'_b$$

Then

$$(m'_a)_{act} = m'_b \left(\frac{28.9}{12}\right) \frac{\text{mol air}}{\text{mol C burned}} = 2.41 m'_b \times \frac{4.76 \text{ mol air}}{3.76 \text{ mol N}_2} \times \frac{\text{mol N}_2 \text{ in air}}{\text{mol C}}$$

$$= 3.04 m'_b \frac{\text{mol N}_2 \text{ in air}}{\text{mol C}}$$

If there were no nitrogen in the fuel, the moles of N_2 in air per mole of C burned would be obtained from the volumetric fractions in the Orsat analysis. That is, we would have

$$(m'_a)_{act} = 3.04 m'_b \frac{(N_2)_{FG}}{(CO)_{FG} + (CO_2)_{FG}}$$

where $(CO)_{FG}$, $(CO_2)_{FG}$, and $(N_2)_{FG}$ are the volume fractions of CO, CO_2, and N_2, respectively, in flue gas. However, the mass fraction of nitrogen in the coal, m'_N (kg N_2/kg C), leads to the following error in $(m'_a)_{act}$:

$$m'_N \frac{\text{kg N}_2}{\text{kg C}} m'_b \frac{\text{kg C burned}}{\text{kg coal}} \left(\frac{4.76 \times 28.9}{3.76 \times 28}\right) \frac{\text{kg air}}{\text{kg N}_2} = 1.3 m'_N m'_b \quad (4.8)$$

Therefore,

$$(m'_a)_{act} = 3.04 m'_b \frac{(N_2)_{FG}}{(CO)_{FG} + (CO_2)_{FG}} - 1.3 m'_N m'_b \quad (4.9)$$

The volume fractions of $(CO)_{FG}$, $(CO_2)_{FG}$, and $(N_2)_{FG}$ in the flue gas are obtained from the Orsat analysis, m'_b is calculated from the refuse analysis, and m'_N is obtained from the ultimate analysis of the fuel. The value of m'_b is obtained from the refuse data by the following series of calculations:

$$m'_{A/r} = 1.0 - \beta = \frac{\text{kg ash}}{\text{kg refuse}} \quad (4.10)$$

$$m'_r = \frac{\text{kg refuse}}{\text{kg coal}} \quad (4.11)$$

$$m'_C = m'_r - m'_a = \frac{\text{kg unburned carbon}}{\text{kg coal}} \quad (4.12)$$

and finally, m_b' is

$$m_b' = \frac{C'}{100} - m_C' = \frac{\text{kg carbon burned}}{\text{kg coal}} \qquad (4.13)$$

This technique can be used to compare the theoretical air-to-fuel ratio with the actual air-to-fuel ratio.

4.2 BURNER DESIGN

Pulverized Coal

Before being supplied to the burner mechanism, coal must be reduced to small particles such that the surface-to-mass ratio is sufficient to ensure complete combustion. In a direct-firing system, the dried, pulverized coal is transported to the burner by the primary air and mixed with the secondary air in the burner. The flow velocity of the coal–air mixture must be high enough to keep the fuel particles suspended in the airstream. This typically requires a minimum flow velocity of 300 ft/min (1.5 m/s) for horizontal burners or 2200 ft/min (11 m/s) for vertical burners. The air in

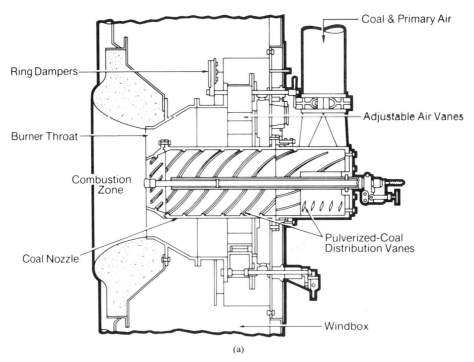

(a)

Figure 4.1 (a) Burner for horizontal firing of coal. (Reproduced with permission from *Combustion—Fossil Power Systems*, Combustion Eng., Inc., Windsor, Conn., 1981.)

Sec. 4.2 Burner Design

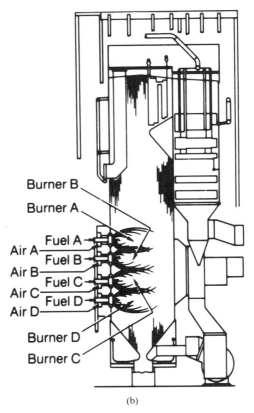

(b)

Figure 4.1 (b) Flow pattern for horizontal (wall) firing. (Reproduced with permission from *Combustion—Fossil Power Systems*, Combustion Eng., Inc., Windsor, Conn., 1981.)

which the coal is suspended is called *primary air*. The additional air, called *secondary air*, required to bring the air-to-fuel ratio to the required level is added at the burner.

The most frequently used burners are circular or cell type in design. Figure 4.1 shows a burner of this design. Note the oil attachment required to get a flare going after burner shutdown. The burners usually fire horizontally or at slight inclinations and are often mounted in the corners of the furnace walls so that the flames can be directed in a circular pattern referred to as *tangential firing*. Another popular arrangement is for the burners to be inserted in the walls at opposite sides of the furnace. This arrangement is called *opposed firing*. Other firing methods are illustrated in Fig. 4.2.

The cyclone firing scheme [see Fig. 4.2(d)] differs substantially from the other approaches. In this scheme, the fuel/air mixture and secondary air are tangentially introduced into a cylindrical combustion chamber so that the material follows a helical path. The chamber length and diameter is designed so that when the exit of

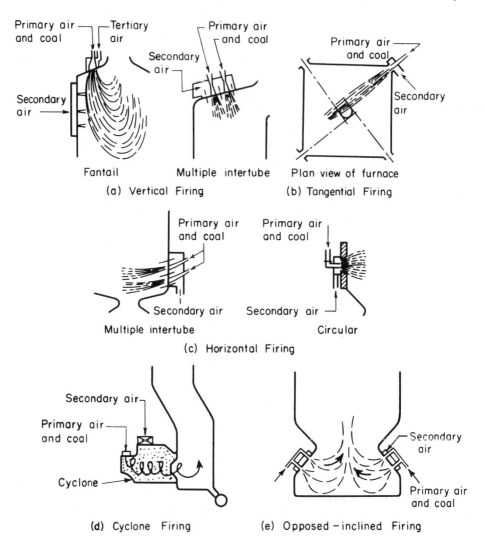

Figure 4.2 Firing methods for pulverized coal: (a) vertical firing; (b) tangential firing; (c) horizontal firing; (d) cyclone firing; (e) opposed-inclined firing. (From *Chemical Engineering Handbook*, by R. H. Perry and C. H. Chilton. Copyright 1973, used with the permission of McGraw-Hill Book Company, New York.)

the chamber is reached, the combustion process is essentially complete. Only hot gases enter the main body of the furnace. Slag (molten ash) forms in the water-cooled ceramic surface of the combustion chamber and is drained.

In most pulverized coal furnaces, the burner direction or attitude can be changed to direct the energy deposition to different regions of the boiler. If the burners are directed downward, the evaporation section of the boiler receives more

Sec. 4.3   Mass and Energy Balances   111

heat. If the burners are directed upward, more heat is transferred to the superheat and reheat sections of the boiler.

Oil-Burning Equipment

Oil burners are also located on the vertical walls of the furnace. The most frequently used burners are the circular type, similar in geometry to the pulverized coal burners. The fuel is fed into the burner in a dense mixture, but the velocity of the air, plus the mixing action of the fuel jets, permits thorough mixing of the fuel and combustion air. Complete combustion requires that the fuel be dispersed into a very fine mist. The process of producing this fine mist is referred to as *atomizing* the fuel. The small particles, with a favorable surface-to-volume ratio, ensure quick ignition and complete combustion. Atomization is accomplished by either mixing the fuel oil with gas (steam or air) or by a mechanical atomizer.

As previously noted, the fuel oil used in large utility boilers generally requires heating to reduce the viscosity and allow for the atomizing process to take place. The fuel oil must be heated to about 57°C (135°F) if No. 4 oil is used or to about 140°C (220°F) if No. 6 oil is used. The oil should not be heated to the point where vapor binding occurs in the pump system. The oil used in these burners must also be free from contamination, such as grit and foreign matter, which tend to clog the burners and control valves. Fine screens or filters to remove small particles are generally used upstream of the burners.

4.3 MASS AND ENERGY BALANCES

Furnace Mass Balance

The calculation of a material balance for a furnace involves the same basic ideas for gas-, coal-, and oil-fired furnaces. The material balance in a simple combustion process accounts for the mass of fuel and air supplied to the furnace (including moisture in the air) and the mass of ash and combustion products produced in the furnace. Figure 4.3(a) illustrates the input and output terms for the mass balance.

The ultimate analysis of the coal is not required, but a proximate analysis, indicating the carbon content, moisture, and ash content, should be available. If the balance is to reflect actual operating conditions, an analysis (such as an Orsat analysis) of the flue gas is necessary. In addition, the refuse from the furnace should be analyzed for ash, unburned carbon, and other matter. The air entering the furnace may be assumed to be of average atmospheric composition and moisture content or pressure and humidity may be measured.

Every material balance requires a *basis* for the calculation. The basis may be a unit mass of fuel or the mass of fuel used in a given cycle of operation, such as a 24-hour period. Alternatively, we may use 1 kg mol of fuel (if the composition is definitely known) or 1 kg mol of a particular component (e.g., carbon).

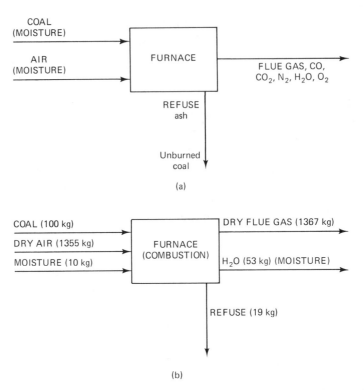

Figure 4.3 (a) Major constituents for a furnace/boiler mass balance; (b) example of a typical mass balance.

First let us consider a mass balance from the designer's point of view.

Example 4.2

Consider a furnace burning a natural gas, consisting of 60 v/o methane and 40 v/o ethane, in 105% of the air required for complete combustion. We are to determine the mass distribution of combustion products.

Solution Basis for solution: 1 g mol of fuel gas. If we assume the fuel gas to be essentially a perfect gas, then each gram mole of fuel gas will contain 0.6 mol of methane and 0.4 mol of ethane. Our basic chemical equation is then

$$0.6(CH_4) + 0.4(C_2H_6) + (1.05)(2.6)(O_2 + 3.76N_2) \rightarrow$$
$$1.4CO_2 + 2.4(H_2O) + (1.05)(2.6)(3.76N_2) + (0.05)(2.6)O_2$$
(4.14)

Note that in balancing the foregoing chemical equation, we determine x, the number of moles of CO_2 from the carbon balance:

$$x = 0.6 + 0.4(2)$$

We then determine the number of moles of H_2O, y, from the hydrogen balance:

$$2y = 0.6(4) + 0.4(6)$$

Sec. 4.3 Mass and Energy Balances

Finally, we find the moles of oxygen, z, that must be added to provide the H_2O and CO_2 formed:

$$2z = 1.4(2) + 2.4$$

With the balanced chemical equation and a knowledge of molecular weights, we determine the masses of the output streams:

$$\text{mass of } CO_2 \text{ produced} = 1.4 \frac{\text{g mol}}{\text{g mol fuel}} \times 44 \text{ g/g mol} = 61.6 \text{ g/g mol fuel}$$

$$\text{mass of } H_2O \text{ produced} = 2.4 \frac{\text{g mol}}{\text{g mol fuel}} \times 18 \text{ g/g mol} = 43.2 \text{ g/g mol fuel}$$

$$\text{mass of } N_2 \text{ remaining} = (1.05)(2.6)(3.76) \frac{\text{g mol}}{\text{g mol fuel}} \times \frac{28 \text{ g}}{\text{g mol}} = 287.4 \frac{\text{g}}{\text{g mol fuel}}$$

$$\text{mass of } O_2 \text{ remaining} = (0.05)(2.6) \frac{\text{g mol}}{\text{g mol fuel}} \times \frac{32 \text{ g}}{\text{g mol}} = 4.2 \text{ g/g mol fuel}$$

Mass-balance computations for a coal-fired furnace are slightly more complex. The complexity is increased when we attempt to perform a mass balance on an operating furnace. In that case, the engineer is generally provided with the following data: analysis of fuel, analyses of refuse from ash pit, and analyses of flue gas, humidity, and pressure of the entering air. The use of this information is illustrated in the following example.

Example 4.3

The following data were obtained from a test conducted on a coal-fired furnace.

Ultimate analysis	w/o
Carbon	66.12
Available hydrogen	3.60
Nitrogen	1.40
Combined water	6.21
Sulfur	1.60
Free moisture	4.48
Ash	16.59
Total	100.00

Total heating value	10,500 $\frac{\text{kcal}}{\text{kg}}$
Total weight of coal fired	45,000 kg
Average temperature of coal fired	23°C

Data on refuse drawn from ash pit:

Ash content	86.23%
Carbon content	13.77%
Average temperature	124°C
Mean specific heat from 23 to 124°C	0.23 $\frac{\text{kcal}}{\text{kg} \cdot \text{°C}}$

Data on flue gas:

Orsat analysis	w/o
Carbon dioxide	11.70
Sulfur dioxide	0.11
Carbon monoxide	0.04
Oxygen	6.37
Nitrogen	81.78
Total	100.00
Average temperature	250°C

Data on air:

Average dry-bulb temperature	23°C
Average wet-bulb temperature	15°C
Average barometric pressure	0.98 bar

Determine the complete material balance.

Solution Basis for solution: 100 kg of coal feed.

1. *Mass of refuse formed*, m_R: The quantity of refuse formed can be weighed directly or can be calculated from the ash content of the coal. By using the ultimate analysis data, we have

$$\text{ash from coal} = 0.1659 \frac{\text{kg ash}}{\text{kg coal}} \times 100 \text{ kg coal} = 16.59 \text{ kg ash}$$

$$\text{ash from refuse} = 0.8623 \text{ kg ash/kg refuse}$$

$$\text{mass of refuse formed} = \frac{16.59}{0.8623} = 19.24 \text{ kg refuse}$$

2. *Mass of the dry gaseous products*, m_{FG}: The weight of the dry gaseous products is calculated from a carbon balance. Carbon is selected because the carbon content of the coal and gaseous products can be accurately determined.

   a. *Carbon balance*: To determine the carbon content of the coal and ash, use 100 kg of coal burned.

$$\text{carbon in the coal} = 100 \text{ kg coal} \times 0.6612 \frac{\text{kg C}}{\text{kg coal}} = 66.12 \text{ kg C}$$

$$\text{kg mol C in coal} = 66.12 \text{ kg C} \times \frac{\text{kg mol C}}{12 \text{ kg}} = 5.51 \text{ kg mol C}$$

$$\text{carbon in refuse, } C_r = 19.24 \text{ kg refuse} \times 0.1377 \frac{\text{kg C}}{\text{kg refuse}}$$

$$= 2.649 \text{ kg C or } 0.22 \text{ kg mol C}$$

Sec. 4.3　Mass and Energy Balances

To determine the carbon in the flue gas, use 1 kg mol of flue gas as the basis for solution.

$$\text{carbon in } CO_2 = C_2$$

$$= 1 \text{ kg mol gas} \times 0.1170 \frac{\text{kg mol } CO_2}{\text{kg mol gas}}$$

$$\times 12 \frac{\text{kg C}}{\text{kg mol } CO_2} \times \frac{1 \text{ kg mol C}}{12 \text{ kg C}}$$

$$= 0.1170 \text{ kg mol}$$

$$\text{carbon in } CO = C_1 = 0.0004 \text{ kg mol}$$

$$\text{total carbon in flue gas} = C_1 + C_2 = 0.1174 \frac{\text{kg mol C}}{\text{kg mol flue gas}}$$

The moles of dry flue gas per 100 kg of coal fired can now be calculated from the results above:

$$\text{number of moles of flue gas} = M_{FG} = \frac{C_{FG}}{C_1 + C_2}$$

$$= \frac{5.29 \text{ kg mol C}}{0.1174 \frac{\text{kg C}}{\text{kg mol flue gas}}} = 45.06 \text{ kg mol}$$

b. *Total dry gaseous products*: Use 100 kg of coal burned as the basis for your calculations.

$$\text{mass of } CO_2 = 45.06 \text{ kg mol gas} \times 0.1170 \frac{\text{kg mol } CO_2}{\text{kg mol gas}} \times \frac{44 \text{ kg } CO_2}{\text{kg mol } CO_2}$$

$$= 232 \text{ kg}$$

$$\text{mass of } CO = 45.06 \times 0.0004 \times 28 = 0.505 \text{ kg}$$

$$\text{mass of } O_2 = 45.06 \times 0.0637 \times 32 = 91.85 \text{ kg}$$

$$\text{mass of } SO_2 = 45.06 \times 0.0011 \times 64 = 3.17 \text{ kg}$$

$$\text{mass of } N_2 = 45.06 \times 0.8178 \times 28.2 = 1039 \text{ kg}$$

$$\text{total mass of dry gaseous products} = 1366.5 \text{ kg}$$

3. *Mass of dry air supplied for combustion*, m_{da}: The weight of dry air supplied for combustion can be calculated from the Orsat analysis data and the carbon content of the refuse, as previously discussed. Another method, used when the carbon content of the refuse is unknown, is based on a nitrogen balance.

   The dry air used for combustion consists not only of oxygen and nitrogen but contains small quantities of rare gases, principally argon. Since both nitrogen and the rare gases are inert, the usual practice is to consider the rare gases as nitrogen and to compensate for this by assigning a molecular weight of 28.2 to "atmospheric nitrogen." Essentially all the nitrogen passes through the furnace and emerges in the flue gas. Any nitrogen in the coal will also be released during combustion and appear in the flue gas. As will be shown, the nitrogen content of the fuel is small compared to the atmospheric nitrogen and can usually be neglected.

Nitrogen balance: Use 100/kg of coal burned as the basis for your solution.

a. *Nitrogen in flue gas*:

$$M_{N_2, FG} = 45.06 \text{ kg mol gas} \times 0.8178 \frac{\text{kg mol N}_2}{\text{kg mol gas}} = 36.85 \text{ kg mol N}_2$$

b. *Nitrogen from coal*:

$$M_{N_2, F} = 100 \text{ kg coal} \times 0.014 \frac{\text{kg N}_2}{\text{kg coal}} \times \frac{1}{28} \frac{\text{kg mol N}_2}{\text{kg N}_2} = 0.0496 \text{ kg mol N}_2$$

c. *Weight of N_2 in flue gas*:

$$\text{weight of N}_2 = 36.85 \text{ kg mol N}_2 \times 28.2 \frac{\text{kg N}_2}{\text{kg mol N}_2} = 1039 \text{ kg N}_2$$

d. *Nitrogen from air*:

$$M_{N_2, a} = M_{N_2, FG} - M_{N_2, F} = 36.8 \text{ kg mol}$$

Dry air is assumed to contain 21% oxygen and 79% nitrogen; thus

$$\text{dry air supplied, } M_{da} = \frac{36.80 \text{ kg mol N}_2}{0.79 \frac{\text{kg mol N}_2}{\text{kg mol air}}} = 46.58 \text{ kg mol}$$

or

$$M_{da} = 46.58 \text{ kg mol} \times 29.1 \frac{\text{kg}}{\text{kg mol}} = 1355 \text{ kg}$$

4. *Weight of moisture in the air*: Basis for solution: 100 kg of coal burned. The weight of moisture in the air can be determined in a number of ways:

(a) by measurement of its dew point and determination of the partial pressure of water vapor from the steam tables for the dew-point temperature,

(b) by hygrometers which measure humidity directly,

(c) by a wet- and dry-bulb psychrometer, or

(d) by chemical analysis by use of desiccants. Based on the wet- and dry-bulb data given for this problem, the molal humidity of the air (moles of water vapor per mole of air) is found to be 0.012. (See Fig. 6.9 for a chart relating wet- and dry-bulb temperature to molal humidity.)

water contained in entering dry air

$$= M_{H_2O} = 0.012 \frac{\text{kg mol H}_2\text{O}}{\text{kg mol air}} \times 46.58 \text{ kg mol air} = 0.559 \text{ kg mol}$$

$$\text{weight of moisture in air} = 0.559 \text{ kg mole} \times \frac{18 \text{ kg}}{\text{kg mol}} = 10.1 \text{ kg}$$

5. *Total volume of wet air introduced*: Use 100 kg of coal as a basis for your solution. Total moles of moisture and dry air, $M_a = M_{da} + M_{H_2O} = 47.14$ kg mol. At standard conditions (1.103 bar, 273°K), 1 kg mol occupies 22.4 m³. Volume corrected to 23°C, 0.98 bar = V_a:

$$V_a = 47.14 \times 22.4 \times \frac{1.013}{0.98} \times \frac{296°K}{273°K} = 1199 \text{ m}^3$$

Sec. 4.3   Mass and Energy Balances   117

6. *Weight of moisture in the flue gas*: The flue gas analysis does not provide data on the quantity of moisture. The moisture content can be calculated from a hydrogen balance which considers the composition of the dry flue gas, the moisture content of the combustion air, and the hydrogen and moisture content of the fuel.

   Hydrogen balance: Basis for solution: 100 kg of coal fired.

   moisture introduced with dry air (calc. 4) = 0.559 kg mol

   water content of coal (combined water) = $\dfrac{6.21}{18}$ = 0.345 kg mol

   moisture on the coal (free water) = $\dfrac{4.48}{18}$ = 0.249 kg mol

   moisture from available hydrogen in the coal (ultimate analysis) = $\dfrac{3.60}{2.016}$ = 1.786 kg mol

   total hydrogen in water of flue gas
   $$= M_{H_2O} = M_{H_2} = 0.559 + 0.345 + 0.249 + 1.786 = 2.939 \text{ kg mol}$$

   moisture in flue gas = 2.939 kg mol $\times$ 18 kg/mol = 52.9 kg

   The partial pressure of the water vapor (p'_{H_2O}) can be found by using the average barometric pressure in the relationship:

   $$p'_{H_2O} = \dfrac{M_{H_2O}}{M_{H_2O} + M_{FG}} \, p = \dfrac{2.939}{2.939 + 45.06} \times 0.98 = 0.06 \text{ bar}$$

7. *Total volume of gaseous products*: Basis for solution: 100 kg of coal.

   total moles of wet gas = $M_{FG} + M_{H_2O}$
   $$= 45.06 + 2.939 = 48.0 \text{ kg mol}$$

   The total volume of flue gas products V_{FG} at 250°C (523°K) and 0.98 bar Hg would be

   $$V_{FG} = 48.0 \times 22.4 \times \dfrac{1}{0.98} \times \dfrac{523}{273} = 2102 \text{ m}^3$$

8. *Summary of material balances*: The final calculations can be checked by means of a balance sheet or a flowchart. The results are illustrated in Fig. 4.3(b). Material balances can also be prepared for nitrogen, oxygen, carbon, and hydrogen. These balances all help verify the accuracy of the calculations. In this example, the results show a small discrepancy resulting from errors in the test data, such as the Orsat and ultimate analyses.

Effect of sulfur on the mass balance. The preceding example did not include sulfur in the coal. If the sulfur content is less than 1%, very little error is introduced if the sulfur is neglected in the calculation. First, it is difficult to include sulfur in these calculations because sulfur is found in a variety of chemical forms in the coal itself. Second, the Orsat analysis does not report SO_2 content of the flue gas, since most of the SO_2 is absorbed in the water of the sampling apparatus. SO_2 can be measured in the flue gas, but it is not done routinely.

It can be shown that neglecting the combustion of even relatively high quantities of sulfur in the fuel will not introduce serious error into either the mass balance calculations or the energy balance calculations. The maximum error introduced is in the order of 3% for the worst case where the sulfur content is high and present in the pyritic form.

Energy Balances

If the combustion process takes place at standard conditions, the heat released may be obtained directly from the standard heats of combustion. Let us again consider the combustion of a fuel gas mixture.

Example 4.4

Assume the combustion of a gaseous mixture at 60% methane and 40% ethane described by Eq. (4.14) and determine the heat released.

Solution If we again take 1 g mol of fuel gas as our basis and find the heat released per gram mole from Table 3.8:

$$Q_0' = 0.6(\text{standard heat of combustion of } CH_4)$$
$$+ 0.4(\text{standard heat of combustion } C_2H_6)$$
$$= 0.6(212,800)\frac{\text{cal}}{\text{g mol}} + 0.4(368,400)\frac{\text{cal}}{\text{g mol}} = \frac{275,000 \text{ cal}}{\text{g mol}} \quad (4.15)$$

Suppose, however, that the combustion process had taken place at some elevated temperature, T_1. The heat of combustion, Q', at T_1 is not identical to the standard heat of reaction, Q_0', at our standard temperature, T_0. We can, however, readily determine Q' from Q_0' by making use of the thermodynamic principle which states that the net energy release in going from state 1 to 2 is independent of the path we take between those states. We may therefore imagine that instead of conducting the combustion isothermally at T_1, we first cool the reactants to T_0 and then burn the fuel at T_0. We finally heat the products of combustion back to T_1. Since the net heat release in the two paths must be identical, we may write

$$Q' = Q_0' + Q_R' - Q_P' \quad (4.16)$$

where Q_R' = heat removed by cooling reactants from T_1 to T_0
Q_P' = heat added by heating products from T_0 to T_1

We may rewrite this in terms of enthalpy changes as

$$Q' = Q_0' + \Delta H_R - \Delta H_P \quad (4.17)$$

where ΔH_R = (enthalpy of reactants at T_1) – (enthalpy of reactants at T_0)
ΔH_P = (enthalpy of products at T_1) – (enthalpy of products at T_0)

In slightly more general form, we have

$$Q' = Q_0' + H_R - H_P \quad (4.18)$$

Sec. 4.3 Mass and Energy Balances

where H_R = enthalpy of reactants relative to T_0
H_P = enthalpy of products relative to T_0

In the latter form we can deal with reactants at differing temperatures.

Since our system is essentially at constant pressure, and provided that there is no change in phase, the enthalpy of component i at temperature T' relative to T_0 is simply

$$H_i = m_i \int_{T_0}^{T'} c_{p_i} \, dT \tag{4.19}$$

where m_i = mass of component i
c_{p_i} = specific heat at constant pressure for component i

If we rewrite Eq. (4.19) by multiplying by $(T' - T_0)/(T' - T_0)$, we obtain

$$H_i = m_i \frac{\int_{T_0}^{T'} c_{p_i} \, dT}{T' - T_0} (T' - T_0) \tag{4.20}$$

where

$$\frac{\int_{T_0}^{T'} c_{p_i} \, dT}{T' - T_0} = \bar{c}_{p_i} = \text{mean value of specific heat of component } i \text{ over the temperature range } (T' - T_0)$$

We simplify our result to

$$H_i = m_i \bar{c}_{p_i} (T' - T_0) \tag{4.21}$$

and can then readily evaluate H_i if $\bar{c}_{p_i}$ is known. Usually, component i values of $\bar{c}_{p_i}$ are obtained from mean molal specific heats $(\bar{c}_p)_{m,i}$, where $\bar{c}_{pi} = (\bar{c}_p)_{m,i}/\text{MW}_i$, and MW is the molecular weight of component i. The polynomial expressions giving $(\bar{c}_p)_{m,i}$ as a function of T' for a $T_0 = 0°C$ have been determined and are listed in Table 4.2 for the common gaseous combustion products.

A complete heat balance also requires consideration of the enthalpy of the ash in addition to that of the combustion gases. Since the exact composition of the ash is generally not known, the exact polynomial expressions for the mean specific heat

TABLE 4.2 MEAN MOLAL SPECIFIC HEATS FOR COMMON GASES PRESENT IN COMBUSTION PRODUCTS (RELATIVE TO 0°C)[a]

Gases	$(\bar{c}_p)_m$ [b]
H_2O	$(\bar{c}_p)_{m,i} = 8.361 + 4.92 \times 10^{-4} T' + 4.46 \times 10^{-7} (T')^2$
CO_2, SO_2, and most triatomic gases	$(\bar{c}_p)_{m,i} = 9.085 + 2.4 \times 10^{-3} T' - 2.77 \times 10^{-7} (T')^2$
N_2, O_2, CO, and most diatomic gases	$(\bar{c}_p)_{m,i} = 6.935 + 3.38 \times 10^{-4} T' + 0.43 \times 10^{-7} (T')^2$

[a] Constants obtained by integrating the equations for instantaneous specific heats provided in *Bureau of Mines Technical Paper 445*, 1929.

[b] $(\bar{c}_p)_{m,i}$ = mean molal heat capacity from 0 to $T'°C$ (cal/g mol · °C).

are not known. The total amount of ash is of significance only when burning coal, and even here the percent ash is small. It is usually adequate to assume that the ash is all SiO_2 and use an average specific heat of 0.26 kcal/kg · °C.

Any water present in the fuel or found in the combustion process will be vaporized. The enthalpy of the water vapor in the combustion gases must therefore include the heat of vaporization. That is,

$$H_w = m_w\left[(\bar{c}_p)_{m,w}(T' - T_0) + h_{fg}\right] \qquad (4.22)$$

where
- H_w = enthalpy of water
- m_w = mass of water
- $(\bar{c}_p)_{m,w}$ = mean specific heat of water between T and T_0
- h_{fg} = heat of vaporization of water at T_0

Example 4.5

We may illustrate the principles involved in computing energy releases by returning to our example in which a mixture containing 60% methane and 40% ethane by volume was burned in 105% of the stoichiometric quantity of air. The fuel gas and air enter the combustion chamber at room temperature. The combustion products leave the chamber at 1000°C. Determine the heat released to the water-cooled combustion chamber walls per gram mole of fuel.

Solution Our balanced chemical equation gave us

$0.6CH_4 + 0.4C_2H_6 + (1.05)(2.6)(O_2 + 3.76N_2) \rightarrow$

$\qquad 1.4CO_2 + 2.4H_2O + (1.05)(2.6)(3.76N_2) + (0.05)(2.6)O_2$

By using 1 g mole of fuel gas as our basis, we had found that the product gases were 61.6 g of CO_2, 43.2 g of H_2O, 287.4 g of N_2, and 4.2 g of O_2. Then

Q' = heat released at 100°C per 1 g mol burned

$\quad = Q'_0 + H_R - H_P$

Now

$Q'_0 = 0.6(CH_4 \text{ heat of combustion/g mol}) + 0.4(C_2H_6 \text{ heat of combustion/g mol})$

$\quad = 0.6 \text{ g mol}\left(212{,}790\,\dfrac{\text{cal}}{\text{g mol}}\right) + 0.4 \text{ g mol}(368{,}400) \text{ cal/g mol}$

$\quad = 275{,}000 \text{ cal}$

Since room temperature and 0°C are close, the heat of reactants relative to 0°C is negligible. Therefore,

$Q' \approx Q_0 - \Delta H_P$

$= 275{,}000 - (1000°C - T_0)\Big[(\bar{c}_p)_{m,O_2}\dfrac{61.6 \text{ g } CO_2}{44 \text{ g/g mol}} + (\bar{c}_p)_{m,H_2O}\dfrac{43.2 \text{ g } H_2O}{18 \text{ g/g mol}}$

$\qquad + (\bar{c}_p)_{m,N_2}\dfrac{281.4 \text{ g } N_2}{28 \text{ g/g mol}} + (\bar{c}_p)_{m,O_2}\dfrac{4.2 \text{ g } O_2}{32 \text{ g/g mol}}\Big]$

$\qquad - 43.2 \text{ g } H_2O \times h_{fg} H_2O$

We determine the values of $(\bar{c}_p)_{m,i}$ for the products from the expressions of Table 4.2 using 1000°C for T. We then obtain Q' from the preceding equation.

Sec. 4.3 Mass and Energy Balances

A situation much more commonly encountered is that in which the fuel is at room temperature but the combustion air has been preheated to some elevated temperature. Under these conditions, the temperature at which the combustion takes place is likely to be unknown, but the temperature of the exit gases is easily measured. In this case, we can again determine the heat released by following a hypothetical path between the initial and final routes. We may assume that we first cool down all reactants to the temperature at which standard heats of reaction are evaluated (18°C), allow the reaction to take place, and then heat the products to the exit temperature. Since the net heat release via this hypothetical path must be identical to that along the actual path, the net heat release per mole of fuel, Q'_1, is given by

$$Q'_1 = Q'_0 + Q'_R - Q'_P \tag{4.23}$$

but in this case

Q'_R = heat removed by cooling reactants from their initial temperature to T_0

Q'_P = heat added by heating products from T_0 to the exit temperature

In almost all cases the fuel entering is nearly at room temperature and hence we may write

$$Q'_1 = Q'_0 + Q'_a - Q'_P \tag{4.24}$$

where Q'_a is the heat removed by cooling air from initial temperature to T_0 and other symbols have their previous meaning. The quantities Q'_a and Q'_P are determined from the change in enthalpy of the product and airstreams. That is,

$$Q'_1 = Q'_0 + \Delta H_a - \Delta H_P \tag{4.25}$$

where ΔH_a = change in enthalpy of air between inlet temperature and T_0

ΔH_P = change in enthalpy of product stream on going from T_0 to exit temperature

We then calculate ΔH_a and ΔH_P from the mean specific heats of the components and the heat of vaporization of H$_2$O. If Q_1 represents the heat released by m_F kilograms of fuel, then if the entering air is dry, we have

$$Q_1 = \frac{m_F}{\mathrm{MW}_F} \left[Q'_0 + (\bar{c}_p)_{m,a}(T_a - T_0)\frac{m_a}{\mathrm{MW}_a} - \sum_i \frac{m_i}{\mathrm{MW}_i}(\bar{c}_p)_{m,i}(T_{\mathrm{exit}} - T_0) - m_w h_{fg} \right] \tag{4.26}$$

where Q_0 = heat of combustion per unit mass of fuel

$(\bar{c}_p)_{m,i}$ = mean molal specific heat of product i between T_0 and T_{exit}

$(\bar{c}_p)_{m,a}$ = mean molal specific heat of air between T_{air} and T_0

m_a = mass of air admitted with m_F kilograms of fuel

m_w = mass of water vapor produced by m_F kilograms of fuel consumed

m_i = mass of component i present in product gas per m_F kilograms of fuel

MW_F, MW_a, MW_i = molecular weight of fuel, air, and component i, respectively

Example 4.6

Let us again apply this approach to the combustion of a mixture containing 60 v/o of methane and 40 v/o of ethane in 105% of the theoretically required air. We will assume that the fuel is admitted at room temperature and the air is preheated to 500°C. We are to determine the heat released in the furnace if the exit gases are at 700°C. In doing these calculations we shall make the simplifying assumption that the mean molal heat capacity from 18°C to T is very little different from that between 0°C and T. This simplification is widely used and is generally satisfactory.

Solution Use 1 kg of fuel as the basis for the solution. We return to the balanced chemical equation that we wrote previously for the combustion of this gaseous mixture. That is,

$$0.4C_2H_6 + 0.6CH_4 + 1.05[2.6(O_2 + 3.76N_2)]$$
$$= 1.4CO_2 + 2.4H_2O + 0.13O_2 + 10.2648N_2$$

We first complete the mass balance. The molecular weight of the fuel mixture is

$$0.4 \times 30 + 0.6 \times 16 = 21.6 \text{ kg/kg mol fuel}$$

We obtain the mass of air admitted per unit mass of fuel from the balanced chemical equation

$$\text{mass of air} = \frac{(\text{mol air/mol fuel}) \times \text{mol wt air}}{\text{wt fuel/mol fuel}}$$

$$= \frac{(1.05 \times 2.6 + 1.05 \times 3.76 \times 2.6) \times 28.97}{0.4 \times 30 + 0.6 \times 16}$$

$$= 17.43 \text{ kg air/kg fuel}$$

We then obtain the weights of the combustion products per kilogram mole of fuel as

$$\text{mass of } CO_2 = \frac{1.4 \times 44}{21.6} = 2.85 \text{ kg/kg fuel}$$

$$\text{mass of } N_2 = \frac{10.26 \times 28}{21.6} = 13.31 \text{ kg/kg fuel}$$

$$\text{mass of } H_2O = \frac{2.4 \times 18}{21.6} = 2 \text{ kg/kg fuel}$$

$$\text{mass of } O_2 = \frac{0.13 \times 32}{21.6} = 0.193 \text{ kg/kg fuel}$$

We next evaluate the mean molal specific heats of air and combustion products from Table 4.2.

$$(\bar{c}_p)_{m,a} = (6.935 + 0.000338T + 0.43 \times 10^{-7}T^2) \text{ cal/mol} \cdot °C$$

Sec. 4.3 Mass and Energy Balances

Evaluating $(\bar{c}_p)_{m,a}$ at 500°C, we have

$$(\bar{c}_p)_{m,a} = 6.935 + 0.000338(500) + 0.43 \times 10^{-7}(500)^2 \text{ cal/g mol} \cdot °C$$
$$= 7.115 \text{ cal/g mol} \cdot °C$$
$$= 7.115 \text{ kcal/kg mol} \cdot °C$$

For CO_2 and H_2O we have

$$(\bar{c}_p)_{m,CO_2} = 9.085 + 0.0024T - 0.277 \times 10^{-6} T^2 \text{ cal/g mol} \cdot °C$$

and

$$(\bar{c}_p)_{m,H_2O} = 8.361 + 0.00442T + 0.446 \times 10^{-6} T^2 \text{ cal/g mol} \cdot °C$$

We evaluate $(\bar{c}_p)_{m,CO_2}$ and $(\bar{c}_p)_{m,H_2O}$ at 700°C and obtain

$$(\bar{c}_p)_{m,CO_2} = 10.629 \text{ kcal/kg mol} \cdot °C$$
$$(\bar{c}_p)_{m,H_2O} = 11.673 \text{ kcal/kg mol} \cdot °C$$

As the following step we evaluate Q_1. We previously determined that combustion of 1 g mol of fuel produced 275,000 cal. Hence combustion of 1 kg mol produces 275,000 kcal. The average mole weight of the fuel gas, MW_{av}, is

$$MW_{av} = 0.6(16) + 0.4(30) = 21.6$$

Hence, for 1 kg of fuel, we have

$$\frac{m_F Q_0'}{MW_F} = \frac{1}{MW_F} Q_0' = 275{,}000 \text{ kcal/kg mol} \times \frac{1}{21.6} \frac{\text{kg mol}}{\text{kg fuel}} = 12{,}730 \text{ kcal/kg fuel}$$

We may now evaluate Q_1 from Eq. (4.26):

$$Q_1 = \frac{m_F}{MW_F}\left[Q_0' + (\bar{c}_p)_{m,a}(T_{air} - T_0)\frac{m_a}{MW_a} - \sum_i \frac{m_i}{MW_i}(\bar{c}_p)_{m,i}(T - T_0) - m_w h_{fg}\right]$$

Note that h_{fg}, the latent heat of evaporation, is evaluated at 18°C, which is the temperature at which the standard heat of reaction is evaluated. From the properties of saturated steam in the Appendix, we find that, at 18°C, $h_{fg} = 2458$ J/kg = 587 kcal/kg or 10,570 kcal/kg mol. Hence, for $m_F = 1.0$,

$$Q_1 = 12{,}730 \text{ kcal/kg fuel}$$
$$+ 7.115 \text{ kcal/kg mol} \cdot °C \,(500 - 18)°C \times \frac{17.43 \text{ kg air/kg fuel}}{28.97 \text{ kg air/kg mol}}$$
$$- \left[10.63 \text{ kcal/kg mol} \cdot °C \times \frac{2.8518 \text{ kg } CO_2/\text{kg fuel}}{44 \text{ kg } CO_2/\text{kg mol}} \right.$$
$$+ 11.67 \text{ kcal/kg mol} \cdot °C \times \frac{2 \text{ kg } H_2O/\text{kg fuel}}{18 \text{ kg } H_2O/\text{kg mol}}$$
$$+ 7.115 \text{ kcal/kg mol} \cdot °C \times \left(\frac{13.31 \text{ kg } N_2/\text{kg fuel}}{28 \text{ kg } N_2/\text{kg mol}} \right.$$
$$\left. \left. + \frac{0.193 \text{ kg } O_2/\text{kg fuel}}{32 \text{ kg } O_2/\text{kg mol}} \right) \right]$$
$$\times (700 - 18)°C - 2 \text{ kg } H_2O/\text{kg fuel}$$
$$\times 587 \text{ kcal/kg } H_2O = 10{,}970 \text{ kcal/kg fuel}$$

A somewhat more complex situation arises in the combustion of coal where all the exit streams are carefully analyzed. If we are to do an accurate computation of the heat release, we must account for unburned coal in the waste (or ash stream) and any carbon monoxide in the flue gas. We do so by subtracting the heating value of the unburned coal and carbon monoxide in the exit streams. Our equation for heat release, Q_1, then becomes

$$Q_1 = (Q_0' + Q_a' - Q_P' - Q_{cw}')\frac{m_F}{MW_F} \qquad (4.27)$$

where Q_{cw}' is the heat of combustion of unburned coal and carbon monoxide in waste and exit streams per mole of fuel and other quantities have their previous meaning. This approach will be illustrated by applying it to the coal combustion example (Example 4.3) discussed previously.

Example 4.7 Furnace Energy Balance

Determine a complete furnace energy balance using the data provided in Example 4.3. This balance is illustrated schematically in Fig. 4.4.

Solution   Basis for solution 100 kg of coal supplied

Energy input calculations:

1. *Heating value of the fuel*: From Example 4.3,

    standing heating value = 10,500 kcal/kg

    energy available = 100 $kg\ coal \times 10{,}500\ kcal/kg = 105 \times 10^4\ kcal$

2. *Enthalpy of the coal*: 0.
3. *Enthalpy of the dry air*: negligible.
4. *Enthalpy of the water vapor in the air* (H_{wa}) (*using data from Example 4.3*): From Tables A.4 and A.5, the heat of vaporization of water at 23°C = 2450 kJ/kg = 585.6 kcal/kg. Hence by using the data of Example 4.3, we have

    water enthalpy = 18 kg/kg mol $\times$ 0.559 kg mol $\times$ 585.6 kcal/kg

    = 5890 kcal

    total energy available = 105.6×10^4 kcal

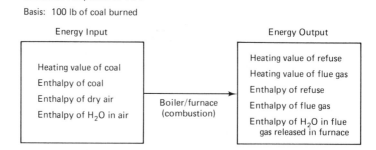

Figure 4.4 Energy balance for a furnace/boiler.

Sec. 4.3   Mass and Energy Balances   125

Energy output calculations:

5. *Heating value of the refuse*, HV_{refuse}:

$$HV_{refuse} = 8055 C_r = 8055 \text{ kcal/kg} \cdot °C \times 2.649 \text{ kg C} = 21,340 \text{ kcal}$$

6. *Heating value of flue gas*, $HV_{flue\ gas}$: The heating value is based on the carbon monoxide content of the flue gas:

$$HV_{flue\ gas} = \frac{m_{CO}}{MW_{CO}}(HV_{CO}) = \frac{0.505}{28}\left(67,620 \frac{\text{k cal}}{\text{kg mol}}\right) = 1220 \text{ kcal}$$

7. *Enthalpy of the refuse*, H_r: The enthalpy of the refuse is found from

$$H_r = m_r(T_r - T_0)(c_p)_r$$

where m_r = mass of refuse, kg
    $(c_p)_r$ = specific heat of the refuse, kcal/kg · °C (given)
    T_r = average refuse temperature
    T_0 = reference temperature (inlet air temperature = 23°C)

$$H_r = 19.24 \text{ kg}(124 - 33)°C \times 0.23 \text{ kcal/kg} \cdot °C = 4447 \text{ kcal}$$

The enthalpies of the CO, CO_2, O_2, SO_2, and N_2 are obtained from their mean specific heats.

$$H_i = \frac{m_i}{W_i}(c_p)_{m,i}(T_{FG} - T_0)$$

where H_i = enthalpy of component i
    mH_i = enthalpy of component i
    MW_i = molecular weight of flue gas component i
    $(c_p)_{m,i}$ = molal specific heat, kcal/kg mol · °C
    T_{FG} = flue gas average temperature, °C

$$\text{enthalpy of } CO_2 = \frac{232}{44}(250 - 23)9.92 = 11,870 \text{ kcal}$$

$$\text{enthalpy of } CO = \frac{0.505}{8}(250 - 23)7.05 = 21 \text{ kcal}$$

$$\text{enthalpy of } O_2 = \frac{91.85}{32}(250 - 23)7.24 = 4714 \text{ kcal}$$

$$\text{enthalpy of } SO_2 = \frac{3.17}{64}(250 - 23)9.92 = 111 \text{ kcal}$$

$$\text{enthalpy of } N_2 = \frac{103.9}{28.2}(250 - 23)7.02 = 58.710 \text{ kcal}$$

$$H_{FG} = \text{total enthalpy of stack gases} = 75,430 \text{ kcal}$$

8. *Enthalpy of water in the stack gas*: The enthalpy of the moisture in the stack gas is based on the heat of vaporization and the degree of superheat for the M_{H_2O} moles of water vapor in the flue gas. By again taking the heat of vaporization as

10,570 kcal/kg mol, we have

specific heat of the water vapor, $(c_p)_{m,H_2O}$ = 8.20 kcal/kg mol · °C

heat of vaporization = $M_{H_2O} \times 10{,}570$ = 2.939 kg mol × 10,542 kcal/kg mol

= 31,065 kcal

superheat enthalpy = $M_{H_2O}(c_p)_{m,H_2O}(T_{FG} - T_0)$

= 2.939 kg mol × 8.20 kcal/kg mol · °C × (250 − 23) °C

superheat enthalpy = 5470 kcal

$H_{H_2O/FG}$ = total enthalpy of water vapor in flue gas = 36,535 kcal

9. *Heat released*: The approach to the final heat balance is summarized in Fig. 4.4. We obtain our heat release, Q_1, by difference:

$$Q_1 = (\text{available energy}) - (HV_{\text{refuse}} - HV_{\text{flue gas}} + H_r + H_{FG} + H_{H_2O/FG})$$

$$= (105.6 \times 10^4) - (21{,}340 + 1220 + 447 + 75{,}430 + 36{,}450)$$

$$= 921{,}100 \text{ kcal}$$

Summary of energy balance (all units kcal):

Input	Output
energy available in fuel = 105×10^4	Q_1 = 921,100
water vapor enthalpy = 5890	HV_{refuse} = 21,340
	$HV_{\text{flue gas}}$ = 1220
	H_r = 447
	H_{FG} = 75,430
	$H_{H_2O/FG}$ − 36,450

4.4 CHEMICAL EQUILIBRIUM

The thermodynamic property that determines whether a chemical reaction will occur spontaneously is called its *thermodynamic potential* or *free energy*. A system will always tend toward the condition with the minimum value of free energy. The total free energy, G, is an extensive property and is defined by

$$G = H - TS \tag{4.28}$$

where H = total enthalpy
S = total entropy
T = absolute temperature

For m kilograms of a pure substance, we obtain G from the intensive values of h and s. That is,

$$G = m(h + Ts) \tag{4.29}$$

Sec. 4.4 Chemical Equilibrium

where h = enthalpy per unit mass (energy/mass)
s = entropy per unit mass (energy/mass $\times T$)

With this notation, the driving force in a chemical reaction is the difference in free energy between the initial and final states, ΔG, where at constant temperature

$$\Delta G = \Delta H - T \Delta S \qquad (4.30)$$

If the free-energy change for a reaction is negative (reaction is *exothermic*—gives off heat) the reaction can occur spontaneously. A positive value for ΔG (reaction is *endothermic*—absorbs heat) indicates that energy must be supplied in order for the reaction to occur.

Free Energy of Formation of a Compound

It is conventional to take the free energy of the chemical elements to be zero at 1 atm pressure and room temperature. The standard free energy of formation of a compound is the free-energy change, ΔG_f^s, when one mole of the compound is formed directly from its constituent elements and reactants and products are at room temperature and have partial pressures of 1 atm. If the reaction takes place at some higher temperature, the free energy of formation, ΔG_f°, will have a different value. Since thermodynamics tells us that the properties of a material in a given end state are independent of the path taken to reach that end state, we can determine ΔG_f° from ΔG_f^s. We imagine that instead of simply reacting the constituents at temperature T', we first cool the reactants to room temperature, allow the reaction to take place and then heat the products back to temperature T'. Since the result of the path must be identical to that obtained by simply conducting the reaction at T', we have for 1 mol of product

$$\Delta G_f^\circ = \Delta G_f^s - \int_{20}^{T'} \left[n_{R_1}(c_p)_{m,i} + n_{R_2}(c_p)_{m,2} + \cdots \right] dT + \int_{20}^{T'} \left[n_p(c_p)_{m,P} \right] dT \qquad (4.31)$$

where n_{R_i} = number of moles of reactant i required
$(c_p)_i$ = molal specific heat at constant pressure of reactant i
$(c_p)_{m,P}$ = molal specific heat at constant pressure of products
ΔG_f^s = molal free energy of formation at room temperature with partial pressures of reactants and products being 1 atm

In most actual situations, the partial pressures of the reactants and products will not be at 1 atm. The effect of such pressure changes on the free energy of a given material are evaluated from the thermodynamic relationship, which states that at constant temperature,

$$dG = v \, dp \qquad (4.32)$$

where v = molal specific volume
p = pressure

In going from standard conditions, p_0, to operating condition, p_1, we have for a material that is an ideal gas,

$$\Delta G = \int_{p_0}^{p_1} v\, dp = RT \ln \frac{p_1}{p_0} \qquad (4.33)$$

where R is the ideal gas constant. The free energy of formation, ΔG_f, of a compound that is an ideal gas and has a partial pressure p_1 at temperature T is then given by

$$\Delta G_f = \Delta G_f^\circ + RT \ln \frac{p_1}{p_0} \qquad (4.34)$$

If we express p in atmospheres and take p_0 as 1 atm, this becomes

$$\Delta G_f = \Delta G_f^\circ + RT \ln p_1 \qquad (4.35)$$

When the material is not an ideal gas, we define a quantity f, called the *fugacity*, such that

$$\int_{p_0}^{p_1} v\, dp = RT \ln \frac{f_1}{f_0} \qquad (4.36)$$

where f_1 and f_0 are the fugacities at p_1 and p_0, respectively. Then

$$\Delta G_f = \Delta G_f^\circ + RT \ln \frac{f_1}{f_0} \qquad (4.37)$$

Equilibrium Constant, K_p

Consider a reaction such as

$$\text{CCl}_4(g) + 2\text{H}_2\text{O}(g) \rightarrow \text{CO}_2(g) + 4\text{HCl}(g) \qquad (4.38)$$

taking place at temperature T. The initial state is defined by a partial pressure of carbon tetrachloride, p_{CCl_4}, and the partial pressure of water vapor, $p_{\text{H}_2\text{O}}$. In the final state, carbon dioxide and hydrogen chloride have partial pressures defined by p_{CO_2} and p_{HCl}.

Using the free energies of formation, the free-energy change for this system, ΔG, can be expressed as

$$\Delta G = \Delta G_{f,\text{CO}_2} + 4\Delta G_{f,\text{HCl}} - \left(\Delta G_{f,\text{CCl}_4} + 2\Delta G_{f,\text{H}_2\text{O}}\right) \qquad (4.39)$$

Each of the free energies of formation in Eq. (4.39) can also be expressed in terms of the free energies of formation at temperature T and the partial pressures of the compounds:

$$\Delta G_{f,\text{CO}_2} = \Delta G_{f,\text{CO}_2}^\circ + RT \ln p_{\text{CO}_2} \qquad (4.40\text{a})$$

$$4\Delta G_{f,\text{HCl}} = 4\Delta G_{f,\text{HCl}}^\circ + RT \ln p_{\text{HCl}}^4 \qquad (4.40\text{b})$$

$$\Delta G_{f,\text{CCl}_4} = \Delta G_{f,\text{CCl}_4}^\circ + RT \ln p_{\text{CCl}_4} \qquad (4.40\text{c})$$

$$2\Delta G_{f,\text{H}_2\text{O}} = 2\Delta G_{f,\text{H}_2\text{O}}^\circ + RT \ln p_{\text{H}_2\text{O}}^2 \qquad (4.40\text{d})$$

Sec. 4.4   Chemical Equilibrium

If Eqs. (4.40a)–(4.40d) are substituted into Eq. (4.39), the result is

$$\Delta G = \Delta G^\circ_{f,CO_2} + 4\Delta G^\circ_{f,HCl} - \left(\Delta G^\circ_{f,CCl_4} + 2\Delta G^\circ_{f,H_2O}\right)$$
$$+ RT\left[\ln p_{CO_2} + \ln p^4_{HCl} - \left(\ln p_{CCl_4} + \ln p^2_{H_2O}\right)\right] \quad (4.41)$$

The first four terms of Eq. (4.41) represent ΔG°, the change in free energy for the reaction when the products and reactants all have partial pressures of 1 atm, while the remaining terms represent the effect of partial pressures on the free-energy change. We may simplify Eq. (4.41) to

$$\Delta G = \Delta G^\circ + RT \ln \frac{p_{CO_2} p^4_{HCl}}{p_{CCl_4} p^2_{H_2O}} \quad (4.42)$$

Observe that in Eq. (4.42) the partial pressures of the products appear in the numerator and the partial pressures of the reactants appear in the denominator. In a chemical reaction, as the reactants are consumed, their pressures are reduced. The logarithm term becomes larger and the value for ΔG more positive (less negative). Similarly, as the reaction continues, the pressures of the products, shown in the numerator, increase. Both effects make ΔG less negative and thus decrease in the reaction driving force.

In reality, there are two competing reactions. The first of these is the forward reaction, in which carbon tetrachloride vapor reacts with water vapor to produce carbon dioxide and hydrogen chloride. The driving force behind this reaction decreases as the reaction proceeds and the reactants are consumed. The second reaction is the reverse reaction, in which carbon dioxide and hydrogen chloride react to form carbon tetrachloride and water vapor. The reverse reaction can be represented by the chemical equation

$$CO_2(g) + 4HCl(g) \rightarrow CCl_4(g) + 2H_2O(g) \quad (4.43)$$

The free-energy change would be

$$\Delta G = (\Delta G)^\circ_{rev} + RT \ln \frac{p_{CCl_4} p^2_{H_2O}}{p_{CO_2} p^4_{HCl}} \quad (4.44)$$

where $(\Delta G)^\circ_{rev}$ is the ΔG° for the reverse reaction.

As the *forward* reaction proceeds, carbon tetrachloride and water are consumed while CO and HCl are produced, thus making ΔG less negative and decreasing the driving force behind this reaction. Equation (4.44) shows that an increase in the pressures of CO_2 and HCl leads to a decrease in the corrective term with a corresponding decrease in ΔG. As the forward reaction proceeds, the driving force behind the reverse reaction increases.

Because of the competing reactions, the system tends to reach an equilibrium condition in which these two driving forces are equal. When this occurs we have

$$CCl_4 + 2H_2O \rightleftharpoons CO_2 + 4HCl \quad (4.45)$$

This is a dynamic equilibrium condition in which both the forward and reverse reactions are occurring simultaneously at the same rate, or $\Delta G = 0$. If ΔG is set equal to 0 in Eq. (4.42), then

$$\Delta G° = -RT \ln \frac{(p_{CO_2})(p_{HCl})^4}{(p_{CCl_4})(p_{H_2O})^2} \qquad (4.46)$$

The parentheses enclosing the pressure terms identify these pressures as equilibrium pressures.

The quantities $\Delta G°$, R, and T are all constant for an isothermal process. This means that the logarithm of the quotient, and hence the quotient, is a constant for equilibrium at a given temperature. This constant is called the *equilibrium constant expressed in partial pressures*, K_p. Equation (4.46) may thus be taken as defining the equilibrium constant, K_p, for a system in which carbon tetrachloride reacts with water vapor to produce carbon dioxide and hydrogen chloride:

$$K_p = \frac{(p_{CO_2})(p_{HCl})^4}{(p_{CC_4})(p_{H_2O})^2} \qquad (4.47)$$

From Eq. (4.46) it is obvious that the equilibrium constant is related to the free-energy change by

$$\Delta G° = -RT \ln K_p \qquad (4.48)$$

In general, the equilibrium constant, K_p, for a reaction in the gaseous phase may be defined as the product of the equilibrium pressures of the gaseous products, each raised to the power of the corresponding coefficient in the chemical equation, divided by the product of the equilibrium pressures of the gaseous reactants, each raised to the power of the appropriate coefficient in the reaction. For the case

$$aA + bB \rightleftharpoons yY + zZ \qquad (4.49)$$

we have

$$K_p = \frac{(p_Y)^y(p_Z)^z}{(p_A)^a(p_B)^b} \qquad (4.50)$$

Note that the units for K_p depend on the exponents a, b, y, and z. Thus for the reaction described by Eq. (4.46), K_p has the units of atm².

When one of the reactants is not an ideal gas, the appropriate partial pressure must be replaced by a fugacity ratio. Assume that in Eq. (4.50), reactant B is not an ideal gas. We then have

$$K_p = \frac{(p_Y)^y(p_Z)^z}{(p_A)^a(f_B/f_{0B})^b} \qquad (4.51)$$

Sec. 4.4 Chemical Equilibrium

In the situation of most concern to us, one of the compounds may be a solid. If that is the case, the fugacity ratio for that component is unity. Thus if compound B is a solid, Eq. (4.50) becomes

$$K_p = \frac{(p_Y)^y (p_Z)^z}{(p_A)^a} \qquad (4.52)$$

Variation of K_p with Temperature

An increase in the temperature of a reacting system at equilibrium through the addition of heat will shift the equilibrium composition in the direction causing an absorption of heat. The effect of increased temperature in exothermic reactions is then to reduce the degree of completion at equilibrium. We may evaluate this effect quantitatively by solving Eq. (4.48) for K_p to obtain

$$K_p = \exp\left(-\frac{\Delta G^\circ}{RT}\right) \qquad (4.53)$$

We then evaluate ΔG° as a function of T using the same approach as that used for ΔG_f°. The results of this computation for typical reactions of interest to the combustion process are shown in Fig. 4.5.

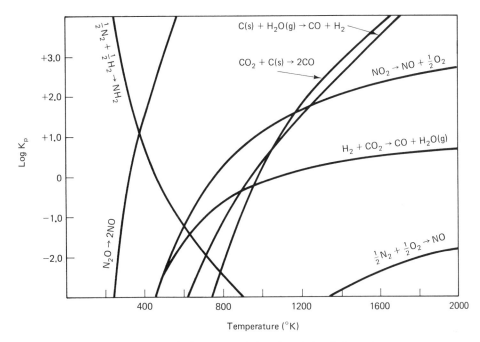

Figure 4.5 Equilibrium constants for selected chemical reactions.

Chemical Equilibrium in the Combustion Process

In complete combustion, carbon, hydrogen, and sulfur react with oxygen to form CO_2, H_2O, and SO_2. Equation (4.54) describes the general form of a complete combustion for a hydrocarbon.

$$C_aH_b + \left(a + \frac{b}{4}\right)O_2 \rightleftharpoons aCO_2 + \frac{b}{2}H_2O \qquad (4.54)$$

Complete combustion of fuel oil containing some sulfur would be written as

$$C_aH_bS_c + \left(a + \frac{b}{4} + c\right)O_2 \rightleftharpoons aCO_2 + \frac{b}{2}(H_2O) + cSO_2 \qquad (4.55)$$

If the oxygen present is not fully adequate for complete combustion of the hydrocarbon, we find that the hydrogen and sulfur are burned H_2O and SO_2, respectively, but that some of the carbon is burned only to carbon monoxide, CO. The amount remaining is determined by balancing the chemical equation. Thus, if only 90% of the stoichiometric amount of oxygen is available, we write

$$C_aH_bS_c + 0.9\left(a + \frac{b}{4} + c\right)O_2 \rightleftharpoons y'CO_2 + x'(CO) + \frac{b}{2}(H_2O) + cSO_2 \qquad (4.56)$$

But from a carbon balance, we have

$$a = y' + x' \qquad (4.57)$$

and from an oxygen balance,

$$0.9\left(a + \frac{b}{4} + c\right) = y' + \frac{x'}{2} + \frac{b}{4} + c \qquad (4.58)$$

By simultaneous solution of the oxygen and carbon balanced, we determine the relative amounts of CO and CO_2 present.

In the combustion of coal, the hydrogen and sulfur present are again always burned to H_2O and SO_2. However, there are several competing reactions which determine the degree of carbon combustion:

$$C(s) + O_2(g) \rightleftharpoons CO_2(g) \qquad (4.59a)$$

$$C(s) + \tfrac{1}{2}O_2(g) \rightleftharpoons CO(g) \qquad (4.59b)$$

$$C(s) + CO_2(g) \rightleftharpoons 2CO(g) \qquad (4.59c)$$

$$CO(g) + \tfrac{1}{2}O_2(g) \rightleftharpoons CO_2(g) \qquad (4.59c)$$

The equilibrium relationship for Eq. (4.59a) is given by

$$K_p = \frac{p_{CO_2}}{p_{O_2}} \qquad (4.60)$$

and for Eq. (4.59c), the equilibrium relationship is

$$K_p = \frac{(p_{CO})^2}{p_{CO_2}} \qquad (4.61)$$

Sec. 4.4 Chemical Equilibrium

In modern coal-burning plants, the gases leaving the flame region may be above 400°C and the equilibrium constant is very large for reactions (4.59a), (4.59b), and (4.59d). It is therefore usual to assume that these reactions go to completion. Reaction (4.59c) is a slower reaction with the relative amounts of CO_2 and CO given by Eq. (4.61). The temperature-dependent value of the equilibrium constant is obtained from Fig. 4.5.

At high temperatures, mixtures of CO and CO_2 are obtained only when less than the stoichiometric quantity of air is present. The reaction

$$CO + \tfrac{1}{2}O_2 \rightleftharpoons CO_2 \qquad (4.62)$$

has a very large equilibrium constant and virtually all the oxygen is consumed. This condition leads to the usual assumption that no oxygen is present in the system at equilibrium if less than the stoichiometric amount of O_2 is supplied. Conversely, when excess oxygen is supplied, it is assumed that no CO is present.

We know that under appropriate conditions carbon can reduce steam to produce CO and H_2. We have previously indicated the importance of this reaction for coal gasification. However, in the presence of oxygen, the CO and H_2 produced will be burned and the reduction of steam need not be considered in normal combustion calculations. Where less than the stoichiometric amount of oxygen is present, the relative amounts of C, CO, and CO_2 are essentially determined by the equilibrium amount of the $C + CO_2 \rightleftharpoons 2CO$ reaction. When excess oxygen is present, complete combustion may be assumed.

Calculation of Equilibrium Composition

From the standpoint of determining the energy release from combustion, chemical equilibrium need be considered only when coal or coke is burned with less than the stoichiometric quantity of air.

Example 4.8

For simplicity, consider the combustion of coke in 80% of the stoichiometric quantity of air at 800°K. The chemical equation for the process may be written as

$$C = 0.8(O_2 + 3.76N_2) \rightleftharpoons x'CO_2 + yCO + z'C + 0.8(3.76N_2)$$

Solution We take 1 g mol of carbon as our basis and write the oxygen and carbon balances as

$$x' + \frac{y'}{2} = 0.8$$

$$x' + y' + z' = 1$$

We solve these two equations simultaneously to obtain y and z in terms of x:

$$y' = 1.6 - 2x' \qquad z' = x' - 0.6$$

Our chemical equation may now be written in terms of x':

$$C + 0.8(O_2 + 3.76N_2) \rightleftharpoons x'CO_2 + (1.6 - 2x')CO + (x' - 0.6)C + 0.8(3.76N_2)$$

To determine x' we make use of our equilibrium relationship for our limiting reaction $C + CO_2 \rightleftharpoons 2CO$:

$$K_p = \frac{(p'_{CO})^2}{p'_{CO_2}}$$

The partial pressures of the component gases may be determined from their mole fractions. Since we are at high temperatures and atmospheric pressure, the gases may be considered ideal and hence for component j, the partial pressure, p'_j, is

$$p'_j = \frac{n_j}{\sum n_i} p = \frac{n_j}{\sum n_i}$$

Note that we may drop the total pressure, p, since it is 1 atm. We can then write

$$n_{CO} = 1.6 - 2x'$$

$$n_{CO_2} = x'$$

$$n_{N_2} = 0.8(3.76)$$

and

$$p'_{CO} = \frac{1.6 - 2x'}{1.6 - 2x' + x' + 0.8(3.76)} \qquad p'_{CO_2} = \frac{x'}{(1.6 - 2x') + x' + 0.8(3.76)}$$

The equilibrium relationship then becomes

$$K_p = \frac{(p'_{CO})^2}{p'_{CO_2}} = \frac{(n_{CO}/\sum n_i)^2}{n_{CO_2}/\sum_i n_i} = \frac{(n_{CO})^2}{n_{CO_2}(\sum_i n_i)} = \frac{(1.6 - 2x')^2}{x'[(1.6 - x' + 0.8(3.76)]}$$

From Fig. 4.5 we find the value of $R \ln K_p$. At 800°K we have $\log K = -2.0$ or $K_p = 0.01$. We then solve the quadratic equation

$$0.01 = 1.6x' - (x')^2 + 3.01x' = (1.6 - 2x')^2$$

for x.

In this simplified approach we have not considered all of the possible chemical reactions simultaneously. It is therefore possible that the value of x' obtained may be unrealistic. That is, the quantity of CO called for by the equilibrium relationship may be in excess of that possible when all the carbon is reacted. We must therefore always determine the maximum possible value for the moles of CO (or maximum value for K_p) and set the CO (or K_p) equal to this value should a larger value be predicted. From our carbon balance equation, we find that when $z' = 0$, we have

$$1 = x' + y'_{max} = (0.8 + y') + y'$$

$$y'_{max} = 0.133$$

Note that once all the carbon has reacted we do not have equilibrium conditions (all constituents are not present) and hence the relationship between p_{CO} and p_{CO_2} required by our equilibrium relationship no longer applies.

In real situations, the combustion temperature is generally unknown. We must then simultaneously determine the flame temperature and degree of equilibrium by an iterative process. We will consider this in the following section.

Examination of Fig. 4.5 shows that at temperatures above 1300°K, the reaction between oxygen and nitrogen in the air can produce observable amounts of NO. Some of this NO is subsequently oxidized to NO_2 after the flue gases are cooled. The mixture of NO_2 and NO produced is generally designated as NO_x since the exact ratio of N to O is variable and depends on a number of factors. The quantity of NO_x formed is so small that it does not affect energy releases, and hence is ordinarily not considered in such calculations. However, the reaction is important, as NO_x is an important atmospheric pollutant. Nitrogen oxides can also be produced by the conversion of some of the organically bound nitrogen compounds in the combustion process. This also has little influence on energy release, but it can be a significant additional source of pollutants.

4.5 THEORETICAL FLAME TEMPERATURES

The theoretical flame temperature is the temperature attained by the adiabatic combustion of a fuel which has been thoroughly mixed with air or oxygen. The maximum theoretical flame temperature of a fuel corresponds to combustion with just the theoretically required quantity of pure oxygen. Similarly, the maximum flame temperature using air corresponds to complete combustion with the theoretically required quantity of normal air. This temperature will be lower than the maximum flame temperature in pure oxygen. In actual combustion chambers, flame temperature is further reduced due to incomplete combustion, excess air requirements, heat losses to surrounding walls, and by heat loss to ash particles suspended in the combustion products.

Flame temperature estimation is required in the design of any combustion chamber, furnace, boiler, or gas turbine. In subsequent sections of this chapter we shall see how estimates of actual flame temperatures are used in the design of the furnaces/boilers of large power plants.

The calculation of the theoretical flame temperature is based on the assumption that the heat released by the combustion process is completely absorbed by reaction products and excess air. The temperature of the product gases is the flame temperature we seek.

In Section 4.3 we had determined that the heat balance could be written as

$$Q_1' = Q_0' + \Delta H_a - \Delta H_P \qquad (4.63)$$

when the fuel entered at room temperature. In the present case, we are assuming no net release of heat in the furnace. That is, all the energy of the reaction is used to heat the product gases. Hence $Q'_1 = 0$ and we have

$$\Delta H_P - \Delta H_a \doteq Q'_0 \qquad (4.64)$$

As before, we evaluate the enthalpy of the air and product streams from the mean molal specific heats of the products and the heat of vaporization of the water vapor. With a slight revision of our previous notation, we have

$$\sum_j n_j(Q'_0)_j = \sum_i \frac{m_i}{\mathrm{MW}_i}(\bar{c}_p)_{m,i}(T_{\mathrm{exit}} - T_0) + m_w h_{fg} - (\bar{c}_p)_{m,a}(T_{\mathrm{air}} - T_0)\frac{m_a}{\mathrm{MW}_a} \qquad (4.65)$$

where n_j represents the number of moles of component j in the fuel mass selected as the basis for solution.

The exit gas, or flame, temperature is unknown and hence the value of $(\bar{c}_p)_{m,i}$ is unknown. We could insert the polynomial expressions for $(\bar{c}_p)_{m,i}$ and attempt to solve the resultant algebraic expression for T_{exit} (the flame temperature). However, analytical solution of cubic equations is difficult and a numerical approach is usually much easier. We simply assume various values of T_{exit} and evaluate both the left- and right-hand sides of the equation. If we then plot the ratio of the two values versus T_{exit}, we readily determine the temperature at which the ratio is unity and the desired equality is attained. This procedure is illustrated in the following example.

Example 4.9

Determine the theoretical (adiabatic) flame temperature obtained from the combustion of coke (taken as pure carbon) with 20% excess air. Assume that the inlet air is at room temperature.

Solution Use 1 kg of C as the basis for the solution. The balanced chemical equation is

$$C + 1.2(O_2 + 3.76 N_2) \rightarrow CO_2 + 0.2 O_2 + 4.512 N_2$$

We then have

$$m_a = \frac{(1.2 + 1.2 \times 3.76) \times 28.97}{12} = 13.80 \text{ kg air/kg C}$$

$$m_{CO_2} = \frac{44}{12} = 3.67 \text{ kg CO}_2/\text{kg C}$$

$$m_{O_2} = \frac{0.2 \times 32}{12} = 0.53 \text{ kg air/kg C}$$

$$m_{N_2} = \frac{4.512 \times 28}{12} = 10.528 \text{ kg N}_2/\text{kg C}$$

The heat-balance equation is

$$\sum_j n_j(Q'_0)_j = \sum_i \frac{m_i}{\mathrm{MW}_i}(\bar{c}_p)_{m,i}(T_{\mathrm{exit}} - T_0) - (\bar{c}_p)_{m,a}(T_a - T_0)\frac{m_a}{\mathrm{MW}_a}$$

We take $T_a \approx T_0$. Therefore,

$$\sum_j n_j(Q'_0)_j = \sum_i \frac{m_i}{\mathrm{MW}_i}(\bar{c}_p)_{m,i}(T_{\mathrm{exit}} - T_0)$$

Sec. 4.5 Theoretical Flame Temperatures

where
$$(\bar{c}_p)_m = 6.935 + 0.000338T + 0.43 \times 10^{-7}T^2 \text{ (kcal/kg mol) for } O_2 \text{ and } N_2$$
and
$$(\bar{c}_p)_m = 9.085 + 0.0024T - 0.277 \times 10^{-6}T^2 \text{ kcal/kg mol} \cdot °C \text{ for } CO_2$$

We know from Table 4.1 that $Q_0' = 97{,}000$ cal/g[†] mole and since $n_F = \frac{1}{12}$,

$$n_F Q_0' = 8080 \text{ kcal/kg C}$$

Therefore, we have

$$8080 \text{ kcal/kg C} = (6.935 + 0.338 \times 10^{-3}T + 0.43 \times 10^{-7}T^2) \text{ cal/kg mol} \cdot °C$$
$$\times (T - 18)°C \times \left(\frac{0.53}{32} + \frac{10.528}{28}\right) \text{ kg mol/kg C}$$
$$+ (9.085 + 2.4 \times 10^{-3}T - 0.2777 \times 10^{-6}T^2) \text{ cal/kg mol} \cdot °C$$
$$\times (T - 18)°C$$
$$\times \frac{3.67}{44} \text{ kg mol/kg C}$$

$$8080 = (3.480 + 0.337 \times 10^{-3}T + 0.02 \times 10^{-6}T^2)(T - 18)$$

We now assume various values of T and search for the correct value where the right-hand side (RHS) of the foregoing equation equals the left-hand side (LHS). We find:

T (°C)	LHS	RHS
1000	8080	3860
2000	8080	8312
1900	8080	7867

By linear interpolation we find that the adiabatic flame temperature is ~1948°C.

A somewhat more complicated situation arises when there is incomplete combustion (less than stoichiometric quantity of oxygen present). In that case we must recognize that the total heat production depends on how much CO_2 and CO are produced and that this in turn depends on the chemical equilibrium achieved. The ratio of CO_2 to CO in the combustion gases was previously shown to be given by

$$K_p = \frac{(p'_{CO})^2}{p'_{CO_2}} \tag{4.66}$$

Consider the general combustion reaction, where the total moles of air are known (or

---

[†] The heat of combustion given in Table 4.1 is negative, as heat is released by the reaction. However, we take Q_0' as positive since it represents heat absorbed by the product gases.

d is known):

$$1C + aH + bS + \left(\frac{a}{4} + b + d\right)(O_2 + 3.764N_2)$$

$$\rightleftharpoons (1x' - y')C + y'CO_2 + x'CO + \frac{a}{2}H_2O$$

$$+ bSO_2 + \left(d + \frac{a}{4} + b\right)(3.76N_2) \qquad (4.67)$$

An oxygen balance will result in an additional equation. Assume that the total number of moles of air supplied is known (d is known). Then

$$y' + \frac{x'}{2} = d$$

or

$$x' = 2(d - y') \qquad (4.68)$$

Equation (4.68) can be utilized to replace x' in terms of y' moles of CO_2. The result is a single unknown x', which can be evaluated from the equilibrium relationship,

$$p'_{CO} = \frac{x'}{x' + y' + \frac{a}{2} + b + \left(d + \frac{a}{4} + b\right)3.76} = \frac{x'}{\phi_0} = \frac{2(d - y')}{\phi_0} \qquad (4.69)$$

$$p'_{CO_2} = \frac{y'}{\phi_0} \qquad (4.70)$$

Equations (4.69) and (4.70) are substituted into Eq. (4.66) for K_p.

$$K_p = \frac{(x'/\phi)^2}{y'/\phi_0} = \frac{(x')^2}{y'\phi_0} = \frac{[2(d - y')]^2}{\phi_0 y'} \qquad (4.71)$$

The flame temperature calculation proceeds in a somewhat similar way to the previous case. A temperature T is assumed in order to calculate K_p and the product gas composition. The heat absorbed by the product gas at this assumed temperature, with the calculated equilibrium gas composition,[†] is compared with the heat of reaction in a heat balance. However, in evaluating Q'_0 we must take account of the fact that only a portion of the fuel is burned to CO_2. With the appropriate value for Q'_0, the ratio of the heat reaction to heat absorbed (left-hand side of heat balance equation over the right-hand side) is again plotted against the assumed flame temperature and the solution found when the ratio is unity.

[†]As noted previously, the analyst must make certain that the limiting CO/CO_2 ratio is not exceeded. If the equilibrium relationship predicts a greater value, the limiting value of the ratio must be used.

Sec. 4.5   Theoretical Flame Temperatures   139

Example 4.10

The product stream of a particular solvent refined coal plant may be considered to contain 0.4 kg of hydrogen for each kilogram of carbon. Other constituents are negligible. Determine the theoretical (adiabatic) flame temperature from combustion of this product with 90% of the stoichiometric quantity of Air. Assume that the inlet air is at room temperature. Also assume that fuel is fed as a water slurry containing 40% fuel by weight.

Solution Basis for solution: 1 kg mol of C. The balance chemical equation is

$$C = aH_2 + bH_2O(l) + 0.9\left(1 + \frac{a}{2}\right)(O_2 + 3.76N_2)$$
$$\rightarrow x'CO + y'CO_2 + z'C + (9 + .45a)3.76N_2 + (a+b)H_2O(g)$$

Since the fuel contains 0.4 kg of hydrogen for each kilogram of carbon, we have for 1 g mol of C:

$$a = \frac{12 \times 0.4}{2} = 2.4 \text{ kg mol } H_2$$

However, the fuel is fed as a slurry: 40 w/o fuel, 60 w/o H_2O; therefore,

$$\frac{\text{wt slurry}}{\text{kg mol C}} = (12 + 2.4 \times 2) \div 0.4 = 42.0$$

$$\frac{\text{mol } H_2O}{\text{mol C}} = 42.0 \times 0.6 \div 18 = 1.4 \text{ kg mol } H_2O$$

or

$$b = 1.4 \text{ kg mol}$$

We then determine the relationship between x' and y' from the carbon and oxygen balance.

Carbon balance: $x' + y' + z' = 1$

Oxygen balance: $0.9 \times 2 \times 2.2 = x' + 2y' + 2.4$

or

$$x' + 2y' = 1.56$$

Hence we write

$$K_p = \frac{(p'_{CO})^2}{p'_{CO_2}} = \frac{[x'/(x'+y'+11.25)]^2}{y'/(x'+y'+11.25)}$$

where $11.25 = (0.9 + 0.45a)3.76 + a + b$. If we assume that all carbon is consumed (i.e., $z' = 0$), then

$$\begin{cases} x' + y' = 1 \\ x' + 2y' = 1.56 \end{cases}$$

Under these conditions, $x' = 0.44$ and $y' = 0.56$, giving

$$(K_p)_{max} = \frac{(0.44/12.25)^2}{0.56/12.25} = 0.0282$$

We now determine K_p and if it is greater than $(K_p)_{\max}$, we will equate it to $(K_p)_{\max}$.

We use the heat-balance equation to determine the flame temperature by taking the gas exit temperature, T_{exit}, as the flame temperature.

$$\sum_j n_j Q_0' = \sum_i \frac{m_i}{\text{WM}_i} (\bar{c}_p)_{m,i} (T_{\text{exit}} - T_0) + m_w h_{fg}$$

where $(\bar{c}_p)_{m,i}$ is obtained from the equations of Table 4.2. We assume various values of T_{exit}. Let $T_{\text{exit}} = 1000°C$ (1273°K). We obtain K_p from Fig. 4.5 and find that $K_p = 160$. Therefore, we set

$$K_p = (K_p)_{\max} = 0.0282$$

and

$$x' = 0.44 \text{ g mol} \qquad y' = 0.56 \text{ g mol}$$

$$(\bar{c}_p)_m = \begin{cases} 7.216 \text{ kcal/kg mol} \cdot °C & \text{for CO, N}_2 \\ 13.227 \text{ kcal/kg mol} \cdot °C & \text{for H}_2\text{O} \\ 11.208 \text{ kcal/kg mol} \cdot °C & \text{for CO}_2 \end{cases}$$

$$\sum_j n_j (Q_0')_j = \begin{pmatrix} \text{heat from combustion} \\ \text{of C} \to \text{CO} \end{pmatrix} + \begin{pmatrix} \text{heat from combustion} \\ \text{of C} \to \text{CO}_2 \end{pmatrix}$$

$$+ \begin{pmatrix} \text{heat from combustion} \\ \text{H} \to \text{H}_2\text{O} \end{pmatrix}$$

LHS = 26,780[†] kcal/kg mol × 0.44 kg mol + 94,400 kcal/kg mol × 0.56 kg mol
 + 68,300 kcal/kg mol × 2.4 kg mol = 228,600 kcal

$$\text{RHS} = \sum_i \frac{m_i}{\text{MW}_i} (\bar{c}_p)_{m,i} (T_{\text{exit}} - T_0) + m_w h_{fg}$$

= [7.216 kcal/kg mol · °C × (00.44 + 7.445) kg mol + 13.227 kcal/kg mol · °C
 × 3.8 kg mol + 11.208 kcal/kg mol · °C × 0.56 kg mol] × (1000 − 18) °C
 + 3.8 kg mol × 10,520 kcal/kg mol

= 151,400 kcal

$$\frac{\text{LHS}}{\text{RHS}} = 1.509$$

Now we let $T = 1500°C$ (1773°K). We again use Fig. 4.5 and find $K_p = 14,600 > (K_p)_{\max}$. Therefore, we have $K_p = (K_p)_{\max} = 0.03187$ and $x' = 0.44$ g mol and $y' = 0.56$ g mol.

$$(\bar{c}_p)_m = \begin{cases} 7.388 \text{ kcal/kg mol} \cdot °C & \text{for CO, N}_2 \\ 15.99 \text{ kcal/kg mol} \cdot °C & \text{for H}_2\text{O} \\ 12.06 \text{ kcal/kg mol} \cdot °C & \text{for CO}_2 \end{cases}$$

The standard heat of reaction remains unchanged and hence LHS = 228,600 kcal.

[†] The heat released when C is burned to CO at standard conditions is obtained by subtracting the heat of combustion of CO from the heat of combustion of carbon.

Sec. 4.6   Calculation of Actual Flame Temperature

Then

$$\text{RHS} = \sum \frac{m_i}{MW_i}(\bar{c}_p)_{m,i}(T_{exit} - T_0) + m_w h_{fg}$$

$$= [7.388 \text{ kcal/kg mol} \cdot °C \times (0.44 + 7.445) \text{ kg mol CO and } N_2$$
$$+ 15.99 \text{ kcal/kg mol} \cdot °C \times 3.8 \text{ kg mol } H_2O$$
$$+ 12.06 \text{ kcal/kg mol} \cdot °C \times 0.56 \text{ kg mol } CO_2] \times (1500 - 18) °C$$
$$+ 3.8 \text{ kg mol } H_2O \times 10{,}520 \text{ kcal/kg mol}$$

$$\frac{\text{LHS}}{\text{RHS}} = 1.003$$

and hence T_{exit} is slightly above 1500°C. We plot the ratio LHS/RHS as temperature and find by extrapolation that $T_{exit} \simeq 1505$°C.

4.6 CALCULATION OF ACTUAL FLAME TEMPERATURES

Actual flame temperatures are always lower than the adiabatic flame temperature since there is always a substantial quantity of heat released to the environment. If the heat release per unit mass of fuel burned, Q_1, is known as a function of temperature, our heat balance becomes

$$\sum_j n_j (Q_0')_j = \sum_i \frac{m_i}{MW_i}(\bar{c}_p)_{m,i}(T_{exit} - T_0) + m_w h_{fg}$$
$$- (\bar{c}_p)_{m,a}(T_{air} - T_0)\frac{m_a}{MW_a} + Q_1 m_F \quad (4.72)$$

If we have a measurement of the flame temperature, we can determine Q_1.

Example 4.11

The fuel stream of Example 4.10 (fuel 0.4 kg of H_2 for each kilogram of carbon suspended in a water slurry containing 40 w/o of fuel) is again burned in 90% of the stoichiometric quantity of air. Assume that the flame temperature from the combustion is found to be 550°C. How much heat is radiated from the flame per kilogram mole of C burned?

Solution By using $T_{flame} = 550$°C, we may obtain K_p as we did in Example 4.10. We find that $K_p = 0.019$. This is less than $(K_p)_{max}$ (see Example 4.10). We therefore need to calculate x', y' and z' using the relationships shown in Example 4.10. Basis for solution: 1 kg mole of C

$$\begin{cases} x' + y' + z' = 1 \\ x' + 2y' = 1.56 \end{cases}$$

$$K_p = 0.01898 = \frac{[x'/(x'+y'+11.25)]^2}{y'/(x'+y'+11.25)}$$

We then obtain

$$x' = 0.371 \text{ kg mol}$$
$$y' = 0.594 \text{ kg mol}$$
$$z' = 0.035 \text{ kg mol}$$

and

$$\sum_j n_j(Q_0') = Q_{C \to CO} + Q_{C \to CO_2} + Q_{H_2 \to H_2O(l)}$$

$$= 26{,}780 \text{ cal/g mol} \times 0.371 \text{ kg mol}$$
$$+ 94{,}400 \text{ cal/g mol} \times 0.594 \text{ kg mol}$$
$$+ 68{,}310 \text{ cal/g mol} \times 2.4 \text{ kg mol}$$
$$= 229{,}950 \text{ kcal per kg mol of C burned}$$

We assume that $T_{\text{exit}} = T_{\text{flame}}$ and for $T_{\text{exit}} = 550°C$,

$$(\bar{c}_p)_m = \begin{cases} 7.079 \text{ kcal/kg mol} \cdot °C & \text{for CO and } N_2 \\ 10.93 \text{ kcal/kg mol} \cdot °C & \text{for } H_2O \\ 10.32 \text{ kcal/mol} \cdot °C & \text{for } CO_2 \end{cases}$$

From the heat balance of Eq. (4.72),

$$\sum_j n_j Q_0' = 229{,}950 = \sum \frac{m_i}{MW_i} (\bar{c}_p)_{m,i} (T_{\text{exit}} - T_0) + m_w h_{fg} + Q_1 m_F$$

or

$$230{,}750 = [7.07 \text{ kcal/kg mol C} \times (0.371 + 7.445) \text{ kg mol} + 10.9269 \text{ kcal/kg mol} \cdot °C$$
$$\times 3.8 \text{ kg mol} + 10.32 \text{ kcal/kg mol} \cdot °C$$
$$\times 0.594 \text{ kg mol}] \times (550 - 18)°C$$
$$+ 3.8 \text{ kg mol} \times 10{,}520 \text{ kcal/kg mol} + Q_1 m_F$$

$$Q_1 m_F = 229{,}950 - 94{,}700 = 135{,}180 \text{ kcal}$$

Since our basis was 1 kg mol of C, heat released = 135,180 kcal/kg mol.

In the absence of flame temperature measurements (e.g., during design) we will need to simultaneously evaluate Q_1 and the flame temperature. We determine the actual flame temperature by essentially the same graphical procedure as described previously. We again plot the ratio of the heat of reaction, left-hand side of Eq. (4.72), to the right-hand side of this equation and determine the temperature at which the ratio is unity.

To evaluate Q_1, let us first consider the interchange of energy between an isothermal black enclosure with area A_T and a gray gas flame filling the enclosure.

Sec. 4.6 Calculation of Actual Flame Temperature

The flame will radiate energy at the rate of q_{rad}, which is given by

$$q_{\text{rad}} = A_T \sigma T_f^4 (\varepsilon_G)_f \tag{4.73}$$

where
σ = Stefan–Boltzmann constant (5.67×10^{-8} W/m² · °K⁴ or 0.17×10^{-8} Btu/h · ft² · °R⁴)
T_f = absolute temperature of gas (°K or °R)
$(\varepsilon_G)_f$ = emissivity of the gas flame at T_f

However, the wall, at absolute temperature T_c, will also radiate energy and a portion of this will be absorbed by the gas. The rate of energy absorption by the gas is obtained from

$$q_a = A_T \sigma T_c^4 (\alpha_G)_c \simeq A_T \sigma T_c^4 (\varepsilon_G)_c \tag{4.74}$$

where $(\alpha_G)_c$ = absorbtivity of gas at T_c, which is approximately equal to $(\varepsilon_G)_c$
$(\varepsilon_G)_c$ = gas flame emissivity at T_c

The net rate of radiation interchange between the gas and the black enclosure is then

$$(q_{\text{rad}})_{\text{net}} = A_T \sigma \left[(\varepsilon_G)_f T_f^4 - (\varepsilon_G)_c T_c^4 \right] \tag{4.75}$$

If we now consider the net rate of interchange between the entire volume of gas and a portion of the enclosure (e.g., area subtended by boiler tubes) having area A_C, we may write

$$q_{\text{TC}} = \sigma \left(A_T \mathscr{F}_{\text{TC}} T_f^4 - A_C \mathscr{F}_{\text{CT}} T_c^4 \right) \tag{4.76}$$

where
q_{TC} = rate of radiative interchange between furnace and gas
$\mathscr{F}_{\text{TC}}, \mathscr{F}_{\text{CT}}$ = combined shape factors accounting for emissivity and the fact that area A_C and gas do not exchange radiation exclusively with each other

In this simple case

$$\mathscr{F}_{\text{TC}} = (\varepsilon_G)_f \frac{A_C}{A_T} \qquad \mathscr{F}_{\text{CT}} = (\varepsilon_G)_c \tag{4.77}$$

where $(\varepsilon_G)_c$ is the emissivity of gas at temperature of A_C and $(\varepsilon_G)_f$ has its previous meaning.

We may extend this concept to the more realistic case of the combustion chamber lined with nonblack, cold tubes which absorb radiation and a refractory that reirradiates all the radiation which falls on it. Equation (4.76) still gives the interchange between the flame and area A_C, which we now take as the effective tube area, but $\mathscr{F}_{\text{TC}}$ and $\mathscr{F}_{\text{CT}}$ must be reevaluated to include the effect of the refractory and the emissivity of area A_c. It may be shown that in this case

$$\mathscr{F}_{\text{TC}} = \frac{1}{\dfrac{A_T}{A_C \varepsilon_t} + \dfrac{1}{(\varepsilon_G)_f} - 1} \qquad \mathscr{F}_{\text{CT}} = \frac{A_T/A_C}{\dfrac{A_T}{A_C \varepsilon_t} + \dfrac{1}{(\varepsilon_G)_c} - 1} \tag{4.78}$$

where A_C = effective tube area
ε_t = emissivity of boiler tubes
A_T = total surface area of enclosure (tube area + refractory)

Note that since the refractory reirradiates all the radiation it receives, its emissivity does not enter into the calculation.

The total rate of heat transfer from the flame in the boiler region, q, is the sum of the radiative heat transfer plus convective heat transfer. That is,

$$q = q_{TC} + h A_C (T_f - T_c) \qquad (4.79)$$

where h is the convective heat-transfer coefficient. In most cases the convective heat transfer is only a small fraction of the radiation heat transfer in the combustion chamber and it is often neglected.

It should be noted that, in calculating the flame temperature in the manner suggested here, we are assuming sufficient turbulence so that exit gas temperature and average flame temperature are equal. In actuality, even in a turbulent furnace, the exit gas temperature is likely to be 100 to 160°C less than the average radiating temperature.

When the absolute temperature of the absorber is less than one-half the flame temperature, the radiation from the cold surface is not a major factor and use of a single emissivity is allowable. Equation (4.76) thus simplifies to

$$q_{TC} = \sigma A_T \mathscr{F}_{TC} (T_f^4 - T_c^4) \qquad (4.80)$$

In some furnace designs (e.g., those using cyclone burners) essentially all the combustion takes place at the burner. The adiabatic flame temperature is attained at the burner chamber exit. The temperature of the hot gases then falls as the gases rise through the furnace. However, if there are cyclone burners at the number of furnace elevations, the approximation of a constant flame temperature may still be reasonable for preliminary design purposes.

Example 4.12

For the condition of Example 4.11, determine the effective furnace area per kilogram mole of C burned per hour if (1) the effective furnace area may be taken at 300°C and (2) the absorbing surface has an emissivity of 0.5 and flame may be taken as black.

Solution Under these conditions, $\mathscr{F}_{TC} = \varepsilon_t (A_C/A_T)$ and Eq. (4.80) simplifies to

$$q_{TC} = \sigma \frac{A_C}{A_T} A_T (T_f^4 - T_c^4) = \sigma A_C \varepsilon_t (T_f^4 - T_c)^4$$

$$= 136{,}000 \text{ kcal/h} = 158{,}000 \text{ W}$$

$$\sigma = 5.669 \times 10^{-8} \text{ W/m}^2 \cdot °K^4$$

$$158{,}000 \text{ W} = 5.669 \times 01^{-8} (\text{W/m}^2 \cdot °K^4) A_C \times 0.5 (823^4 - 573^4) °K^4$$

$$A_C = 14.52 \text{ m}^2$$

4.7 FLAME EMISSIVITY

To evaluate flame emissivity appropriately, it must be recognized that thermal radiation emitted by flames is due to three major phenomena:

1. Infrared radiation from combustion product gases
2. A continuum of thermal radiation from high-temperature soot and ash particles
3. Chemiluminescence, a form of nonequilibrium radiation

Chemiluminescence contributes very little to the overall radiative heat transfer and will be neglected.

Let us first consider the radiation emanating from a hot gas. When a gas is heated, it emits radiation in certain specific sections of the infrared spectrum. Conversely, when blackbody radiation traverses a hot gas mass, radiation is absorbed in the same specific regions. Some gases do not show significant absorption–emission bands in the infrared region at the temperatures of interest in combustion chambers. This is true for gases such as oxygen and nitrogen with symmetric molecules. However, gases such as CO, SO_2, CO_2, and water vapor do show absorption bands of importance. Since CO and SO_2 are present in only small amounts in industrial flames, we are concerned primarily with radiation from CO_2 and water vapor.

The quantity of radiation emitted by a gas flame containing CO_2 and H_2O depends on the temperature of the gas flame, the mass of gas in the flame, and the geometry of the gas mass. At a given total pressure, the mass of gas in a hemispherical flame can be expressed in terms of the product of the partial pressure of the radiating gas and L, the radius of the hemisphere. It is convenient to express the quantity of radiation emitted as the ratio of the radiation received from the gas mass at the center of the hemispherical base to the radiation that would be received at the same position form a hemispherical blackbody, having the same radius. This ratio is nothing more than the gas emissivity.

The emissivities of CO_2 and water vapor at 1 atm total pressure are functions of T for various values of the product pL. For other geometries we replace L by the average mean beam length, L_m. We may define L_m so that it applies to radiation over the entire envelope (entire surface of gas volume). For common geometries of interest, L_m is obtained from the ratio L_m/D, where D is a characteristic dimension. Table 4.3 lists this ratio and the characteristic dimension, D, for several common geometries. For shapes not listed in the table, the approximation $L_m = 0.88 \times 4(V/A_T)$, where V is the gas volume and A_T is the total surface area that encloses the gas volume, is often suggested.

For the practical case where water vapor and carbon dioxide are present together in the combustion product gas, the total radiation will be less than the

TABLE 4.3 MEAN BEAM LENGTHS FOR VOLUME RADIATION TO ENTIRE ENCLOSURE SURFACE

Shape	Characteristic dimension, D	L_m/D
Sphere	Diameter	0.63
Right circular cylinder		
Height = $\frac{1}{2}$ diameter	Diameter	0.45
Height = diameter	Diameter	0.6
Height = 2 × diameter	Diameter	0.73
Height = ∞	Diameter	0.90
Rectangular parallelepiped		
Cube	Edge	0.6
1 × 1 × 4	Shortest edge	0.81
Infinite parallel planes	Clearance	1.8
Space between tubes on square pitch in tube bank	Clearance	3.5

contribution of each component taken separately because of the spectral shadowing. There is also some spectral shadowing by sulfur dioxide, which further reduces the emissivity of the mixture. However, the SO_2 effect is small and is usually neglected.

Although obtaining the exact value of emissivity for water vapor–CO_2 mixtures is complicated, a reasonably good approximation can be obtained easily. Use is made of the fact that for these mixtures, ε_G decreases with rising temperature in such a manner that the product $\varepsilon_G T$ depends almost exclusively on the product $(p'_{CO_2} + p'_w)L_m$. Figure 4.6 shows this relationship. The value of $\varepsilon_G T$ varies slightly with the ratio p'_w/p'_{CO_2}, but as may be seen in the figure, over the ratio range 1 to 2

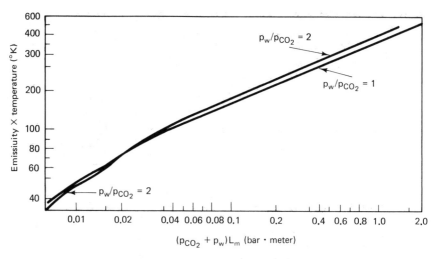

Figure 4.6 Emissivity of combustion products.

Sec. 4.7 Flame Emissivity

little variation is seen. Combustion gases containing both CO_2 and water vapor are generally in this range.

The approximation of Fig. 4.6 is best in the range 850 to 1700°K. However, the inaccuracy introduced by using this approximation outside this range is probably smaller than the uncertainty in the other components of furnace gas emissivity.

Emissivity of Soot and Fuel Particles

Soot particles form during combustion when there is insufficient air for complete combustion or the air is not well mixed with the fuel. The soot particles tend to be small, approximately 0.02 to 0.06 μm in size. The concentration of soot particles in a flame cannot be predicted by analysis. Actual soot concentrations have to be determined by measurement. Soot radiation is often referred to as *luminosity*.

The emissivity at any particular wavelength λ, is called the *monochromatic emissivity*, ε_λ, and is independent of the soot particle size. The monochromatic emissivity depends primarily on the total particle volume per unit volume of space, v_s. ε_λ can be estimated from

$$\varepsilon_\lambda = 1 - e^{-K'v_s L_m/\lambda} \tag{4.81}$$

where L_m represents the mean beam length and K' is a constant. Integration over all wavelengths leads to the expression for the soot emissivity, ε_s, as

$$\varepsilon_s = 1 - \left(1 + \frac{K'Tv_s L_m}{c_2}\right)^{-4} \tag{4.82}$$

where c_2 is the second Planck radiation constant. Values for K'/c_2 are obtained experimentally from the complex refractive index of soot. These values depend on the hydrogen-to-carbon ratio of the coal used. Typical K'/c_2 values range from

$$480 \text{ m}^{-1}/°\text{K} < \frac{K'}{c_2} < 980 \text{ m}^{-1}/°\text{K} \quad \text{for } 0 < \frac{\text{H}}{\text{C}} < 0.4$$

A typical practical value would be 500 m^{-1}/°K (85 ft^{-1}/°R).

The conversion of carbon to soot in coal or oil systems where a very visible, luminous flame is present can be as high as 2%. This value corresponds to a soot emissivity of 0.5 to 0.75. However, under the usual conditions coal flames are only moderately luminous and addition of 0.1 to the nonluminous gas emissivity is usually sufficient when calculations are based on a mean flame temperature.

The presence of burning, nearly black, fuel particle clouds in flames can substantially add to flame emissivity. A cloud of particles may be characterized by a mean beam length L_m and a projected area per unit volume of space, A/V. The emissivity, ε_p, of the cloud due to the particles is then given by

$$\varepsilon_p = 1 - e^{-(A/V)L_m} \tag{4.83}$$

Alternatively, the cloud of particles may be characterized by a number density of N

with a mean projected area per particle of A_p. The cloud emissivity is then given by

$$\varepsilon_p = 1 - e^{-NA_p L_m} \qquad (4.84)$$

As a typical example, a cloud with 200-μm fuel oil particles, $L_m = 3$ m, and combustion at stoichiometric conditions results in a cloud-particle contribution to emissivity of 0.30. This will be reduced as the fuel particles are burned off.

Total Flame Emissivity

The combined flame emissivity due to radiation from the gases, soot, and particles may be estimated from the component emissivities if they are known. The combined emissivity, ε_{comb}, is approximated by

$$\varepsilon_{comb} = 1 - (1 - \varepsilon_G)(1 - \varepsilon_s)(1 - \varepsilon_p) \qquad (4.85)$$

where ε_G = emissivity due to gas radiation occurring alone
ε_s = emissivity due to soot radiation occurring alone
ε_p = emissivity due to particle radiation occurring alone

4.8 FLUIDIZED-BED COMBUSTION SYSTEMS

When a gas is blown upward through a granular bed of particles, fluid friction produces a pressure drop which increases with increasing gas velocity. If the gas flow rate is continuously increased, a velocity will be reached at which the gas pressure drop just equals the weight of particle bed per unit area plus the friction of the particle bed against the walls. If the gas flow is maintained at this rate, and the particles are free flowing, the particles are separated from each other and the bed has a much more open arrangement. The suspended mass of particles, which resemble a boiling liquid, is referred to as a *fluidized bed*.

The pressure drop, Δp, through a packed bed of small, uniform particles may be calculated from

$$\Delta p = \frac{f_p l u^2 v_s \rho_g}{g_c D_p (1 - v_s)^3} \qquad (4.86)$$

where l = depth of packed bed
u = gas velocity based on total cross-sectional area of bed
D_p = particle diameter
f_p = packed-bed friction factor = $[150/(D_p u \rho_g / \mu_g v_s)] + 1.75$
g_c = gravitational conversion factor (one in SI units)
v_s = volume of solid particles/total volume
ρ_g = gas density
μ_g = gas viscosity

If we assume that friction along the wall is small, we obtain the fludization velocity

Sec. 4.8 Fluidized-Bed Combustion Systems

by equating Δp to the bed weight per unit area; that is,

$$\Delta p = \rho_s v_s l \qquad (4.87)$$

where ρ_s is the density of solid particles.

In a fluidized-bed combustion system, crushed coal, ash, and limestone are mixed together in a bed levitated by incoming combustion air. The combustion air enters the bottom of the furnace and flows upward through the bed, causing it to be suspended in the furnace. The boiler tubes are immersed in the fluidized bed. This results in direct contact between the burning particles of coal and the boiler tubes. Very high rates of heat transfer are obtained, thereby reducing the furnace area and size. The resulting combustion temperature is considerably lower than in a regular furnace/boiler.

The major attraction of the fluidized-bed combustion process is the potential for a direct reduction in emission of pollutants. The somewhat lower combustion temperatures, 900°C (1952°F), reduce the formation of nitrogen oxides. Most important, however, the limestone introduced with the pulverized coal reacts with the sulfur dioxide in the fluidized bed to produce calcium sulfites or sulfates. These

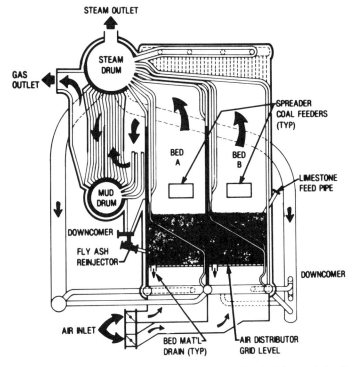

Figure 4.7 Small fluidized-bed furnace/boiler. (Reproduced with permission from *Proceedings of the American Power Conference*, Vol. 41, 1979.)

calcium salts are solids and remain trapped in the combustion chamber. The result is a reduction in the need for pollution control and abatement equipment, even with the combustion of high-sulfur coal.

Figure 4.7 illustrates a small fluidized-bed steam generator. Pebble size coal is fed at the top of the bed by a spreader-stoker. Crushed limestone is fed at a single point at the top of each bed (larger units might require multiple limestone feeders). Spent bed material is removed at the bottom of the bed. The bulk of the heat transfer occurs within the bed itself, but the hot gases leaving the bed (bed operating temperature $\simeq 900°C$) are cooled by radiation to tubes lining the furnace and then by convection to tube banks. The larger fly ash particles, which contain considerable unburned carbon, are collected by a cinder trap and then are reinjected into bed A. Fine particulates are removed by a baghouse dust collector.

Note that the tubes in the fluidized bed see severe service, and adequate tube lifetime is of concern. Severe failure of a single tube will shut down the boiler. To avoid this difficulty, some advanced designs replace the boiler tubes in the fluidized bed by heat pipes containing sodium which transmit the heat to the boiler. Since each heat pipe is an isolated system, an individual failure does not require shutdown of the entire system.

SYMBOLS

A	area, m²
A_C	effective tube area in furnace, m²
A_T	total enclosure area, m²
c_p	specific heat at constant pressure, J/kg·°K or kcal/kg·°C
$(\bar{c}_p)_{m,i}$	mean molal specific heat of component i at constant pressure, kJ/kg mol·°K or kcal/kg mol·°K
c_2	second Planck radiation constant
C'	percent carbon in coal
C_1	carbon in the CO, kg mol/kg mol flue gas
C_2	carbon in the CO_2, kg mol/kg mol flue gas
C_{FG}	carbon in the flue gas, kg mol
C_r	carbon in the refuse, kg per mole carbon
d	dilution coefficient
D	diameter or other characteristic dimension, m
D_p	particle diameter, m
f	fugacity
f_p	packed-bed friction factor
$\mathscr{F}_{CT}$	Combined shape factor accounting for emissivity and view effects based on A_C

Chap. 4 Symbols

$\mathscr{F}_{TC}$	combined shape factor accounting for emissivity and view effects based on A_T
g_c	gravitational conversion factor [unity in SI units, 32.2 ft/s² (lb$_m$/lb$_f$ in British units)]
G	total free energy, J
ΔG	change in free energy, J
ΔG_f	free energy of formation, J
h	specific enthalpy, J/kg or kcal/kg
h_{fg}	latent heat of vaporization of water at 1 bar, J/kg or kcal/kg
$\hbar$	convective heat-transfer coefficient, W/m² · °K
H	enthalpy, J or kcal
H'	percent hydrogen in coal, w/o
ΔH	enthalpy change, J or kcal
K	constant
K_p	equilibrium constant
l	depth of packed bed, m
L	radius of gas hemisphere, m
L_m	mean beam length, m
m'_a	(mass of air required for complete combustion)/(mass of fuel)
$(m'_a)_{act}$	mass of air supplied per unit mass of fuel (actual air-to-fuel ratio), kg air/kg fuel
$m'_{a/C}$	mass of air required for complete combustion/mass of carbon
$m'_{A/r}$	mass of ash/mass of refuse
m'_b	mass of carbon burned/mass of coal
m'_C	mass unburned carbon/mass of coal
m_i	mass of component i
m'_N	mass of nitrogen/mass of coal
m'_r	mass of refuse/mass of coal
M	number of kilogram moles
$M'_{a/F}$	moles air/moles fuel
M_{da}	dry air supplied for combustion, kg mol
MW	molecular weight
n	number of moles of a substance or compound
n_{Rj}	moles of reactant j required
N	number density, number/m³
N'	percent nitrogen in coal, w/o
O'	percent oxygen in coal
p	pressure, Pa or bar

p'	partial pressure, Pa, bar, or atm
Δp	pressure drop, Pa (N/m^2)
q_a	rate of energy absorption by the gas, W
q_r	rate at which the flame radiates energy, W
q_{rad}	rate of radiative interchange between flame and black enclosure, W
$(q_{\text{rad}})_{\text{net}}$	net rate of radiation interchange of energy between gas and black enclosure, W
q_{TC}	rate of interchange of heat between flame and cool surface in furnace due to radiation, W
Q	heat released or added to system, J or kcal
Q_1	net heat released by m_F kg fuel, J or kcal
Q'	heat of combustion at nonstandard conditions per mole of fuel, kcal/kg mol or J/kg mol
Q'_0	standard heat of combustion per mole fuel, kcal/kg mol or J/kg mol
Q'_1	net heat release per mole of fuel, kcal/kg mol or J/kg mol
Q'_a	heat removed by cooling air from initial temperature to T_0 per mole of fuel, kcal/kg mol or J/kg mol
Q'_{cw}	heat of combustion of unburned coal and CO in waste and exit streams per mole fuel, kcal/kg mol or J/kg mol
Q'_p	heat added by heating products from T_0 to T_1
Q'_R	heat removed by cooling reactants from T_1 to T_0 per mole of fuel, kcal/kg mol or J/kg mol
R	ideal gas constant
s	specific entropy, J/kg · °K
S	total entropy, J/°K
S'	percent sulfur in coal
ΔS	entropy change
T	temperature, °K (unless specified otherwise)
T_c	absolute temperature of furnace wall, °K
T_f	absolute temperature of the flame or gas, °K
u	gas velocity, m/s
v	molal specific volume, m^3/kg mol
v_s	volume of solid particles/total volume
V	gas or cloud volume, m^3
V_a	volume of wet air for combustion, m^3
V_{FG}	total volume of flue gas products, m^3
Z_i	atoms per mole of fuel

Greek Symbols

$(\alpha_G)_c$	absorbtivity of gas at T_c
β	mass fraction of carbon in the refuse
ε_G	emissivity of gas
$(\varepsilon_G)_c$	emissivity of the gas at T_c
$(\varepsilon_G)_f$	emissivity of the gas flame at T_f
ε_p	emissivity of particles
ε_s	emissivity of soot
ε_t	emissivity of boiler tubes
ε_λ	monochromatic emissivity
λ	wavelength
μ	viscosity, Pa · s
ρ	density, kg/m^3
σ	Stefan–Boltzmann constant, 5.67×10^{-8} W/m^2 · °K^4

Superscripts

°	state at which reactant and product partial pressures are 1 atm
s	standard conditions

Subscripts

a	air
0	initial or standard condition
1, 2, ···	state or condition 1, 2 of the system
av	average
C	carbon
CO	carbon monoxide
CO$_2$	carbon dioxide
da	dry air
exit	exit conditions
F	fuel
FG	flue gas
g	gas
H$_2$O	water
i	component i
max	maximum
N$_2$	nitrogen

O_2	oxygen
op	actual quantity used
P	product
r	refuse
R	reactant
rad	radiant
rev	reverse
s	solid
S	sulfur
SO_2	sulfur dioxide
theo	theoretical
w	water

PROBLEMS

4.1. Determine the higher heating value of coal having the ultimate analysis shown in Table 3.12 by **(a)** the Dulong–Berthelot formula and **(b)** the heating values of its constituents.

4.2. The coal described in Table 3.12 is burned in a boiler from which flue gas exits at 300°C. The Orsat analysis of the flue gas indicates that it contains 12% CO_2, 7% oxygen, and negligible amounts of CO. The ash in the ash pit contains 10 w/o carbon, is withdrawn at 200°C, and has a specific heat of 0.2 kcal/kg. Estimate the heat supplied to the boiler per kilogram of coal burned. Assume that the combustion air is dry and at 20°C, and that heat loss to the environment from the furnace wall is negligible.

4.3. A fuel has the following elemental analyses:

	w/o
C	82.1
S	0.2
H_2	17.7

If the fuel is burned in 20% excess air, determine the percent by volume of each constituent in the flue gas.

4.4. A gaseous fuel consisting of 60% ethane and 40% methane is burned in air. Determine the adiabatic flame temperature when it is burned with **(a)** the stoichiometric quantity of air and **(b)** 20% excess air.

4.5. Methane gas is to be used as a fuel for a gas turbine. Turbine blade properties limit the inlet gas temperature to 600°C. Determine how much excess air must be supplied so that the inlet temperature limitation is not exceeded.

4.6. Determine the adiabatic flame temperature when coal with the following ultimate analysis is burned with 30% excess air:

	w/o
Carbon	65.0
Sulfur	0.5
Nitrogen	1.5
Available hydrogen	4.0
Combined water	9.0
Moisture	10.0
Ash	10.0
Total	100.0

4.7. Determine the flame temperature when the coal described in Problem 4.6 is burned with 93% of the stoichiometric quantity of air and 55% of the heat generated is radiated to the walls.

4.8. Use the equilibrium constants given in Fig. 4.5 to estimate the mole fractions of nitrogen oxides in the flue gas of Problem 4.6. (Only three simultaneous equations, obtained from the equilibrium relationships, must be solved if the N_2 mole fraction is assumed to be known.)

4.9. Estimate the emissivity of the flame obtained under conditions of Problem 4.4(b) assuming that the combustion takes place in a 4 m × 4 m × 4 m enclosure.

4.10. The fuel oil described in Problem 3.1 is burned in an enclosure which may be considered to be a 3.5-m-diameter cylinder, 4 m high. The walls may be considered to be 50% refractory and 50% steel tubes at a temperature of 350°C. Optical pyrometer readings indicate that the flame temperature is 1000°C. Estimate the radiant heat transfer from the flame to the tubes lining the walls.

4.11. Determine the velocity of room-temperature air required to fluidize a bed of coke particles. The coke particles may be taken to have a density 80% of that of carbon, to have an average diameter of 1 cm, and to form a bed of 40% porosity.

BIBLIOGRAPHY

BAUMEISTER, T. (ed.), *Marks' Standard Handbook for Mechanical Engineers*, 8th ed. New York: McGraw-Hill, 1981.

BERKOWITZ, N., *An Introduction to Coal Technology*. New York: Academic Press, 1979.

Combustion, 3rd ed. Windsor, Conn.: Combustion Eng., Inc., 1981.

CULP, A., JR., *Principles of Energy Conversion*. New York: McGraw-Hill, 1979.

HOUGEN, O. A., K. M. WATSON, and R. A. RAGATZ, *Chemical Process Principles—Part I*: *Material and Energy Balances*, 2nd ed. New York: Wiley, 1958.

KRENZ, J., *Energy: Conversion and Utilization*. Boston: Allyn and Bacon, 1976.

PERRY, J. H. (ed.), *Chemical Engineers Handbook*, 4th ed. New York: McGraw-Hill, 1981.

SPALDING, D., "Mathematical Models of Continuous Combustion," *Proc. Conference on Emissions from Continuous Combustion Systems*, Sept. 27, 1981.

5

Fossil-Fueled Steam Power Plants: Primary Systems

5.1 INTRODUCTION TO POWER PLANT SYSTEMS AND COMPONENTS

This chapter presents an overview of the primary system of a modern fossil-fueled power plant. Emphasis will be placed on the basic functions and design features required for coal-fired power plants since these are the predominant fossil-fueled plants.

The block diagram of Fig. 5.1 shows the major segments of a modern coal plant. After crushing and pulverizing, the coal is fed to the furnace, where it is burned. In the flame region, much of the heat generated is transferred to boiler tubes which line the walls. The steam–water mixture formed in the boiler tubes proceeds to a steam drum, where the vapor and liquid are separated. The liquid is returned to the boiler tubes and the vapor proceeds to a superheater. There the hot gases leaving the flame region superheat the saturated steam produced in the boiler section. The superheated steam proceeds to the turbine, where its thermal energy is converted to mechanical energy.

The steam leaving the high-pressure stage of the turbine is reheated in the furnace prior to being sent to the lower-pressure sections of the turbine. The steam exiting from the low-pressure turbine proceeds to a condenser. The liquid from the condenser is reheated in regenerative feedwater heaters, using steam bled from the turbine, prior to being returned to the furnace. The feedwater then proceeds to the furnace, where it is brought to near saturation in the economizer. The gases leaving the economizer preheat the entering air and are then discharged through the stack. Particulates in the gas stream are removed by electrostatic precipators prior to discharge. In newer plants, the exit gases are also scrubbed to remove SO_2.

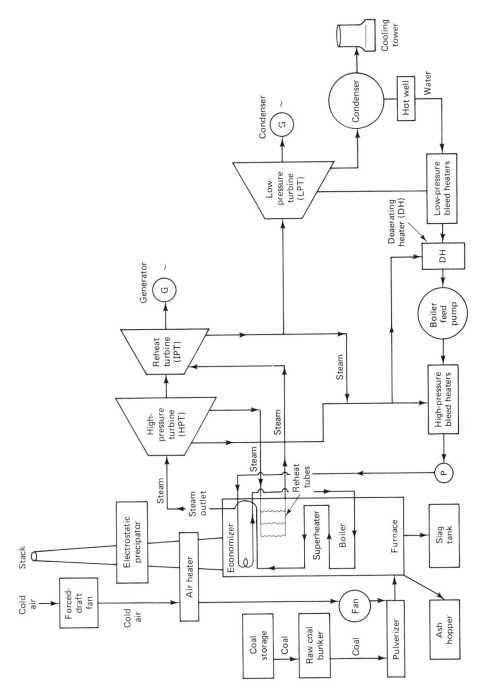

Figure 5.1 Major components of a coal-fired power station.

157

The major factor in establishing the size of the power plant components is the required net power output. However, the design is significantly affected by the characteristics of the coal being burned, coal delivery methods, combustion technique selected, and constituents of the coal ash. The most significant coal characteristics are:

Moisture content
High heating value
Grindability index
Ash content
Ash fusion temperature
Sulfur content

These specific coal characteristics are far more meaningful in the design of the equipment and the station than is the rank of the coal being burned. Each specific coal characteristic can affect the design of certain plant equipment and components.

As an example, consider the effect of moisture on power plant components. The general effect is to increase most boiler components sizes and to reduce boiler efficiency. A high moisture content will require a larger air heater. This, in turn, requires an increase in the temperature of the inlet air to the pulverizers since a greater volume of air is required to dry the coal in the pulverizer. This increases the initial primary air fan size and reduces the forced draft fan size. The burning of coal with a high moisture content results in a higher flue gas dew point, and raises the stack temperature needed to prevent corrosion. The net result of a high-moisture-content coal is to reduce the overall power plant efficiency.

The high heating value of the coal is another property that significantly affects plant design. The high heating value (HHV) establishes the rate at which coal is fed into the furnace (firing rate). Since the HHV can vary from 19,000 kJ/kg (8000 Btu/lb) to well over 28,500 kJ/kg (12,000 Btu/lb) and since the firing rate for a given furnace or boiler output is inversely proportional to the HHV, it is easy to see that mass firing rates can vary by over 50%. Thus the HHV will affect the design of the boiler combustion zone, burners, pulverizers, coal feeders, silo storage requirements, coal-handling equipment, and the size of the active and inactive coal storage.

5.2 FUEL HANDLING AND PREPARATION FOR BURNING

Coal

Handling systems. The coal-handling systems at a large modern power plant unload the coal, handle it in the yard, and prepare the coal for burning in the furnace. Figure 5.2 is a block diagram which shows the major steps in the coal-handling operation at a power plant. Since most coal plants burn pulverized coal, this section focuses on the preparation of pulverized coal.

Sec. 5.2   Fuel Handling and Preparation for Burning

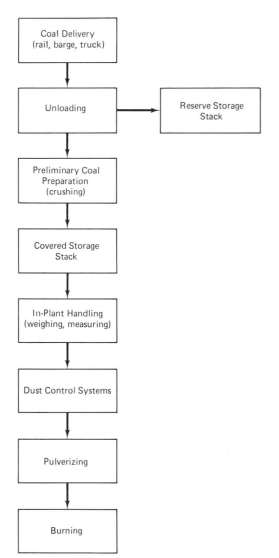

Figure 5.2   Coal-handling systems.

Shipment to power plants.   Rail shipment of coal accounts for nearly three-fourths of the coal delivered to U.S. power plants. Unit trains supply the coal for large power plants. Shipments of 10,000 tons or more are quickly unloaded by rapid-discharge bottom-dump railroad cars. A 100-ton load of coal can be discharged from a hopper car in 20 seconds.

Some western surface-mined coal and some eastern coal from Pennsylvania, West Virginia, and Ohio is shipped by barge. Barge shipment of coal accounts for approximately 10% of the coal used by U.S. utilities. The barges used to ship coal are generally not self-unloading, except for a few barges on the Great Lakes. The power plant must therefore have its own equipment for unloading barge coal. A typical unloader consists of a tower housing a trolly boom with a clamshell bucket. The bucket swings over the barge to load and dumps the coal in hoppers on shore.

The remainder of the coal used by utilities is shipped by either truck or overland conveyer. Overland conveyers are used where power plants are located near large mines.

Coal stored at a power plant may be placed in either active storage or reserve storage. The active storage supplies the normal power operation while the reserve storage pile is kept for emergencies and delays in regular shipment. Current utility practice is to keep large inactive reserves, up to 500,000 tons, to ensure a reliable fuel supply for 60 to 90 days of plant operation.

Since reserve storage piles are usually outdoors, frozen coal is a problem often faced by power plants in the north during the winter months. Surface moisture on the coal causes the coal particles to freeze together. The difficulties caused depend on the coal particle size. Large chunks of coal have weaker ice bonds than small fine particles because the surface-to-volume ratio is much lower. Severe winter weather conditions can cause entire train loads or reserve stacks of coal to freeze solid.

There are three methods of dealing with frozen coal: heat, mechanical devices, and freeze-control agents (FCAs). The heating operation is performed in an enclosed thawing shed that is heated by steam, electric, or gas heaters. This method is becoming increasingly expensive. Alternatively, mechanical devices can be used to shake, crush, or vibrate in order to break up the coal–ice mass. Freeze-control agents act as antifreeze to weaken the crystalline structure of the ice bond. The most widely used FCA is ethylene glycol. It is applied at a few pints per ton of coal when the coal is loaded into the car or silo. Ethylene glycol has proved highly successful as a freeze-control agent.

Coal in active storage is generally kept under cover to protect it from moisture and freezing. Large concrete bunkers immediately adjacent to the boiler plant are usually used. In some cases, both the active and inactive storage is under cover. This is particularly true for western coal, much of which is subbituminous, as it tends to produce large quantities of small particles, known as *fines*. With these coals, dust and moisture can become serious problems, making closed storage attractive. Large concrete silos are often used for this purpose. The silos are normally kept under a slight negative pressure to minimize dust problems.

Handling equipment. Very large quantities of coal are handled daily at modern coal stations. A 600-MWe unit, at full power operation, handles and burns approximately 300 tons of coal per hour, or 7200 tons of coal per day. Some of the coal delivery and receiving systems must be designed to handle up to 4000 tons of coal per hour.

One of the major tasks is building the active and reserve "stacks" of coal. Although there are a number of techniques and a wide variety of equipment that can be used to stack and reclaim coal, stacking conveyors are widely used. A particularly

Sec. 5.2 Fuel Handling and Preparation for Burning   161

useful device is the bucket wheel stacker/reclaimer, which is designed to perform both the stacking and reclaiming operations. Such a device would typically be used to build long triangular piles by taking coal from a conveyer running parallel to the pile. When the pile of coal reaches a maximum height, the stacker moves a few feet and continues to build another pile. Reclaiming is done by reversing the direction of the wheel and conveyor.

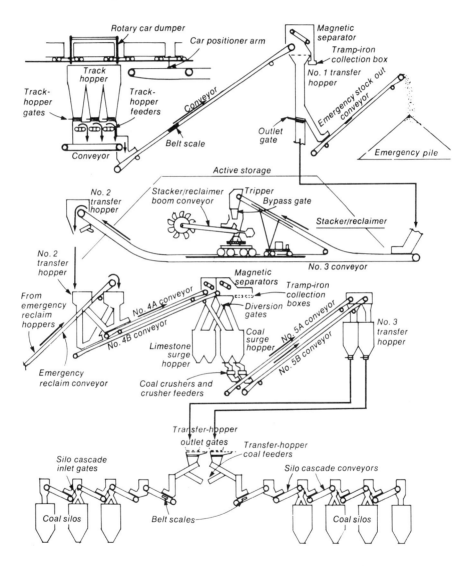

Figure 5.3 Coal handling at a large power station. (Reprinted with permission from *Power*; copyright 1979 McGraw-Hill, Inc., New York.)

Other types of in-plant handling systems are lowering wells, telescoping chutes, and belt conveyors. Belt conveyors can transfer the coal from the unloading area to various desired locations in the plant. These conveyors are effective on horizontal runs and inclines up to 20°. Discharge is usually from the end of the conveyor. Belt conveyors and feeders are rated on their volumetric capacity, which must be sufficient to meet coal requirements.

Figure 5.3 illustrates the coal-handling system at a large modern plant. This system is designed to handle up to 3500 tons/h of coal delivery, stacking, and reclaiming.

Crushing. Most coal for stoker furnace operation is bought in the required size. Crushers are usually required for pulverized coal furnace operation since unsized coal is normally purchased. The crushing can be done in the yard before the coal is stored or in the plant before it is used. Coal that is stored outdoors is normally crushed after it is reclaimed from the pile and before it is placed in the boiler silos or bins. Several types of crushers can be used: rotary breakers, single- and double-roll crushers, and hammermills. The particle size of the crushed coal is still relatively large and the coal can be safely stored in bunkers prior to pulverizing.

Inside storage and handling. The inside storage and handling systems must ensure a continuous, dependable supply of coal to the firing system. The bunkers, silos, and chutes must be sized properly. The condition of the coal is critical to a smooth flow of material. Wet, small-particle coal will adhere or stick to handling equipment and cause disruption in flow. If the surface moisture content exceeds 5%, coal handling becomes potentially difficult.

Coal chutes should be carefully designed to permit the smooth flow of coal from bunker to pulverizer. All chute angles should be made steep and with few reductions in cross sections. Tapered or uniform enlargements should be used in chute design.

The coal bunker is generally an integral part of the power plant building. The bunkers are designed to provide sufficient coal storage adjacent to the furnace so that there is always a ready supply for the pulverizers. Bunker capacity is generally set for storage of a 30-hour supply of coal at full power operation. Bunkers for a typical large plant would contain 10,000 to 20,000 tons of coal. Bunkers have been built of many different materials: stainless steel cladding on low-carbon steel plate, reinforced concrete, and even tile. The bunker shape is designed to enhance coal flow and cross sections are generally rectangular, parabolic, or cylindrical. The bunkers are often suspended from beams above the firing floor to save space. They may have multiple outlets feeding a number of pulverizers.

Dust control systems. Fugitive dust control is part of the total coal-handling system. Such control is required to guard against dust explosions as well as to meet state and federal health regulations. Dust is produced in the unloading of the coal, during the building of storage piles, while conveying coal, and at belt unloading and discharge areas.

Dust is usually controlled by dust collection systems with appropriate exhaust volumes at each transfer point. Alternatively, dust generation may be suppressed by using a conveyor arrangement with a wet suppression. However, this approach is expensive. Dry collection systems utilizing fabric filters can be effective in hot and cold climates. When a dry collection system is used, the disposal of the collect dust becomes a problem. One solution is to combine the collected dust with oil or some other type of binder and pelletize the mixture. These pellets are reintroduced into the conveyor system for subsequent firing in the furnace.

Dust collection and the reuse of the collected dust has received increased attention with increased power plant sizes. In large stations, as much as 50% of the coal may be in fines, especially with plants burning low-rank coal. Hence any loss of a substantial fraction of the fines leads to a marked increase in fuel cost. The best solution appears to be in the designing the plant to minimize coal handling and transfer points.

Preparation for firing: pulverizers. Most large coal-burning power plants now utilize suspension firing of coal. That is, fine coal particles suspended in air are blown into the furnace, where they are burned. The coal must be pulverized into small enough particles to be burned completely before reaching a cool surface. In most power plants, the coal is fired immediately after the pulverizing operation, although some stations store the pulverized coal for short periods before firing. The advantage to some storage capacity is that the coal can be pulverized at an average rate rather than peak rates dictated by load conditions.

The coal preparation system pulverizes the coal and delivers air for complete combustion. The primary air dries the coal and transports it through the pulverizer. The burner mixes the primary pulverizer air with an outside source of air to achieve complete combustion.

There are two types of pulverizers that are widely used. A medium-speed pulverizer (75 to 225 rpm) can be used for all types of bituminous coal. These pulverizers have a low power requirement, quick response to load change, and produce a fine coal powder. This class of pulverizer is available in a ball race type of design, or ball and roller type of design. High-speed pulverizers (above 225 rpm) are also used for all types of bituminous coal. This type of pulverizer is usually an impact mill. The pulverizer consists of a crushing section, a grinding section, and an exhauster. High-speed pulverizers require very little storage but are easily damaged by foreign material.

A typical medium-speed pulverizer is shown in Fig. 5.4. The coal is ground to a fine powder by crushing. The lower race, turned by a motor, causes the balls to rotate. The coal entering the center of the mill is thrown out against the moving balls, which crush it against the races. Primary pulverizer air transports the coal to the burner.

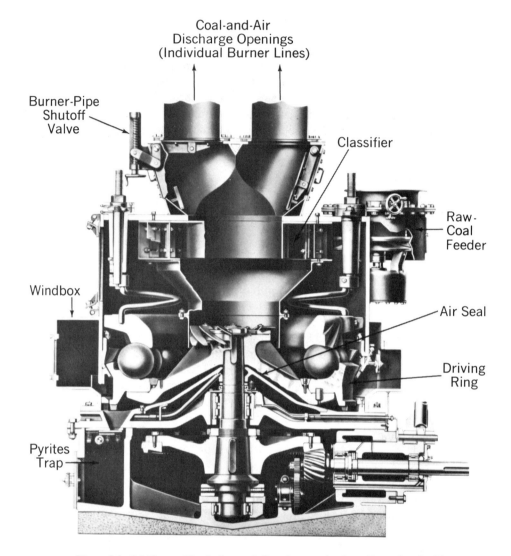

Figure 5.4 B&W type EL single-row ball-and-race pulverizer. (Reproduced with permission from *Steam, Its Generation and Use*, The Babcock & Wilcox Company, Lynchburg, Va., 1978.)

The degree of pulverization required varies with the type of coal and furnace design. Typically, from 65 to 80% of the pulverized coal would be required to pass through a 200-mesh screen (200 openings per linear inch, ~ 75 μm). The pulverizer capacity factor of any given pulverizer (ratio of actual capacity to capacity under some standard condition) increases with an increase in the coal's grindability index and decreases with the fineness required.

Petroleum

Transportation, handling, and storage. Oil is transported by tanker, barge, rail, truck, or pipeline. The relative costs vary sharply. Shipping over water is the least expensive, with truck shipment being the most costly.

Oil, being a liquid, is easily handled. After the shipment of oil arrives at the plant site, it is pumped into storage tanks which protect the oil from contamination and evaporation. Each surface storage tank is surrounded by a large cofferdam. In case of tank failure or fire, this cofferdam must contain the entire contents of a full storage tank. The National Fire Protection Association has prepared a standard set of rules for the safe storage and handling of petroleum. These rules are the basis of many local and state regulations and should be consulted as a guide to safe design procedures.

Pumping equipment for heavy oils must include heating facilities. In addition, the pipes and tanks must be cleaned at regular intervals due to the sludge formation that is characteristic of these heavy oils.

Preparation for firing varies with the fuel characteristics. The heavy oils used for utility applications must be heated to reduce the viscosity before the oil will atomize for burning. The heavy oil is pumped with either a rotary gear or reciprocating pump. The oil is usually heated by steam in a conventional shell-and-tube heat exchanger.

Estimation of Fuel Requirements

We may determine the fuel requirements from a knowledge of the total steam requirements and the heat released in the furnace per unit mass of fuel. We determine the latter quantity from the furnace energy balance calculations described in Chapter 4.

We may state our problem as: Determine the mass of fuel required per unit time given (1) the higher heating value of the fuel, (2) ultimate analyses of the fuel, (3) excess air used and entering temperature of air, (4) temperatures of exit gases and ashes, (5) steam exit conditions and production rate, and (6) feedwater temperature.

This problem may be solved by making a heat balance around the entire furnace. We may then write

enthalpy of entering streams + heat of reaction = enthalpy of exit streams

$$\dot{m}_a h_a + \dot{m}_F Q_0'' + \dot{m}_1 h_1 = \dot{m}_1 h_{ss} + \dot{m}_F \left[\sum_i \frac{m_i'}{MW_i} (\bar{c}_p)_{m,i} (T_{exit} - T_0) \right]$$

$$+ m_w' h_{fg} \dot{m}_F + m_A' (\bar{c}_p)_A \dot{m}_F (T_{ash} - T_0) \quad (5.1)$$

where $\dot{m}_a$ = mass of air required per hour
h_a = specific enthalpy of entering air
$\dot{m}_F$ = mass of fuel required per hour

Q_0'' = higher heating value (standard heat of reaction with product water as liquid per unit mass of fuel)
$\dot{m}_1$ = mass of feedwater required per hour
h_1 = specific enthalpy of entering feedwater
h_{ss} = specific enthalpy of superheated steam
m_i' = mass of product i in flue gas per unit mass of fuel
m_A' = mass of ash per unit mass of fuel
$(\bar{c}_p)_{m,i}, MW_i$ = mean molal specific heat of product i and molecular weight of product in flue gas, respectively
$(\bar{c}_p)_A$ = mean specific heat of ash, kJ/kg · °C or kcal/kg · °C
m_w' = mass of water vapor discharged in flue gases per unit mass of fuel
h_{fg} = heat of vaporization of water at 1 atm, kcal/kg
T_0 = base temperature (temperature at which Q_0' evaluated)

Since the preheating of the air is done within the furnace, the air enters the furnace at room temperature and its enthalpy is negligible. We may therefore eliminate the $\dot{m}_a h_a$ term and simplify to obtain

$$\dot{m}_1(h_{ss} - h_1) = \dot{m}_F \left\{ Q_0' - \left[\sum_i \frac{m_i'}{MW_i} (\bar{c}_p)_{m,i} (T_{\text{exit}} - T_0) \right] \right.$$
$$\left. - m_w' h_{fg} - m_A'(\bar{c}_p)_A (T_{\text{ash}} - T_0) \right\} \quad (5.2a)$$

or

$$\dot{m}_F = \frac{\dot{m}_1(h_{ss} - h_1)}{Q_0'' - \sum_i \frac{m_i'}{MW_i} (\bar{c}_p)_{m,i} (T_{\text{exit}} - T_0) - m_w' h_{fg} - m_A'(\bar{c}_p)_A (T_{\text{ash}} - T_0)}$$

(5.2b)

If steam reheat takes place in the boiler, the numerator of the right-hand side of Eq. (5.2a) must be modified to account for the heat released in the steam reheater.

5.3 STEAM GENERATOR CONFIGURATION

The Early Fire Tube Boilers

A fire tube boiler is a water-filled vessel containing a number of tubes which serve as channels for the flow of the combustion product gases. Heat is transferred from the hot gas through the tube walls to the pressurized water in the vessel.

The very early designs were based on a single tube. As power requirements increased, multiple tubes were utilized to increase the heat-transfer area (see Fig. 5.5). The fire tube boiler was widely used for power production up to the latter part

Sec. 5.3   Steam Generator Configuration

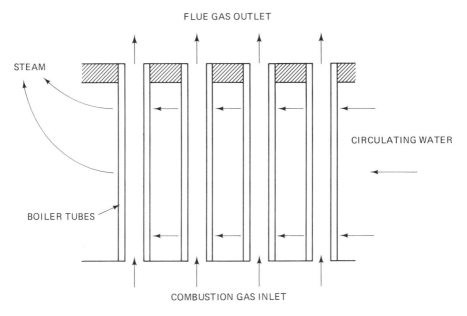

Figure 5.5 Early-design fire tube boiler.

of the nineteenth century. It is still encountered in some small units producing steam for industrial processes or heating of commercial buildings.

The fire tube boiler system was limited in size, steam pressure, and degree of operating safety. In the event of a major tube failure, high-pressure water can be forced directly into the furnace area, rapidly creating large quantities of steam which have the potential for major boiler explosion.

Development of the Water Tube Boiler

The safety concerns and limitations of the fire tube boiler led to the development of the water tube boiler. In this design, the water flows through the inside of the tubes and the combustion gases flow on the outside. Heat is again transferred from the hot gas through the tube wall to the water, which is now on the inside of the tube.

The first water tube boiler was built and patented in England by Sir William Blakey in 1766. John Stevens, American inventor and founder of Stevens Institute of Technology, patented and built a water tube steam boiler in 1803. Stevens' use of his device to power a steamboat on the Hudson River near New York City was the first practical application of the water tube boiler.

The next major innovation was the development around 1870 of the inclined tube boiler (see Fig. 5.6). The inclined water tubes connected water spaces at the front and rear, with a steam space above. The design provided better water circulation and additional heat-transfer area. Twenty years later, Stirling developed a design connecting the steam generating tubes directly to a steam separating drum.

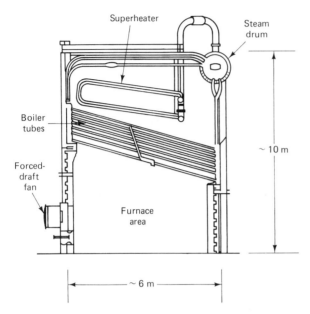

Figure 5.6 Inclined or straight tube-type boiler.

Because Stirling's boilers had bent steam tubes, the design became known as the bent-tube boiler.

Subsequent developments in the early twentieth century consisted of improvements in boiler tube and vessel materials which allowed for higher temperature and pressure operation. Steam conditions were improved by 1925 to 1200 psi and 700°F.

Water-Cooled Integral Furnace/Boiler Design

Major changes in boiler concepts occurred in the late 1920s. As power station electrical outputs were increased, it became necessary to build larger boiler units. The stoker furnace and boiler reached their design limits and new coal-burning methods were needed. The solution to this design limitation was provided by the pulverized coal-fired burner and the water-cooled furnace/boiler design. The first designs simply incorporated water cooling of the furnace into existing boiler designs, with the furnace cooling system separate from the boiler circulation. This arrangement was shortly abandoned for an integral arrangement in which the furnace water-cooled surface and the boiler surface were arranged as a single unit.

Water-cooled furnaces are now used in virtually all utility boilers. The water tubes are located on the furnace walls and on the top of the furnace. The water in these tubes serves as a coolant for the furnace walls and as the boiler section of the power plant. The water-cooled furnace design reduces the transfer of heat to the structural members, thus allowing for larger furnace volumes with increased steam temperature and pressure conditions. In addition, the water-cooled structure leads to improved mechanical design of the furnace as well as reduced external heat losses.

Sec. 5.3 Steam Generator Configuration

The heat-absorbing surfaces in the furnace receive the heat generated by the products of combustion. The primary mechanisms of heat transfer in the furnace are radiation from the fuel bed or fuel particles and nonluminous radiation from the combustion products. Convective heat transfer from the furnace gases plays a significant but much less important role. Such boilers are therefore often called *radiant boilers*. Any deposits of ash or slag reduce the effective absorption of heat by the furnace wall surfaces.

All early radiant boiler designs were combined with the "natural circulation" concept, as are many modern radiant boiler designs. In the natural-circulation boiler, the difference in specific gravity between water and steam provides the driving force for circulation. The principle is illustrated in Fig. 5.7. Saturated water flows from the steam drum via the downcomer to the lower drum. The burner heat input to the user area pipe reduces the fluid density in this region, resulting in an upward flow to the steam drum.

Figure 5.8 shows a schematic of an early pulverized coal unit which combined the radiant boiler and natural circulation concepts. Water flows downward from the

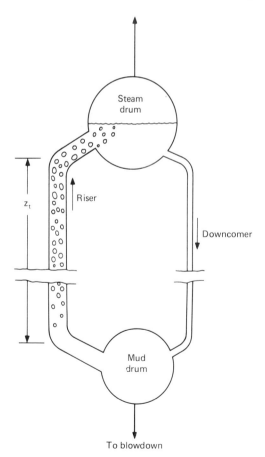

Figure 5.7 Principle of the natural circulation boiler (From *Principles of Energy Conversion*, by A. Culp; copyright 1979—used with the permission of McGraw-Hill Book Company, New York.)

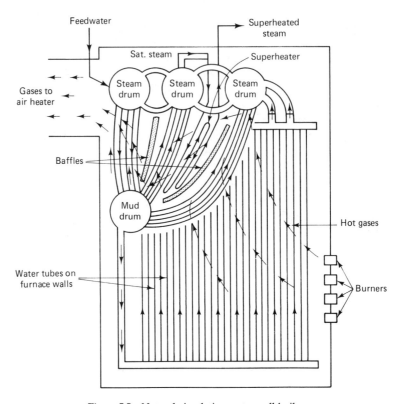

Figure 5.8 Natural-circulation water wall boiler.

mud drum to headers feeding the tubes lining the walls of the radiant furnace. The low-density steam–water mixture in these tubes rises to the steam drum at the upper right. The steam is separated and flows to the central drum, where it is removed (the central drum, being relatively quiescent, allows droplet carryover to settle out). Feedwater enters the drum at the left and mixes with the saturated liquid in the left and central drum. The cooler liquid flows down to the mud drum.

The hot gases leaving the radiant zone still contain a substantial amount of energy which must be utilized. Some of this energy is transferred by convection from the hot gases to the boiler tubes running from the mud drum to the two steam drums at the right. Additional energy is transferred by convection to the steam flowing inside the superheater tube banks. Further cooling takes place by convective heat transfer to the feedwater in the "economizer" tube bundle running from the mud drum to the steam drum at the upper left.

Modern Steam Generators

The quantity of steam that can be separated by surface separation per unit surface area is quite limited. Surface separation in a simple steam drum is inadequate for the large steam flows produced by modern boilers. Here centrifugal separators

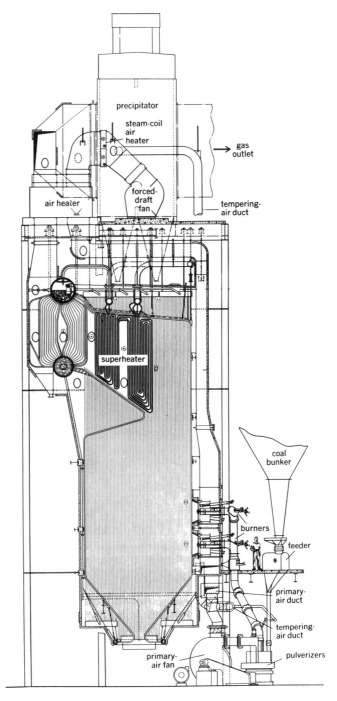

Figure 5.9 Modern natural-circulation steam generator. (From *Encyclopedia of Science and Technology*; copyright 1977—used with the permission of McGraw-Hill Book Company, New York.)

contained within the steam drum (see Section 5.5) are required. Although the centrifugal separators add significantly to the head loss, the long length of the riser and downcomer tubes lead to a sufficient head differential for adequate circulation.

A modern natural-circulation steam generator is shown in Fig. 5.9. As in the earlier design shown in Fig. 5.8, the bulk of the steam is generated in the radiant boiler tubes along the furnace walls. The steam–water mixture flows upward to the main steam drum (upper drum), where a centrifugal separator separates the steam and water. After passing through a drier, the steam proceeds to the two convective superheaters. In the superheater, the hot gases leaving the furnace superheat the steam to the desired temperature.

The liquid in the steam drum is mixed with feedwater and then flows downward through the tubes at the far left to the lower drum. This drum feeds liquid to the radiant tubes. Some of the liquid from the lower drum flows upward in the right hand portion of the tube banks connecting the upper and lower drums and is partially vaporized there. The left-hand tube bank acts as an economizer (feedwater heater). The exiting gases then proceed to an air heater, where they preheat the incoming combustion air.

A number of modern boilers used forced circulation rather than natural circulation. This scheme is illustrated in Fig. 5.10. Operation of such steam generators is basically similar to that of a natural-convection unit. Such units are particularly useful at very high pressures (e.g., above 160 bar), where the reduced density difference between liquid and vapor limits the natural circulation heat available. Steam generators of this type are often called *pump-assisted circulation* units since

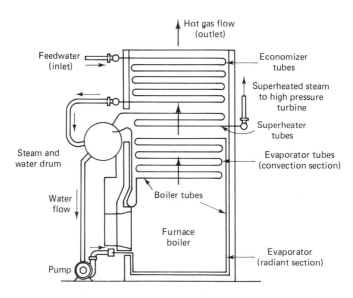

Figure 5.10 Modern forced-circulation steam generator.

the natural-circulation head is still significant and the pump is called upon to supply only a portion of the frictional head loss.

Assisted circulation boilers can supply between 130,000 and 3,000,000 kg steam per hour (300,000 to 7,000,000 lb/h), depending on the design. Steam pressures are below the critical pressure and generally range from 115 to 165 bar (1700 to 2400 psi).

Once-Through Boilers (Universal Pressure or Benson Boilers)

Both the natural circulation and assisted circulation steam generators are designated as recirculating units since only a fraction of the water flowing in the steam generating tubes is evaporated. The remainder of the water is recirculated. In the once-through design, all of the water entering the boiler is evaporated.

In this system, water is pressurized to about 340 bar (5000 psi) in the main feed pump. The compressed water is directed to the economizer section of the boiler (Fig. 5.11) and passes upward to the economizer outlet header. The water is then piped down to the lower furnace area, where it turns and flows upward, cooling the furnace walls. The fluid is then used to cool the front and rear furnace roof. The fluid is next forced to flow through the convection enclosure space to a lower header system. From this header, tubes transport the fluid to the superheater inlet header.

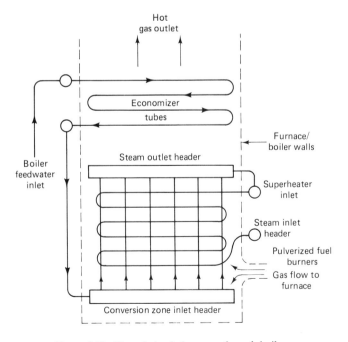

Figure 5.11 Forced-circulation once-through boiler.

The boiler capacity in the once-through boiler ranges from 130,000 to 4,500,000 kg/h (300,000 to 10,000,000 lb/h). The operating pressure is usually subcritical, but these boilers can also operate in the critical pressure region up to 240 bar (3500 psi). This system requires extremely pure water with a maximum impurity level of a few parts per billion.

Boiler Rating and Performance

Steam generating units are designed for specific operating conditions and sold with a guarantee of performance. The boiler rating is usually defined in terms of steam output, kilograms per hour (pounds per hour), at a specified pressure and temperature. Reheated steam is included in this requirement in terms of the quantity of reheat steam at specific inlet and outlet steam pressures and temperatures. Performance characteristics also include the identification of major losses from the boiler system and finally a calculation of the boiler efficiency.

The index that best describes the quality of a boiler is the boiler efficiency, η_B.

$$\eta_B \equiv \frac{\text{total energy (thermal) transferred to the boiler system coolant}}{\text{energy released in the combustion of the fuel}} \times 100 \quad (5.3)$$

The total heat added to the coolant includes the heat added in the boiler tubes, economizer, evaporation, superheater, and reheat sections of the boiler system.

If the mass flow rates of fuel and boiler and reheat coolant are known or measured, the boiler efficiency can be calculated from the equation

$$\eta_B = \frac{\dot{m}_s(h_2 - h_1) + \dot{m}_{rh}(h_4 - h_3)}{\dot{m}_F(\text{HHV})} \times 100 \quad (5.4)$$

where h_1 = specific enthalpy of the incoming boiler feedwater, kJ/kg
 h_2 = specific enthalpy of the steam leaving the superheater, kJ/kg
 h_3 = specific enthalpy of the steam entering the reheater, kJ/kg
 h_4 = specific enthalpy of the steam leaving the reheater, kJ/kg
 HHV = high heating value of the coal (or fuel), kJ/kg
 $\dot{m}_F$ = coal (or fuel) consumed, kg/h
 $\dot{m}_s$ = steam flow rate from superheater, kg/h
 $\dot{m}_{rh}$ = reheat steam flow rate, kg/h

If these data (mass flow rates) are not available, then an alternative method of calculating boiler efficiency can be employed. We may obtain η_B from

$$\eta_B = \frac{(\text{HHV}) - \sum L_i^\circ}{\text{HHV}} \times 100 \quad (5.5)$$

The HHV again represents the higher heating value of the coal (or fuel) and the $\sum L_i^\circ$ term is the summation of the major losses from the boiler system based on a unit mass of fuel input.

Sec. 5.3   Steam Generator Configuration

The major losses from the boiler system are due to:

Incomplete combustion, L_1^o 2.5-3 %
Unburned carbon, L_2^o 1-2%
Sensible heat loss to dry gas, L_3^o 10%
Moisture, L_4^o 5-6%
Moisture in the combustion air, L_5^o .5% to .8%
Thermal radiation, L_6^o 2%

The losses above are evaluated by utilizing data from the ultimate analysis of the fuel, the Orsat analysis of the flue gas, the coal refuse analysis, and boiler pressure and temperature data. Since these parameters can be measured with more accuracy than the mass flow rates, Eq. (5.5) will give a more accurate estimate of the boiler efficiency. Each of these loss terms can be estimated through the use of the calculational procedures that follow.

The first loss, due to incomplete combustion, is the energy loss due to the formation of carbon monoxide instead of carbon dioxide. This loss is obtained by using the Orsat analysis data and the refuse data:

$$L_1^o = 24{,}000 m_b' \frac{\text{CO}'}{\text{CO}' + \text{CO}_2'} \quad \text{kJ/kg} \tag{5.6}$$

where m_b' = mass of carbon burned per unit mass fuel = $(C'/100) - m_C'$
C = percent carbon from the as-burned ultimate fuel analysis
CO′ = percent of CO in the flue gas
CO_2' = percent of CO_2 in the flue gas

This loss is normally about 2.5 to 3.0% of the energy supplied.

The second loss, the unburned carbon loss L_2^o, also utilizes the refuse data and is simply

$$L_2^o = m_C' (\text{HHV})_{\text{carbon}} \quad \text{kJ/kg} \tag{5.7}$$

where m_C' in the mass of unburned carbon in refuse per mass of fuel. Values of L_2^o range between 1 and 2% of the available energy.

The dry gas loss, L_3^o, is due to the combustion air supplied to the boiler. It is a sensible heat loss calculated from the actual dry air-to-fuel ratio and the gas temperatures:

$$L_3^o = m_{\text{gas}}' \bar{c}_p \left(T_{\text{gas out}} - T_{\text{gas in}} \right) \quad \text{kJ/kg} \tag{5.8}$$

where m_{gas}' = kilograms of dry flue gas/kilogram of fuel
$\bar{c}_p$ = average specific heat of the flue gas, kJ/kg·°C
$T_{\text{gas in}}$ = gas temperature at the inlet, °C
$T_{\text{gas out}}$ = gas temperature at the outlet, °C

The weight of dry flue gas must be corrected for refuse, moisture, and hydrogen mass fractions obtained from the refuse and ultimate analysis. The specific heat can be estimated by using air data. The dry gas loss can be as large as 10% of the energy supplied.

The two moisture losses are due to the moisture in the fuel L_4^o and the moisture in the combustion air, L_5^o. The fuel moisture loss per kilogram of fuel burned, L_4^o, is

$$L_4^o = \left(m_m' + 9m_{H_2}'\right)(h_s - h_w) \quad \text{kJ/kg} \tag{5.9}$$

where m_m' = mass fraction of moisture in the coal (as-burned analysis)
m_{H_2}' = mass fraction of hydrogen in the coal (as-burned analysis)
h_s = specific enthalpy of steam at product gas outlet temperature, kJ/kg
h_w = specific enthalpy of water at air inlet temperature, kJ/kg

The partial pressure of the water vapor may have to be estimated in order to evaluate h_s. The approximate partial pressure of the water vapor at the usual product gas outlet conditions is 1 to 2 psia. Moisture in the coal can reduce boiler efficiency by 5 or 6%.

The second moisture loss is an order of magnitude smaller than L_4^o. Values are typically 0.5 to 0.8% loss in boiler efficiency. Loss L_5^o is due to the moisture that enters the boiler (as a vapor) with the combustion air and leaves the superheater region with the combustion products (also as a vapor). The loss is a sensible heat loss given by

$$L_5^o = 1.926\left(m_a'\right)_{\text{act}}(X)\left(T_{\text{gas out}} - T_{\text{gas in}}\right) \quad \text{kJ/kg} \tag{5.10}$$

Terms not previously defined are:

$X \equiv$ specific humidity of combustion air, kg water/kg dry air

$(m_a')_{\text{act}} \equiv$ actual air to fuel ratio (mass of air supplied per unit mass of fuel), kg air/kg fuel

The final loss term that is usually included in boiler efficiency calculations is the heat loss due to thermal radiation from the boiler, L_6^o. It is a small loss amounting to only 0.2% loss in efficiency for typical boiler operating conditions. The radiation loss cannot be calculated simply but can be estimated from Fig. 5.12.

$$L_6^o = (\text{HHV})_{\text{coal}} \times R_L \quad \text{kJ/kg} \tag{5.11}$$

where R_L is the radiation loss factor from Fig. 5.12.

After estimation of the boiler losses, the boiler efficiency is calculated from Eq. (5.5). Although the method described herein provides a reasonable estimate of boiler efficiency, it may not be accurate enough for contract specification evaluation. The American Society of Mechanical Engineers (ASME) has prepared a Power Test Code which contains the detailed procedures and methods that should be used for evaluating boiler performance.

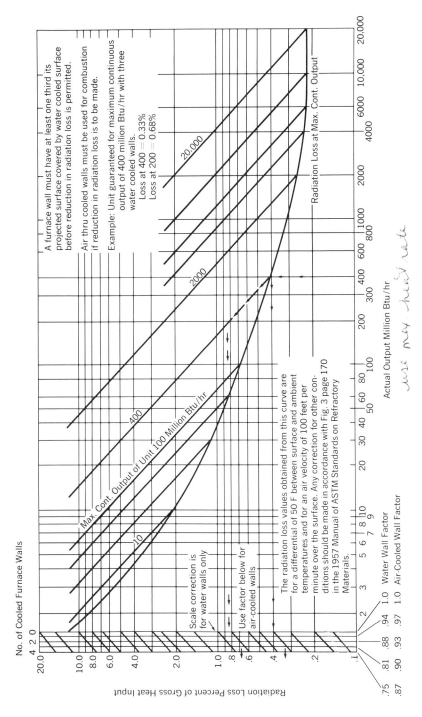

Figure 5.12 Radiation loss as a percent of gross heat input. (Reproduced with permission from *Steam, Its Generation and Use*, The Babcock & Wilcox Company, Lynchburg, Va., 1978.)

5.4 STEAM GENERATOR ANALYSIS

Computation of Thermal Loads

The heat loads on the various segments of the furnace may be determined by simple heat balances on the water or air streams. To illustrate this, let us consider a numerical example.

Example 5.1

We are to produce 100,000 kg/h of steam at 160 bar and 200°C superheat from water entering the furnace at 50°C below saturation. Air is fed at a rate of 1.1×10^5 kg/h to be preheated 200°C. Estimate the heat that must be transferred to the boiler superheater, economizer and air heater. Assume that the steam cycle contains no steam reheater.

Solution

Economizer: The water enters the furnace, and hence the economizer, at 50°C below saturation. If we assume that we will heat the water close to saturation, say 15°C below saturation, in the economizer, then q_e, the rate of heat transfer in the economizer, is

$$q_e = \dot{m}_1 (h_{eo} - h_{ei}) \tag{5.12}$$

where $\dot{m}_1$ = feedwater flow, kg/h
h_{eo} = specific enthalpy of water at economizer outlet
h_{ei} = specific enthalpy of water at economizer inlet

At 160 bar, $T_{sat} = 347.3°C$; therefore,

T_{ei} = feedwater inlet temperature = 297.3°C

T_{eo} = feedwater outlet temperature = 332.3°C

We obtain the feedwater enthalpies at T_{eo} and T_{ei} by assuming that the enthalpy of liquid water is unaffected by pressure, and hence we then take them as equal to those of saturated liquid at T_{eo} and T_{ei}, respectively. From the steam tables, we have

$h_{eo} = 1544.5$ kJ/kg $h_{ei} = 1329.7$ kJ/kg

$q_e = 100,000$ kg/h $(1544.5 - 1329.7)$ kJ/kg $= 21.5 \times 10^6$ kJ/h

Boiler: In the boiler, the water is heated from the economizer outlet temperature to saturation and then the required steam flow is produced. Our heat balance then gives us q_b, the rate of heat transfer in the boiler:

$$q_b = \dot{m}_1 (h_g - h_{eo})$$

where h_g is the specific enthalpy of saturated steam at 160 bar and the other symbols

Sec. 5.4 Steam Generator Analysis

have their previous meanings. From the steam tables, we have

$$h_g = 2582 \text{ kJ/kg}$$

$$q_b = 100{,}000 \text{ kg/h } (2582 - 1544.5) \text{ kJ/kg} = 103.7 \times 10^6 \text{ kJ/h}$$

Superheater: The heat load in the superheater, q_s, is obtained from the enthalpy change that occurs when the fluid is superheated by the required amount. Hence

$$q_s = \dot{m}_1 (h_{ss} - h_g)$$

where h_{ss} is the specific enthalpy of steam at 160 atm and 200°C superheat (532.3°C). We now use the superheated steam table to determine $h_{ss} = 3209$ kJ/kg.

$$q_s = 100{,}000 \text{ kg/h } (3209 - 2582) \text{ kJ/kg} = 62.7 \times 10^6 \text{ kJ/h}$$

Air heater: From the heat balance on the airstream, q_{ah}, the heat transferred to the air is

$$q_{ah} = \frac{\dot{m}_a (\bar{c}_p)_m}{\text{MW}_a} (T_a - T_{\text{in}})$$

where MW_a = molecular weight of air
 $\dot{m}_a$ = mass of air required per hour
 T_a = temperature of exit air
 T_{in} = temperature of inlet air

Since T_{in} is close to T_0, we can use $(\bar{c}_p)_m$ based on 0°C with little error. From Table 4.2 we obtain the equation for the mean molal specific heat of air.

From 0°C to T°C:

$$(\bar{c}_p)_m = 6.935 + 3.38 \times 10^{-4} T + 0.43 \times 10^{-7} T^2 \quad \text{kcal/kg mol}$$

For T = 200°C:

$$(\bar{c}_p)_m = 7.0 \text{ kcal/kg mol} \cdot °C = 29.2 \text{ kJ/kg mol} \times \frac{\text{kg mol}}{28.9 \text{ kg}} = 0.975 \frac{\text{kJ}}{\text{kg} \cdot °C}$$

$$q_{ah} = 1.1 \times 10^5 \text{ kg/h} \times 0.975 \text{ kJ/kg} \cdot °C \times (200°C - 20°C) = 19.2 \times 10^6 \text{ kJ/h}$$

Heat Transfer in the Radiant Boiler

In Chapter 4 we developed the equations describing, q_{TC}, the rate of heat transfer from the flame to the boiler tubes lining the refractory surface. We had, for the case when the absorber was less than half the flame temperature, that

$$q_{\text{TC}} = \sigma A_T \mathscr{F}_{\text{TC}} (T_f^4 - T_c^4) \tag{5.13}$$

where

$$\mathscr{F}_{\text{TC}} = \left[\frac{A_T}{A_C \varepsilon_t} + \frac{1}{(\varepsilon_G)_f} - 1 \right]^{-1}$$

and

A_T = total enclosure area

A_C = effective area of cold surface

$(\varepsilon_G)_f$ = emissivity of hot gases at flame temperature T_f

ε_t = emissivity of boiler tubes at temperature T_c

σ = Stefan–Boltzmann constant $(5.67 \times 10^{-8} \text{W/m}^2 \cdot {}^\circ\text{K}^4)$

Means for estimating $(\varepsilon_G)_f$ were discussed at length in Section 4.7. The total area of the refractory enclosure, A_T, may be readily estimated for any assumed geometry. The effective area, A_C, is that portion of the boiler tube area which effectively receives the radiation from the flame. The ratio of A_C/A_T may be obtained from Fig. 5.13.

The total radiative heat transfer may then be determined if T_f and T_c are known. The value of T_c is generally set at approximately the saturation temperature of the boiling fluid inside the tube. Since the boiling heat-transfer coefficient is very

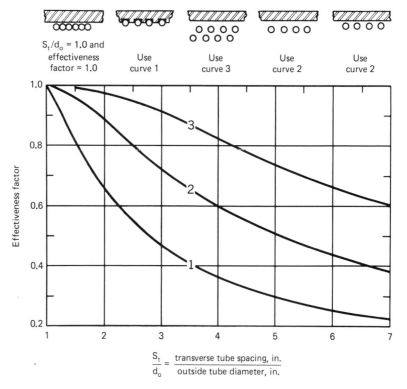

Figure 5.13 Furnace-wall area-effectiveness factor. (Reproduced with permission from *Steam, Its Generation and Use*, The Babcock & Wilcox Company, Lynchburg, Va., 1978.)

high and the tube wall temperature drop is small, the difference in temperature between the inner and outer tube walls can be ignored.[†] The value of T_f is generally unknown and must be determined iteratively as described in Chapter 4.

If the designer is attempting to meet a given thermal load in the boiler, he or she will have to assume a value of A_T. The designer then must select a geometric arrangement for the boiler tubes lining the wall to get A_C/A_T. With these quantities and information on the coal heating value, composition, firing rate, air rate, and air temperature, the designer can iteratively determine q and T_f. If the value of q is not satisfactory, a revised value of A_T must be assumed.

For approximate calculations, it is usually assumed that convection heat transfer to the boiler tubes in the radiant section may be ignored.

Heat-transfer tube banks directly above the flames. In addition to the boiler tubes embedded in the refractory walls, there are usually two or three rows of boiler tubes suspended in the space immediately above the flame region. These tubes are generally called *slag-screen* or *boiler screen* tubes. These rows of tubes receive radiation directly from the flame and screen subsequent rows from the high radiation flux. We may calculate the radiative heat transfer from the flame to these first rows in exactly the same manner as used for the radiative transfer to the tubes lining the refractory walls. The value of A_C is again obtained using Fig. 5.13.

Although the majority of the radiation from the flame will be absorbed by the first rows, some direct radiation will be absorbed by the succeeding rows, which may contain superheater tubes. We may consider that the value of A_C/A_T for one row of tubes of a given spacing provided by Fig. 5.13 represents the fraction, x, of the net direct radiation heat transfer to area A_T. If a fraction x of the net saturation to A_T is absorbed by the first row, the second row will absorb the same fraction x of the remaining $(1-x)$. Similarly, the third row will absorb a fraction x of that remaining after the first two. For example, if $A_C/A_T = 0.7$ for the first row, $A_C/A_T = 0.21$ for the second row, and 0.063 for the third row curve 3 of Fig. 5.13 is obtained from curve 2 on this basis. When a deep tube bank is present, A_C/A_T for the entire bank $\simeq 1.0$.

In addition to the direct radiation from the flame, the tubes in the screen area also receive radiation from the hot gases flowing through the tube bank itself. We know from Chapter 4 that q, the net interchange between a hot gas at temperature T_G and with emissivity ε_G, and black area A at T_c which is seen completely is given by

$$q = A\sigma\left[(\varepsilon_G)_G T_G^4 - (\varepsilon_G)_c T_c^4\right] \qquad (5.14)$$

[†] This approximation remains reasonable as long as there is nucleate boiling on the inside of the tube. At high steam fractions, nucleate boiling will occur as long as the wall is covered with a thin liquid film. At very high qualities we have dry-out of this film and a pronounced deterioration in heat transfer. Satisfactory boiler operation requires that we stay below the dry-out quality (steam fraction) unless a once-through boiler is being designed. Heat fluxes must then be kept low above the dry-out location or excessive tube wall temperatures result.

When $T_c \leq \frac{1}{2}T_G$, this may be simplified to

$$q = \sigma A (\varepsilon_G)_G (T_G^4 - T_c^4) \tag{5.15}$$

If the cold surface is gray rather than black, multiplication of the foregoing equation by ε_1, the emissivity of the cold surface, allows for the reduction of primary radiation beams but secondary reflections are ignored. Hence multiplication by ε_1 overcorrects. For most calculations it is adequate to use $(1 + \varepsilon_1)/2$. Hence q_{ir}, the radiation from the hot gases in the bundle, is approximated by

$$q_{ir} = \sigma A (\varepsilon_G)_G \frac{1 + \varepsilon_1}{2} (T_G^4 - T_c^4) \tag{5.16}$$

The value of ε_1 is known for any given tube material and ε_G may be obtained for a given gas composition if the mean beam length is known. We obtain the mean beam length from Fig. 5.14.

The hot gases flowing across the tube bank also transfer heat to the tubes by convection. The total heat transfer, q_t, to a given row of tubes in tube bank is then given by

$$q_t = q_r + q_{ir} + h(T_G - T_c)A \tag{5.17}$$

where q_r = direct radiation heat transfer from flame region
q_{ir} = radiation heat transfer from hot gases flowing through tube bank (intertube radiation)
h = convective heat-transfer coefficient, energy/(area)(time)(unit temp. diff.)

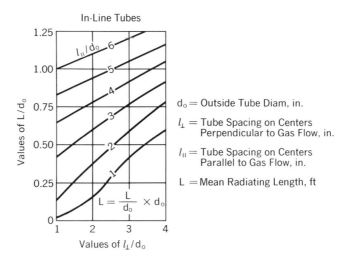

Figure 5.14 Mean radiation length, L, for various tube sizes and arrangements. (Reproduced with permission from *Steam, Its Generation and Use*, The Babcock & Wilcox Company, Lynchburg, Va., 1978.)

Sec. 5.4 Steam Generator Analysis

The direct radiation from the flame is significant only in the first two or three rows. In this short distance the gas temperature does not change markedly and Eqs. (5.16) and (5.17) may be used, with A representing the tube area in the screen rows.

In subsequent rows, direct radiation from the flame may be ignored. The total heat transfer rate, q_t in any row, is given by

$$q_t = q_{ir} + h A_1 (T_G - T_c) \qquad (5.18)$$

with A_1 representing the tube area in that row. To obtain the heat transfer across the entire bundle of tubes, we would write

$$q_t = q_{ir} + h A_h (\Delta T)_{LM} \qquad (5.19)$$

where A_h = bundle heat-transfer area
$(\Delta T)_{LM}$ = log mean temperature difference between the gas and tube surface across the bundle
q_{ir} = rate of intertube radiation heat transfer from hot gases to entire bundle

This is often rewritten as

$$q_t = (U_{ir} + h) A_h (\Delta T)_{LM} \qquad (5.20)$$

where U_{ir} is the equivalent intertube radiation conductance, energy/(area)(time)(unit temp. diff.). The value of U_{ir} will depend on the temperature of the receiving surface, $(\Delta T)_{LM}$, and the emissivities of the gas and tube. Since tube emissivities are virtually constant under usual industrial conditions, we may write

$$U_{ir} = (\varepsilon_G)_G U_r \qquad (5.21)$$

where U_r, the basic radiation conductance, is only a function of $(\Delta T)_{LM}$ and T_c. Values of U_r are provided in Fig. 5.15. (Note that conversion from British units is required.) Values of $(\varepsilon_G)_G$ are obtained by using the mean beam length provided by Fig. 5.14 together with the graphs of Chapter 4.

Convective heat transfer. We indicated previously that the high-boiling heat-transfer coefficient on the inside of the tubes in the boiler screen allowed this resistance to be neglected. In the superheater region, the resistance to heat transfer of the steam flowing inside the tubes is of a similar order of magnitude to the resistance on the outside of the tube and may no longer be ignored. However, the resistance to heat flow of the gas and steam films is considerably larger than the tube wall resistance. The metal wall resistance may therefore be neglected and the overall coefficient of heat transfer, U, may be approximated by

$$\frac{1}{U} = \frac{1}{h_{Tc}} + \frac{1}{h_{Ti}}$$

or

$$U = \frac{h_{Tc} h_{Ti}}{h_{Tc} + h_{Ti}} \qquad (5.22)$$

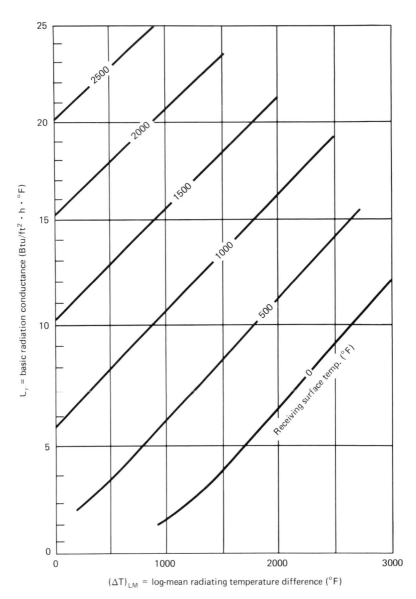

Figure 5.15 Radiation conductance as a function of ΔT_{LM} and T_c. (Reproduced with permission from *Steam, Its Generation and Use*, The Babcock & Wilcox Company, Lynchburg, Va., 1978.)

Sec. 5.4 Steam Generator Analysis

where h_{Tc} = film conductance (heat-transfer coefficient) on the outside of the superheater tubes (hot gas side)

h_{Ti} = film conductance (heat-transfer coefficient) on the inside of the superheater tubes (steam side)

For superheater tubes in the region where intertube radiation heat transfer is still significant, we generally will not know the surface temperature of the metal but will know the temperature of the steam inside the tubes. We may conservatively evaluate the area required to obtain the desired degree of superheat by rewriting Eq. (5.20) as

$$q_t = (U_{ir} + U) A_{\hbar} (\Delta T)_{LM} \tag{5.23}$$

and base $(\Delta T)_{LM}$ on the total log mean temperature difference between the gas and steam streams (see Fig. 5.16). We obtain the required temperature change in the gas stream from a heat balance across the bundle, that is,

$$\dot{m}_s (h_1 - h_2) = \sum_j \frac{\dot{m}_j}{MW_j} \left[(\bar{c}_p)_{m,j} \right]_2 T_2 - \left[(\bar{c}_p)_{m,j} \right]_1 T_1 \tag{5.24}$$

where
$\dot{m}_s$ = mass flow rate of steam
h_2, h_1 = specific steam enthalpy at temperatures T_2 and T_1, respectively
$\dot{m}_j$ = mass flow rate of jth component in gas stream
MW_j = molecular weight of jth component
$[(\bar{c}_p)_{m,j}]_1, [(\bar{c}_p)_{m,j}]_2$ = mean molal specific heats of jth component in gas evaluated at T_1 and T_2, respectively

Various empirical or semiempirical correlations have been devised to evaluate the film coefficient for the hot gases flowing across the superheater tubes and the steam inside the tubes. For flow of fluids inside tubes and conduits, the Dittus–Boelter correlation in widely used:

$$\frac{h D_e}{k} = 0.023 \left(\frac{D_e G'}{\mu} \right)^{0.8} \left(\frac{c_p \mu}{k} \right)^n \tag{5.25}$$

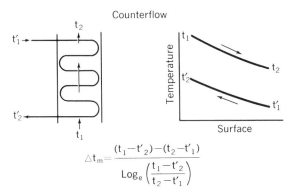

Figure 5.16 Log mean temperature difference and counterflow. (Reproduced with permission from *Steam, Its Generation and Use*, The Babcock & Wilcox Company, Lynchburg, Va., 1978.)

where h = heat-transfer coefficient, W/m$^2 \cdot$°C
D_e = equivalent diameter, m
G' = mass velocity, kg/h$\cdot$m^2
k = thermal conductivity of fluid, W/m$\cdot$°C
c_p = specific heat of fluid, kJ/kg$\cdot$°C
μ = fluid viscosity, kg/m$\cdot$s
n = 0.4 for heating, 0.3 for cooling

The physical properties of the fluid are evaluated at the fluid bulk temperature. This correlation would be used to determine the heat-transfer coefficient to the steam inside the superheater tubes. It may also be applied to flow parallel to banks of tubes.

Similar relationships have been developed to evaluate the film coefficient for banks of tubes in cross flow. For staggered banks of tubes, Weisman obtained

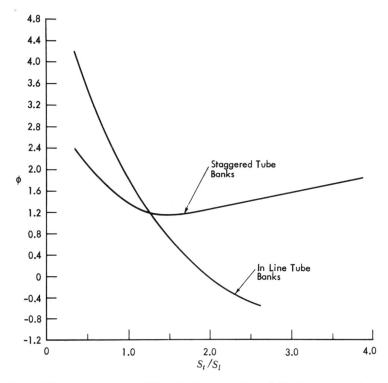

Figure 5.17 Arrangement coefficients for heat transfer to fluids flowing normal to tube banks. (From L. S. Tong and J. Weisman, *Thermal Analysis of Pressurized Water Reactors*, 1970; reprinted with the permission of the American Nuclear Society, LaGrange Park, Ill.)

$$\frac{hD_o}{k}(a_f)^{\psi b} = 0.38\left(\frac{D_oG'}{\mu}\right)_f^{0.61}\left(\frac{c_p\mu}{k}\right)_f^{0.4} \quad \text{for } 300 < \frac{D_oG'}{\mu} \leq 40{,}000 \quad (5.26)$$

and

$$\frac{hD_o}{k} = 0.051\left(\frac{D_oG'}{\mu}\right)_f^{0.8}\left(\frac{c_p\mu}{k}\right)_f^{0.52} \quad \text{for } 80{,}000 < \frac{D_oG'}{\mu} < 800{,}000 \quad (5.27)$$

where a_f = fraction of the total cross-sectional area in an infinite array taken up by the fluid
G' = mass velocity (in this case, based on the average flow area of bundle)
$b = 0.175(D_oG'/\mu)^{-0.07}$
ψ = is a function of S_t/S_l, the ratio of the tube pitch transverse to the flow to the pitch parallel to the flow

The value of ψ is obtained from Fig. 5.17 (usually $1.2 < \psi < 1.6$). The subscript f indicates that the fluid properties for Eqs. (5.26) and (5.27) are evaluated at an average film temperature given by $(T_{\text{fluid}} + T_{\text{surface}})/2$.

While Eq. (5.26) or (5.27) may be used to evaluate the film coefficient for the boiler region tubes exposed to the gases leaving the radiant section of the furnace, expressions of the form

$$\frac{hD_o}{k} = C\left(\frac{D_oG'}{\mu}\right)_f^n\left(\frac{c_p\mu}{k}\right)^m \quad (5.28)$$

where D_o = tube outside diameter
C, m, n = coefficients determined from the tube bank geometry

are also commonly used. Note that in this case G' is based on the minimum flow area between the tubes. The subscript f retains its previous meaning.

5.5 STEAM SEPARATION AND PURIFICATION

All recirculating boilers operating below the critical pressure require a steam drum for separation of steam from the steam–water mixture coming from the boiler tubes. The separated water is returned, along with feedwater, to the boiler region of the furnace. For boiler operation above the critical pressure, there is no distinction between the water and steam phases. The mixture is simply considered a "fluid" and steam separation is not possible.

In addition to providing steam separation, the steam drum is designed to (1) contain the water–steam mixture during load changes that result in significant steam volume changes in the system, (2) serve as a location for addition of water treatment chemicals if desired, and (3) remove solid and particulate matter from the steam.

High-purity steam is required for use in modern high-pressure turbines since if the steam contains as little as 0.6 ppm of solid matter, deposits form on turbine blades.

The desired steam separation is accomplished in two steps. The first, or primary separation, is the removal of the major part of the water from the mixture. This prevents the vapor phase from being recirculated to the heating tubes.

The second step (or secondary separation) is the removal of the entrained droplets from the steam. This step is sometimes referred to as *steam scrubbing* or *steam drying*. Both steps are performed by the steam drum.

In the early boilers, steam separation was accomplished by simple separation by gravity at the steam–water interface. However, for such separation to work, the steam velocity at the separating surface must be below the "carryover" velocity. At steam velocities above this critical value, there is very substantial entrainment of liquid droplets. As pressure increases, the surface tension decreases sharply. This, together with the decreased difference between gas and liquid densities, leads to a rapid decrease in the size of the droplets formed. This smaller droplet size in turn causes a sharp decrease in the carryover velocity (see Fig. 5.18). As pressure is increased above atmosphere, the increased gas density increase first causes an increase in the mass that can be separated per unit area per unit time despite the decrease in carryover velocity. However, further increase in pressures causes the mass separation rate to decrease again. At the high pressures of modern boilers, the surface are that would be required to meet the demand of a large boiler would be excessive.

Improved performance can be obtained by appropriate baffling (see Fig. 5.29). However, in high-pressure boiler units, particularly large units with long boiler tubes, part of the circulating head can be used to provide a separating force many times greater than gravity. This is the principle behind the centrifugal separating devices included in modern drums such as the one shown in Fig. 5.19. Such drums usually contain both cyclone separators and scrubber elements.

The cyclone separators are spaced along the longitudinal axis of the drum and the steam–water mixture enters tangentially from the riser pipes. The centrifugal force throws the water outward against the cylinder walls. The less dense steam is directed toward the core of the cylinder and then flows upward. The water flows downward, discharging through pipes located in the cylinder bottom. This water, now free of bubbles, returns through the downcomers to continue the natural-circulation process. The steam, containing some droplets, flows upward in the drum through the scrubbing elements.

Droplet carryover is most likely when wide variations in load occur. Since all utility boilers will see such variation at some time, drying or scrubbing is required to maintain steam purity. Any droplets carried outside the steam drum will be evaporated in the superheater and then any solids dissolved in the droplets would appear in the steam. Droplet removal is accomplished by large corrugated metal plates installed in the top section of the steam drum. These scrubbers provide an additional large surface area upon which the droplets impinge as the steam flows

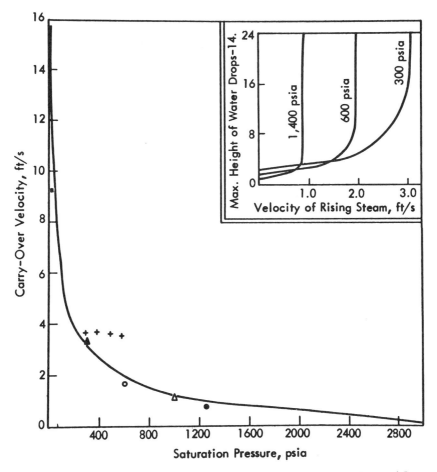

Figure 5.18 Surface disengagement carryover velocities. (From L. S. Tong and J. Weisman, *Thermal Analysis of Pressurized Water Reactors*, 1979; reprinted with the permission of the American Nuclear Society, LaGrange Park, Ill.)

between the corrugated plates. The water collected is drained to the pool below the scrubber. In modern high-capacity boilers, the steam leaving the steam drum has less than 1.0 ppm of total solids.

One of the most troublesome contaminants of steam is silica (SiO_2) since silica will distribute itself between the steam and water phases even in the absence of droplet carryover. The ratio of the equilibrium concentration (by weight) of silica in the steam to the silica concentration in the water is called the *distribution ratio*. At low pressures this ratio is very low, but at the saturation temperature corresponding to 2000 psi, it has risen to about 0.03. It then climbs very sharply, so that at the saturation temperature 3000 psi it is at 0.22. Since silica deposits can form in the

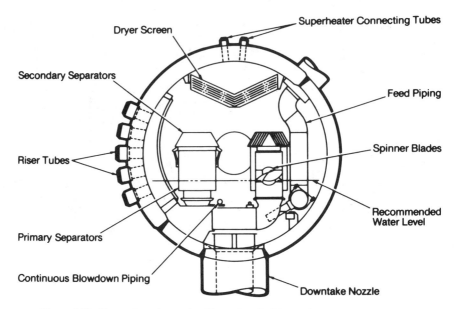

Figure 5.19 Steam drum internals. (Reproduced with permission from *Combustion—Fossil Power Systems*, Combustion Eng., Inc., Windsor, Conn., 1981.)

steam turbine when the silica concentration in steam exceeds 0.025 ppm, this would require that boiler water be kept at an unrealistically low solids level. To avoid this problem, steam washing is used. Relatively cold feedwater is distributed across the steam scrubbers. Since the silica vapor concentration is higher than that in equilibrium with the cold feedwater, the silica vapor is absorbed in the feedwater and the concentration in the steam is reduced to acceptable levels.

5.6 PRIMARY SYSTEM HEAT EXCHANGERS

Feedwater System and Economizer

In utility boilers, the gas temperature leaving the boiler superheat/reheat region can be as high as 1000°F. If these combustion gases were simply exhausted to the atmosphere at this temperature, the boiler and overall plant system efficiency would be low. In addition, the feedwater at the boiler inlet could be well below saturation temperature. It is for this reason that the boiler exhaust gases are used for the last stage of boiler feedwater heating. Since the device designed to preheat the feedwater

Sec. 5.6 Primary System Heat Exchangers

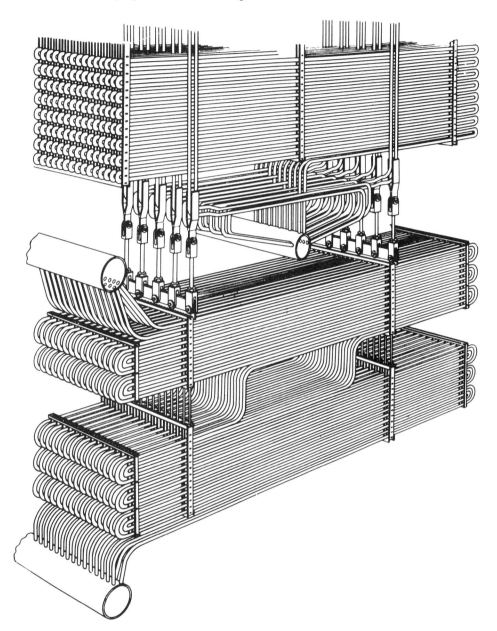

Figure 5.20 Economizer for a Babcock & Wilcox boiler. (Courtesy of The Babcock & Wilcox Company, Lynchburg, Va., 1983.)

improved the economy (or efficiency) of the system, it became known as the economizer. Steam generator efficiency rises approximately 1% for each 5.5°C (10°F) rise in feedwater temperature produced by the economizer.

The feedwater to the economizer, which is the last feedwater heater, is delivered by the boiler feed pump. An economizer is a forced-flow, once-through, convective heat-transfer device. It usually consists of an array of steel tubes with the pressurized feedwater on the inside and the combustion gases on the outside. Since gas heat-transfer coefficients are usually low, extended surface design heat exchangers are common (finned tube design).

Economizers are classified as *horizontal* or *vertical tube* types, depending on the geometrical arrangement; and as *longitudinal* or *cross flow*, depending on the direction of gas flow with respect to the tube geometry. Other design characteristics may include such features as a return bend or continuous tube, staggered or in-line tube arrangements, and steaming or nonsteaming. In a steaming economizer, steam would begin to form in the economizer section before the boiler inlet. The steam formed is limited to about 20% of the boiler feed at full output. A typical economizer is shown in Fig. 5.20.

The temperature of the economizer tube metals is generally close to the temperature of the boiler feedwater flowing through the tubes. If the feedwater temperature is too low, condensation can form on the external side (gas side) of the economizer tubes. This condensate is strongly acidic and corrosive because of the sulfur dioxide and sulfur trioxide in the flue gas. These two impurities also raise the dew point of the exhaust gases, thereby increasing the potential for condensation. Figure 5.21 shows the limiting metal temperatures to avoid external corrosion in economizers for coal and oil systems.

To avoid internal corrosion of the economizer tubes, the feedwater inlet temperature should be above 100°C (212°F). Internal corrosion and pitting may occur if the feedwater contains more than 0.007 ppm of dissolved oxygen. Therefore, it is important to provide deaeration of the feedwater for the removal of oxygen. Since internal corrosion is accelerated at low value of pH in the feedwater, it is desirable to maintain a pH of 8 or 9 for the water flowing through the economizer.

Closed Feedwater Heaters

It has been noted previously that the train of feedwater heaters is used to improve the thermodynamic efficiency of the power plant. They also serve to limit or prevent thermal transients that might otherwise occur in the system. The number and size of heaters vary with the type and size of the power plant. A large fossil-fueled plant may have from two to four closed low-pressure heaters, a deaerator heater (open heater), and two or three closed high-pressure heaters (see arrangement in Fig. 5.22).

Closed feedwater heaters are essentially tube-and-shell heat exchangers with the pressurized feedwater flowing through the tubes and steam condensing in the

Sec. 5.6   Primary System Heat Exchangers   193

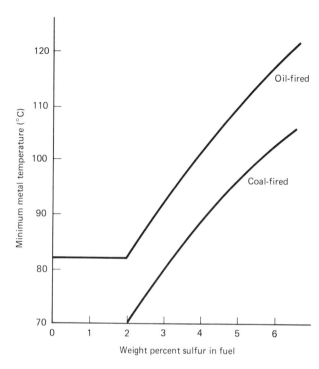

Figure 5.21 Minimum metal temperature to avoid external metal corrosion. (From *Mark's Standard Handbook for Mechanical Engineers*, by T. Baumeister, E. Avallone, and T. Baumeister III, copyright 1978—used with the permission of McGraw-Hill Book Company, New York.)

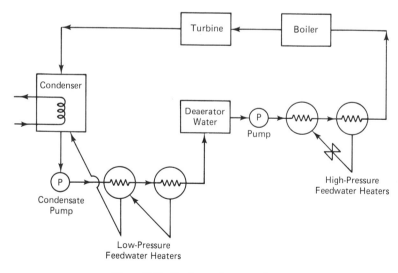

Figure 5.22 Feedwater heater train.

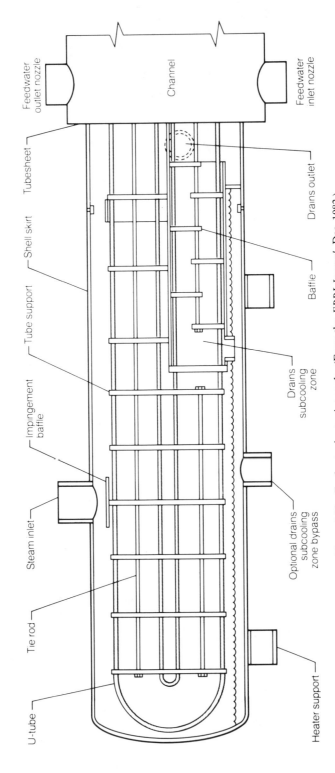

Figure 5.23 Feedwater heater internals. (From the *EPRI Journal*, Dec. 1982.)

shell. Low-pressure heaters are located upstream of the boiler feedpump and are usually designed for pressures of 900 psi or less. High-pressure heaters, located on the downstream side of the boiler feed pump, operate above 1500 psi up to critical water pressures. Steam and water pressures may differ. A single pump may be used to circulate the water through a series of feedwater heaters. A modern version of a feedwater heater is shown in Fig. 5.23.

The shell-and-tube arrangements vary. The tube arrangements may consist of a nonremovable straight tube bundle or a 180° bend tube bundle. Heaters may be used in a horizontal or vertical attitude. In some designs, the high-pressure heater contains a desuperheating section which may heat the feedwater above saturation temperature. All heaters must be provided with vents and drums. Periodic venting is required to remove any accumulation of noncondensable gases. A steam trap usually allows discharge of only liquid condensate through the drain line. The condensate is flashed across a throttling valve to the pressure of the next feedwater heater in the train. The resulting steam–water mixture joins the bleed stream from the turbine in heating the next feedwater air.

Deaerators (Deaerating Heaters)

The deaerating feedwater heater is an open feedwater heater that operates at a pressure slightly higher than atmospheric pressure. The main function of this heater is to remove dissolved gases, especially oxygen and CO_2, from the boiler feedwater, thereby reducing corrosion levels throughout the system. The feedwater must be heated to saturation temperatures where the gas solubility is near zero. By being part of the feedwater train, the deaerator also helps to improve the system thermodynamic efficiency. The removal of oxygen and carbon dioxide from the boiler feedwater is an essential aspect of the boiler feedwater conditioning process.

The deaerating heater shown in Fig. 5.24 uses steam bled from the high-pressure turbine to heat the feedwater to saturation temperature. The water is mechanically agitated to enhance the release process. The released gases are removed by a steam ejector and released to the atmosphere. A modern deaerator will reduce oxygen content to less than 0.005 cm^3/liter and carbon dioxide content to a negligible amount.

Condensers

The power plant condenser receives the low-pressure exhaust from the steam turbine, and hence it precedes the feedwater heaters previously described (see general arrangement in Fig. 5.22). The primary function of the condenser is to produce a vacuum or desired back pressure at the turbine exhaust for the improvement of the power plant heat rate. The condenser also serves to condense the wet turbine exhaust steam into an easily pumped liquid. Finally, the condenser deaerates

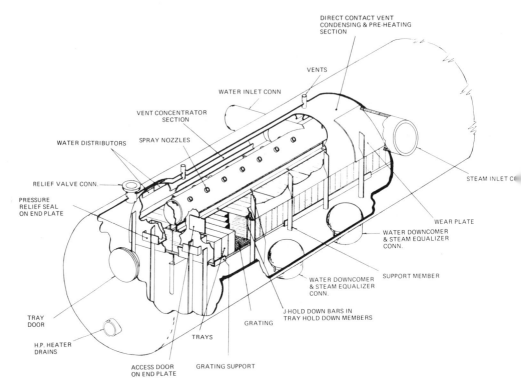

Figure 5.24 Deaerating heater. (Courtesy of Crane Co., Cochrane Environmental Systems Division, King of Prussia, Pa.)

the condensate. Figure 5.25 shows a large condenser unit typical of those used for power plant operation. In the past, most power plants were generally located on a river, lake, or bay such that access was available to cooling water. This cooling water from these sources removed the heat from the condensers. Although this practice is still common, in many cases environmental regulations prohibit such operation. Many newer units are built with large cooling towers which provide atmospheric cooling of the cooling water circulated through the condenser.

The condenser is a shell-and-tube heat exchanger, with the cooling water on the tube side and turbine exhaust steam, heater drains, steam dumps, and so on, on the shell side. Most manufacturers utilize tube bundle configurations that are unique to their particular design philosophy. The design objectives are to minimize pressure losses between the turbine exhaust and the air offtake and tubes are arranged to promote optimum heat-transfer rates. Large power plant condensers are rectangular in shape to conserve space. The turbine normally exhausts downward into the condenser. The condenser tubes are connected to large chambers, or water boxes, at each end of the condenser. Because of the increased pressure resulting from the need to circulate condenser water through cooling towers, water boxes are now generally made with curved surfaces as shown in Fig. 5.25. With cooling towers, pressures

Sec. 5.6 Primary System Heat Exchangers

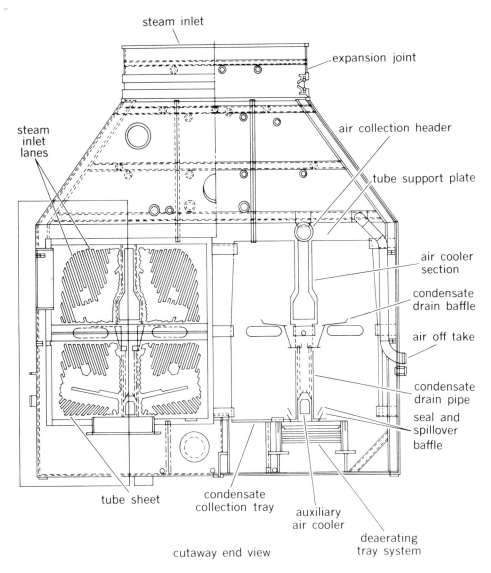

Figure 5.25 Turbine steam condenser. (Courtesy of Foster Wheeler Energy Corp., Livingston, N.J.)

range from 4 to 5.5 bar (60 to 80 psi), whereas with straight-through river cooling, pressures are below 2 bar (30 psi).

The water boxes divide and separate the flow path in the condenser. This design feature permits one half of the unit to be shut down for repair or cleaning while the other half is operating. The turbine can thus be kept running at reduced load.

When ocean, river, or lake water is used, thermal performance and corrosion resistance of the condenser tubing may be seriously affected by biofouling. Biofouling may appear as slime and algae deposition on tube surfaces or tube blockage by barnacles, mussels, seaweed, and debris. A number of methods are available to control biofouling and provisions must be made to perform some or all of these operations. These methods include backwashing or flushing, warm-water recirculation, manual or automatic tube cleaning with rubber plug or brushes, and chlorine or chemical dosing. Recent environmental regulations have curtailed the use of chlorine.

A relatively recent development is the use of air-cooled condensers in place of the standard water-cooled units. Such condensers are located out of doors and are cooled by forced convection to a rapidly flowing airstream. Although such a unit may not be able to maintain as high a vacuum as a water-cooled condenser, it may still be economically attractive when it eliminates the need for a water cooling tower.

Auxiliary equipment for condenser operation is necessary to remove air from the steam space and drain the condensate from the hot well. The equipment typically includes circulating pumps, condensate pumps, steam jet ejectors (for gas removal), and relief valves.

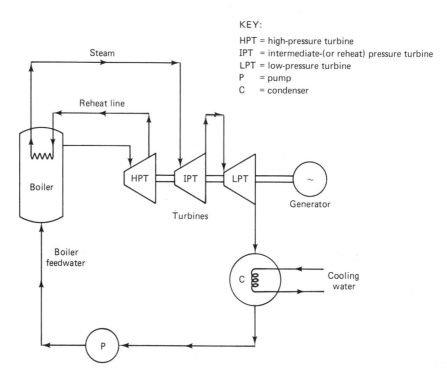

Figure 5.26 Single-stage reheat turbine.

Reheaters

Reheating is normally accomplished by passing steam from an intermediate turbine stage to a flue-gas-heated reheater located in the main boiler section of the plant. The reheated steam is fed back to the next turbine stage. High-pressure steam can also be used for reheating intermediate pressure steam. Figure 5.26 is a schematic representation of a reheat cycle.

As can be determined from the thermodynamic considerations that reheating increases the overall cycle efficiency. More important, however, use of reheat limits the end-point moisture (moisture at last stage of turbine). To keep turbine blade corrosion and erosion at acceptable limits, the moisture content of saturated steam is limited to 10 to 15%.

The T-S diagram illustrates the effect [see Fig. 2.6(a)] of a simple reheat stage. The cycle without reheat would have end point (b) while the cycle with reheat has end point (c). Note that the end point is much closer to the saturation line with reheat and this means less moisture. Also note that the cycle efficiency has improved. Although modern plants have several reheat stages, the general effect is similar.

Studies have been performed to determine the optimum number of reheat cycles for a power plant. Cycle efficiency can be increased by several percentage points through the use of six or seven reheat cycles. Beyond this point the capital cost of the reheaters exceeds the small additional incremental gain in performance.

Heat Exchanger Analysis

Heat-flow analysis for most of the shell-and-tube heat exchangers follows the basic form

$$q = UA_h(\Delta T)_{\text{LM}} \tag{5.29}$$

where U = overall coefficient of heat transfer for conduction and convection resistances, W/m²·°C
 A_h = heat-transfer area, m² (based on outside surface area)
 $(\Delta T)_{\text{LM}}$ = log mean temperature difference between the fluid streams, °C

For the simple case of a cylindrical tube, such as shown in Fig. 5.27, the sum of the heat flow resistances, Σ/R, would be

$$\frac{1}{U} = \frac{1}{h_i}\frac{r_o}{r_i} + \frac{1}{h_o} + \frac{r_o \ln(r_o/r_i)}{k} + \frac{1}{(h_f)_o} \tag{5.30}$$

where r_o, r_i = outer and inner tube radii, respectively
 h_o, h_i = convective film coefficients at outer and inner surfaces, respectively
 k = conductivity of the tube wall
 $(h_f)_o$ = fouling coefficient (heat conductance of tube wall deposit)

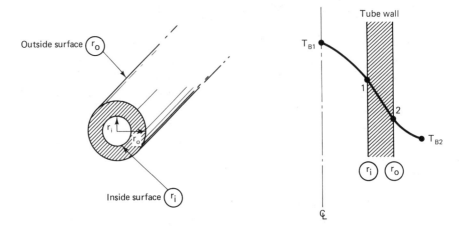

Figure 5.27 Temperature profile across a cylindrical heat-exchanger tube.

The principal challenge in the heat-transfer analysis centers in methods for calculation of the convective film coefficients, h_1 and h_2, and evaluating the effects of fouling and scaling (deposition of soot and ash outside of tube and deposition of insoluble scale on inside of tube). The determination of the fouling coefficient is beyond the scope of the text (see the Bibliography at the end of this chapter). Techniques for evaluating the convection heat-transfer coefficients required will be presented in this section.

Heat-transfer coefficients in water-cooled surface condensers. Although there are several types of steam condensers used in power plants, the most efficient and widely used is the water-cooled surface condenser previously described.

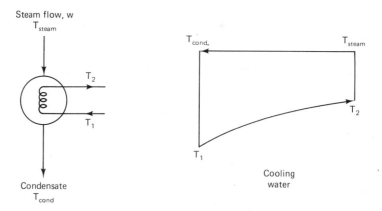

Figure 5.28 Temperature profile in a water-cooled surface condenser.

Sec. 5.6 Primary System Heat Exchangers

Usually it is a single-pass radial flow design. Steam enters from the top and condensate is collected in the hot well at the bottom of the cylindrical shell. Cooling water inlet and outlet are through separate water boxes. As may be seen from Fig. 5.28, heat flows from the wet steam outside the tubes (shell of exchanger) across the tube wall to the cooling water on the inside of the tubes in the shell region. The total heat removed would be calculated by Eq. (5.29) and the overall coefficient of heat transfer is of the form of Eq. (5.30).

The resistance to heat transfer on the shell side is due to collection of a thin film of liquid on the tube. The heat-transfer resistance, h, is usually estimated based on *Nusselt's equation* for laminar film condensation on a single horizontal tube:

$$h = 0.725 \left[\frac{\rho_l (\rho_l - \rho_v) g h_{fg} k^3}{\mu D_o (T_s - T_w)} \right]^{-1/4} \quad (5.31)$$

where ρ_l, ρ_v = density of liquid and vapor, respectively
k = thermal conductivity of liquid
μ = viscosity of liquid
D_o = outside diameter of tube
g = acceleration of gravity
T_s, T_w = temperature of steam and wall, respectively

For a bank of N_0 tubes in a vertical row, the mean condensation coefficient $\bar{h}_2$ is related to the coefficient h for the top tube by

$$\bar{h}_2 = h N_0^{-1/4} \quad (5.32)$$

The values obtained by this approach are conservative (low). Note that since the average wall temperature is unknown until the heat-transfer rate is calculated, an iterative calculation is required. The resistance to heat transfer on the water side is readily determined using the Dittus–Boelter equation [Eq. (5.25)] with the properties of the cooling water.

Reheater heat-transfer coefficients. In these units, wet saturated steam is reheated to superheated steam. Hot flue gases on the outside of the tubes transfer heat through a convective gas film, through the tube wall, and finally to a convective steam film on the inside of the reheater tube. If the tube wall is neglected, the overall coefficient of heat transfer, U, is

$$U = \frac{h_{\text{steam}} \times h_{\text{flue gas}}}{h_{\text{steam}} + h_{\text{film gas}}} \quad (5.33)$$

The convective film coefficient for steam may be conservatively estimated from the Dittus–Boelter equation with the fluid properties evaluated at dry-steam conditions. In the wet-steam region the actual heat transfer will be higher than that estimated.

The flue-gas-side convective film coefficient would be evaluated using a relationship for cross flow across the tube bank [Eq. (5.26), (5.27), or (5.28)]. The properties are, of course, evaluated at flue gas conditions.

Closed feedwater heater heat-transfer coefficient. The arrangement of the closed feedwater heater is essentially the same as that found in the condenser. Hence the same relationships apply.

Economizer heat-transfer coefficient. The convective heat-transfer coefficient to water inside the tubes is obtained by the Dittus–Boelter equation (5.25). Since the gas flows across the tube bank, we again use Eq. (5.26), (5.27), or (5.28) for computation of the heat-transfer coefficient on the flue gas side. When extended surfaces are used on the flue gas side, an effective surface area must be used. The effective surface area allows for the fact that the fins will be at a lower temperature than the tube wall. The effective surface area is obtained by multiplying the total area of the tubes and extended surface by a effectiveness factor. The value of this factor will depend on the particulars of a given design.

5.7 WATER AND STEAM FLOW IN THE PRIMARY SYSTEM: NATURAL-CONVECTION SYSTEMS

Water enters the furnace at the bottom header of the economizer. It then flows upward through the economizer and exits through the outlet heater and into the piping that connects to the steam drum. Natural circulation carries the water from the steam drum through downcomer pipes to the distributor supplying the boiler tubes. The fluid boils in the furnace tubes and a two-phase mixture of steam and water flows to a furnace wall header and then through riser tubes to the steam drum. This circulation is shown schematically in Fig. 5.29.

In a natural-circulation system, the driving head is the head difference developed because of the density difference between the fluid in the riser and downcomer tubes. That is, the pressure difference Δp, available to drive the flow is

$$\Delta p = \left(\rho_l z_t - \int_0^{z_t} \rho_{tp}\, dz \right) \frac{g}{g_c} \qquad (5.34)$$

where z_t = total height between lower header and steam drum
ρ_l = density of liquid in downcomer line
ρ_{tp} = density of two-phase mixture in furnace tubes
g = gravitational acceleration (9.8 m/s² in SI units, 32.2 ft/s² in British units)
g_c = gravitational conversion constant [unity in SI units, 32.2 ft/s² (lb$_m$/lb$_f$ in British units)]

Sec. 5.7 Water and Steam Flow in the Primary System

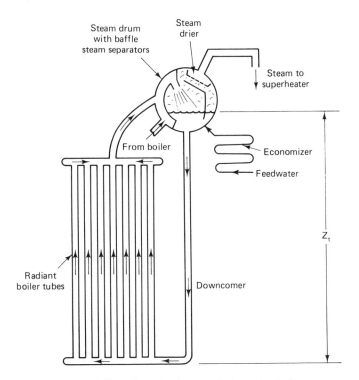

Figure 5.29 Schematic of a natural-circulation boiler.

In the furnace region, the density of the fluid will be continuously changing due to vaporization. Since heat input per unit length may not be constant, it is often easiest to treat the density variation in a stepwise fashion and make the approximation.

$$\int_0^{z_t} \rho_{tp} \, dz - \sum_j (\bar{\rho}_g \Delta z_j) \qquad (5.35)$$

where $\bar{\rho}_j$ = average density of two-phase mixture in jth segment of furnace tube or riser
Δz_j = vertical height of jth segment

Although the vapor will be traveling at a somewhat higher velocity than the liquid, at the high pressure of modern boilers, the ratio of vapor to liquid velocity does not depart greatly from unity. With this assumption, the average density, $\bar{\rho}$, is readily computed from the quality, x. We define

$$\alpha = \text{void fraction} = \frac{\text{volume of vapor in mixture}}{\text{total volume of liquid–vapor mixture}}$$

and recognize that

$$\bar{\rho}_j = \alpha_j \rho_v + (1 - \alpha_j)\rho_l \qquad (5.36)$$

where ρ_l and ρ_v are the densities of saturated liquid and vapor, respectively. Further, it may be shown readily that when vapor and liquid velocities are equal,

$$\alpha = \frac{1}{1 + [(1-x)/x](\rho_v/\rho_l)} \qquad (5.37)$$

where x is the quality, that is, the mass of vapor flowing per total mass flowing. Once x is obtained by a heat balance, the value of $\bar{\rho}_j$ is easily computed.

The pressure loss, Δp_L, which must be equated to Δp, is obtained by summing the losses in all the components. The calculation is somewhat complicated by the presence of a two-phase mixture in the risers. In the most accurate way of treating such systems, the frictional pressure drop is calculated as if the liquid phase were flowing alone and the result multiplied by a two-phase friction factor multiplier, ϕ^2. The value of ϕ^2 is a function of pressure, x, and flow rate. A relatively simple correlation may be obtained by defining the parameter $(\bar{X})^2$ as

$$\bar{X}^2 = \frac{(dp/dz)_l}{(dp/dz)_g} \qquad (5.38)$$

where $(dp/dz)_l$ = frictional pressure drop if liquid phase were flowing alone
$(dp/dz)_g$ = frictional pressure drop if vapor phase were flowing alone

Then we may use Chisholm's representation of the Lockart and Martinelli correlation where ϕ^2, the two-phase friction multiplier, is given by

$$\phi^2 = 1 + \frac{20}{\bar{X}} + \frac{1}{\bar{X}^2} \qquad (5.39)$$

The total frictional loss, Δp_f, is the summation of the losses in all components. That is,

$$\Delta p_f = \sum_j \frac{f' \Delta z_j (G'_l)_j^2}{2g_c D_e \rho_l} \phi_j^2 + \sum_k \frac{K_k (G'_k)^2}{2g_c \rho_k} \qquad (5.40)$$

where f' = single-phase friction factor evaluated as if liquid alone flowing in duct
$(G'_l)_j$ = liquid mass velocity in jth pipe length (mass/time area)
G_k = total mass velocity in restriction or area change k
ρ_k = average fluid density at restriction or area change k
K_k = pressure loss coefficient for restriction or area change k
ρ_l = liquid density

The pressure drop through the primary steam separator would be treated by assigning an appropriate value of K.

Since vapor is being generated in the furnace tubes, the mixture is being accelerated. The total pressure loss Δp_L is therefore given by

$$\Delta p_L = \Delta p_f + \Delta p_a \qquad (5.41)$$

when Δp_a is the acceleration pressure drop. By a simple force–momentum balance it may be shown that

$$\Delta p_a = \frac{(G')^2}{g_c \rho_l} \left[\frac{(1 - x_{\text{exit}})^2}{1 - \alpha_{\text{exit}}} + \frac{x_{\text{exit}}^2}{\alpha_{\text{exit}}} \frac{\rho_l}{\rho_v} - 1 \right] \qquad (5.42)$$

where G' is the total mass velocity (mass/area time) and α_{exit} and x_{exit} represent the values of α and x at the exit of the furnace tubes.

It should be recognized that an iterative calculation is required. Since the total flow circulation is unknown, a value must be assumed. The values of Δp_L and Δp are then calculated. If, as is most likely, they fail to agree, another total flow must be chosen and the calculation repeated. When agreement between Δp and Δp_L is obtained, the exit void fraction (α) must be checked against the maximum recommended void fraction (α). These recommended void fractions are generally based on keeping the quality sufficiently low so that the dry-out quality will not be exceeded. At about 100 bar, one boiler manufacturer requires $\alpha \leq 0.67$; at 33 bar, $\alpha \leq 0.77$.

Forced-Circulation Systems

The general procedure is essentially as described above. However, since a pump is to be used, the desired flow may be selected and the total differential pressure loss, Δp_L, and total driving pressure, Δp, may be calculated directly. The difference ($\Delta p_L - \Delta p$) is the pressure rise that must be supplied by the circulating pump.

5.8 STEAM GENERATOR CONTROL

The objective of the steam plant control system is to provide the steam flow required by the turbine at design pressure and temperature. Further, it is desired that the steam plant respond rapidly to condition changes without any significant oscillations or hunting. The variables that can be controlled to provide the desired operations are fuel firing rate, airflow, gas-flow distribution, feedwater flow, and turbine valve setting. The key measurements that describe the plant performance are steam flow rate, steam pressure, steam temperature, primary and secondary airflow rates, fuel firing rate, feedwater flow rate and steam-drum level, and electrical power output. The control system must act on the measurement of these plant parameters so as to maintain plant operation at the desired conditions.

Most modern control instrumentation utilizes *closed-loop control*. In this mode, the actual output of the system is measured and compared to some demand signal (set point). The difference between the measurement and demand, called the *error signal*, is then used to reduce the difference between measurement and demand (set point) to near zero.

Proportional control is the simplest type of closed-loop control. In this mode, the controller output is proportional to the error signal. The control signal will be either directly or inversely proportional to the error signal, depending on the control action required.

To ensure stable behavior of a proportional controller, there is a dead band around the set point in which no controller output is provided. With a simple proportional controller, the final stable value of the control variable will then be "offset" from the set point by some small amount. This offset can be eliminated by incorporating *integral* or *reset control* action within the controller. In integral control, the control action is based on the integral of the deviation between the controlled variable and the set point over the time period in which the deviation occurs. By incorporating this action the offset produced by simple proportional control is eliminated while maintaining system stability. This use of the terminology *reset control* for this control mode comes about since the band of proportional action is shifted or reset so that the controlled variable operates about a new base point.

Almost all power plant control instruments combine both proportional and integral control actions. In some cases, derivative control is also added. A derivative control action is determined by the rate of change of the controlled variable. If a variable begins to change rapidly, a large control signal is provided. Once the rate of change is reduced, the proportional and integral actions take over and do the final positioning. The combination of all three modes of control action provides more rapid response and more stable operation than possible with simpler systems.

Steam Demand

In modern central generating systems, when the load on the system changes, the load dispatch system determines which turbine–generator is going to accept the load change. The load dispatch system then provides a signal to the appropriate turbine–generator indicating the revised electric load to be carried. A controller on the turbine–generator compares the new electrical load requirement (new "set point") with the actual electrical load being generated and produces an error signal. This error signal activates the turbine admission valves, and more or less steam is admitted to the turbine.

By changing the steam flow to the turbine, the mechanical work produced by the turbine changes and the electrical load (MWe) changes proportionally. The mechanism by which this occurs is discussed in Chapter 8.

Firing Rate

If an increase in electrical load is to be produced, the turbine admission valves open wider admitting more steam to the turbine. This sudden increase in steam flow is not met by an immediate increase in steam generation and hence the steam pressure will fall. Similarly, a decrease in load leads to reduced steam flow which produces a rise in steam pressure.

The controller, which senses steam pressure, adjusts the fuel firing rate in response to a steam pressure increase or decrease. In a pulverized-coal fired boiler, the signal goes to a pulverizer master controller, which provides signals to the coal feeders providing fuel to the individual pulverizers. Since burners do not operate properly below a specified fuel firing rate, as the load falls, the pulverizer master controller terminates feed to upper burners so that the feed rates to those remaining in service are adequate.

The primary and secondary airflow rates are adjusted in accordance with the changed fuel firing rates. At the same time the firing rate is increased, the uptake draft is increased by opening the damper at the furnace exit. This decreases the pressure in the furnace, which is sensed by a unit controlling the dampers on the forced draft fans (blower) supplying air to the furnace. The positions of the dampers are adjusted to maintain the furnace pressure essentially constant with load. A proportional controller with automatic reset is generally used for this purpose. Note that at low loads, some blowers will have to be completely shut down.

Feedwater Control

In a recirculating boiler, one might consider feedwater control in terms of the maintenance of a safe water level in the steam drum. In the early steam plants, the level control approach was used. The level in the steam drum was monitored and feedwater was added as required to maintain the desired level. This "single-element" level control does not work well if large load fluctuations occur. When there is a sudden increase in steam demand, there will be a sudden reduction in boiler pressure before the combustion control system can respond with increased steam generation. The volume of water displaced by steam bubbles from flashing water causes an apparent increase in drum water level (*level swell*). A single-element level control would respond by closing the feedwater inlet just at the time more feedwater flow is required.

The problems caused by level swell may be avoided by using what is known as three-element feedwater control. In this scheme, the steam flow to the turbine and the feedwater flow to the furnace provide the primary signals (two elements of the control system). The feedwater flow is balanced against the steam flow by a proportional controller. Since the metering of the steam and feedwater flow cannot be 100% accurate, the average level in the boiler drum may slowly rise and fall with time. Integral action based on the level signal (third element of control system) trims the feedwater flow so that the average drum level does not drift.

In once-through boilers, the control scheme must be modified since no steam drum is provided. The feedwater flow is balanced against steam flow or boiler demand based on a load signal. A reset signal is provided using feedwater temperature error. A deviation from the desired feedwater temperature necessitates more or less feedwater to compensate for the change in the extraction flow to the feedwater heaters.

Steam Superheat Control

It is necessary to maintain steam temperature at or near the design point. Excessive temperatures will lead to excessive turbine corrosion, while low temperatures will cause a significant decrease in plant efficiency.

The temperature of the superheated steam reaching the turbine depends on a number of factors: load on the plant, amount of excess air used in combustion, feedwater temperature, heating surface cleanliness, distribution of heat input among the burners at different positions, variation in fuel characteristics, use of saturated steam for soot blowing, or auxiliaries. The most important of these factors is load variation.

Both the volumetric flow and temperature of combustion gases increase as load is increased. Thus in convective superheaters the steam temperature increases as load increases. In a radiant type of superheater, the radiant heat input does not increase much with load; therefore, the steam temperature decreases with load. By using a radiant superheater in parallel with a convective superheater it is possible to maintain a more constant steam temperature. However, since most of the heat is usually transferred in the convective superheater, an overall rise in temperature with load is still seen. Additional control and adjustment techniques are therefore required.

An additional technique for limiting the variation in steam temperature with load is through burner control. Here the effective boiler furnace volume is varied in response to load variations. The effective boiler volume is reduced by not firing the upper burners at low loads. In addition, the burners can, in some designs, be tilted downward at low loads and upward at high loads. By changing the effective boiler area, the heat absorption at low loads is reduced and the gas temperature entering the convective superheater is kept near the desired level. This decreases the variation of steam temperature with load.

The effect of load variation may be minimized further by use of controlled gas recirculation. In this method (see Fig. 5.30) some gas from the boiler, economizer, or air heater outlet is reintroduced near the furnace inlet by means of suitable blowers and ductwork. The primary function of gas recirculation is to reduce the heat absorption in the furnace (boiler region) by diluting the furnace gases and thus lowering the temperature in the radiant zone. This reduces the heat absorption in the boiler zone while increasing the gas velocity going through the convective superheater and increasing its heat absorption. Thus, by recirculating gas at low loads, higher steam temperatures are possible at low loads.

In some gas recirculation systems recirculated gas is admitted near the furnace exit. This is called *gas tempering*. In this approach the furnace exit temperature is reduced without affecting furnace heat absorption. In once-through boilers, a combination of gas tempering and recirculated gas addition at the furnace inlet is

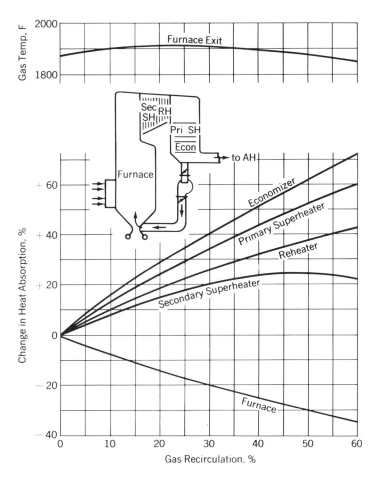

Figure 5.30 Effect of gas circulation on the heat absorption pattern at a constant firing rate. (Reproduced with permission from *Steam, Its Generation and Use*, The Babcock & Wilcox Company, Lynchburg, Va., 1978.)

common. Furnace exit-gas temperature and furnace heat absorption can then be independently controlled.

In general, gas recirculation and burner control can provide a nearly constant superheat temperature for a given furnace design from somewhat over 50% of full load to over 90% of full load. However, such control is generally not adequate to maintain the desired steam temperature at nearly full load. If design temperature conditions are maintained at the part loads, when all burners are on and tilted upward and there is no gas recirculation, the steam temperatures tend to exceed the design point as full load is approached (see Fig. 5.31). At these highest loads it is necessary to reduce the steam temperature. This is called *attemperation*. Such attemperators may be of the surface type, in which a heat exchanger is used, or of

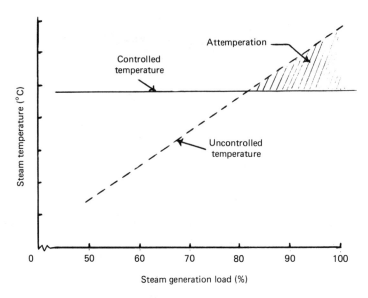

Figure 5.31 Steam temperature variation with load.

the direct contact type. The most common form of the direct contact attemperator is the spray attemperator. In this unit, feedwater is sprayed into the steam through a spray nozzle discharging into a venturi which provides mixing.

In controlling steam temperature via gas recirculation, a two-element control is generally used. Since the response to damper motion is slow, primary air flow usually governs the initial damper position. Final adjustment comes from integral action on the steam temperature. When beyond the gas recirculation range, the attemperator is controlled by balancing feedwater flow to the spray against steam flow or load with final integral action (reset) determined by the steam temperature.

It should be noted that the presence of a reheater will complicate the designer's task. The methods used for superheat control will also affect the temperature of the reheated steam. This reduces the designers' freedom of action. In a recirculating unit, reducing the effective furnace area will increase both superheat and reheat temperature. The latter action may be undesirable.

SYMBOLS

a_f	fraction of total cross-sectional area in an infinite array which is taken up by the fluid
A	area, m^2
A_C	effective area of cold surface, m^2

Chap. 5 Symbols

A_h	heat-transfer area, m^2
A_T	total enclosed area, m^2
c_p	specific heat, kJ/kg · °C
$\bar{c}_p$	mean specific heat of flue gas, kJ/kg · °C
$(\bar{c}_p)_{m,i}$	mean molal specific heat, kJ/kg mol · °C
C'	percent carbon from ultimate analysis
CO'	percent of CO in flue gas
CO_2'	percent of CO$_2$ in flue gas
D_e	equivalent diameter, m
D_o	outside tube diameter, m
f'	single-phase friction factor
$\mathscr{F}_{TC}$	combined shape factor accounting for emissivity and view effects based on A_T
g	acceleration of gravity, m/s^2
g_c	gravitational conversion constant (unity in SI units)
G'	mass velocity, kg/m^2 · s
h	fluid enthalpy, kJ/kg
h_a	specific enthalpy of the entering air, kJ/kg
h_{ei}	specific enthalpy of water at economizer inlet, kJ/kg
h_{eo}	specific enthalpy of water at economizer outlet, kJ/kg
h_f	specific enthalpy of feedwater entering furnace, kJ/kg
h_{fg}	heat of vaporization of water at 1 atm, kJ/kg
h_g	specific enthalpy of saturated steam, kJ/kg
h_s	specific enthalpy of steam at product gas outlet temperature, kJ/kg
h_{ss}	specific enthalpy of superheated steam, kJ/kg
h_w	specific enthalpy of water at air inlet temperature, kJ/kg
h	convective heat-transfer coefficient (film coefficient), W/m^2 · °K
$\bar{h}$	average heat-transfer coefficient, W/m^2 · °K
h_{Tc}	film conductance on outside of superheater tubes, W/m^2 · °K
h_{Ti}	film conductance on inside of superheater tubes, W/m^2 · °K
$(h_f)_o$	fouling coefficient, W/m^2 · °K
HHV	high heating value of coal, kJ/kg or kcal/kg
k	thermal conductivity, W/m · °C
K_k	pressure loss coefficient, dimensionless
L_1°	boiler loss due to incomplete combustion, kJ/kg
L_2°	boiler loss due to unburned carbon, kJ/kg

L_3^o	boiler loss due to dry gas, kJ/kg
L_4^o	boiler loss due to moisture, kJ/kg
L_5^o	boiler loss due to moisture in combustion air, kJ/kg
L_6^o	boiler loss due to thermal radiation, kJ/kg
$(m_a')_{act}$	mass of air supplied per unit mass of fuel (actual air-to-fuel ratio), kg air/kg fuel
m_A'	mass of ash per unit mass of fuel
m_b'	mass of carbon burned per unit mass of fuel
m_C'	fraction of unburned carbon in refuse
m_{gas}'	mass of dry flue gas per unit mass of fuel
m_{H_2}'	mass fraction of hydrogen in coal
m_m'	mass fraction of moisture in coal
m_w'	mass of water vapor discharged in flue gas per unit mass of fuel
$\dot{m}_a$	mass of air required per hr, kg/h
$\dot{m}_F$	mass of fuel required per hr, kg/h
$\dot{m}_1$	feedwater flow, kg/h
$\dot{m}_{rh}$	reheat steam flow rate, kg/h
$\dot{m}_s$	steam flow rate, kg/h
MW	molecular weight
N_0	number of tubes in row
Δp	pressure difference, Pa
Δp_a	acceleration pressure drop, Pa
Δp_f	frictional pressure loss, Pa
Δp_L	total pressure loss, Pa
q	rate of heat transfer, W
q_{ah}	rate of heat transfer to airstream, kJ/h
q_b	rate of heat transfer in the boiler, kJ/h
q_e	rate of heat transfer in the economizer, kJ/h
q_{ir}	rate of radiation heat transfer from hot gases in tube bank, W
q_r	rate of direct radiation-heat transfer from flame, W
q_{rad}	rate of heat transfer in the radiant boiler, W
q_s	rate of heat transfer in the superheater, W
q_t	rate of total heat transfer to a given row of tubes, W
q_{TC}	rate of heat transfer to a given row of tubes, W
Q_0''	higher heating value (standard heat of reaction with product water as liquid), kJ/kg or kcal/kg
r_o, r_i	outer and inner tube radii, respectively, m

Chap. 5 Symbols

R_L	radiation loss factor (see Fig. 5.12)
S_l	tube pitch parallel to flow, m
S_t	tube pitch transverse to flow, m
T	temperature
T_0	base temperature
T_a	temperature of exit air from air heater, °C
T_c, T_f	tube and flame temperatures, respectively, °K
T_{ei}	feedwater inlet temperature, °C
T_{eo}	feedwater outlet temperature, °C
$T_{\text{gas in}}$	gas temperature at inlet to boiler, °C
$T_{\text{gas out}}$	gas temperature at boiler outlet, °C
T_{in}	inlet air temperature to air heater, °C
T_G	hot gas temperature, °K
T_s	steam temperature, °K
T_w	wall temperature, °K
$(\Delta T)_{\text{LM}}$	log mean temperature difference—gas to tube, °C
U	overall coefficient of heat transfer, W/m² · °C
U_{ir}	equivalent intertube radiation conductance, W/m² · °K
U_r	basic radiation conductance, W/m² · °K or Btu/h · ft² · °F
x	quality, mass of vapor/total mass of vapor plus liquid
X	humidity of combustion air, kg H₂O/kg dry air
$\overline{X}^2$	ratio of liquid-phase pressure drop to gas-phase pressure drop
z	height, m
z_t	total height between lower header and steam drum, m

Greek Symbols

α	void fraction
ε	emissivity of the cold surface
ε_1	emissivity of cold surface
ε_G	emissivity of the hot gases
$(\varepsilon_G)_f, (\varepsilon_G)_c$	emissivity of gases at flame and tube temperatures, respectively
ε_t	emissivity of boiler tubes
η_B	boiler efficiency, %
μ	fluid viscosity, kg/m · s
ρ	density, kg/m³

$\bar{\rho}$	average density of two-phase mixture, kg/m³
σ	Stefan–Boltzmann constant, 5.67×10^{-8} W/m² · °K⁴
ϕ^2	two-phase function multiplier
ψ	function of the ratio S_t/S_l

Generally Used Subscripts

1, 2, 3, ...	locations 1, 2, 3, ...
a	air
f	evaluated at film temperature
G	at gas temperature
i	inner
j, k	segment or unit number
l	liquid
o	outer
v	vapor

PROBLEMS

5.1. A power plant produces 250,000 kg of steam per hour at a pressure of 170 bar. The steam is superheated to 200°C above saturation. The flue gases leave the furnace at 275°C and feedwater enters at 225°C. The fuel (bituminous coal) has a lower heating value of 32,000 kJ/kg and is burned with 13 kg of air per kilogram of fuel. Determine the total hourly fuel requirements if heat losses to walls and to discharge ash are 6% of the energy supplied.

5.2. Calculate the "boiler efficiency" for
 (a) the furnace described in Problem 5.1.
 (b) the furnace described in Problem 4.2.
 (c) the furnace described in Problem 4.2 if radiation losses from the furnace walls were not negligible.

5.3. The radiant boiler of a small furnace is to provide 100,000 kg/h of saturated steam at 150 bar. The furnace has a 3 m × 3 m square cross section. The furnace walls are lined with a single row of 1-in.-O.D. water tubes placed on 1.5-in. centers. The flame is at 800°C and contains 15 v/o steam, 20 v/o CO_2, and no particles. Estimate the total length of water tubing required.

5.4. A power plant is to produce 300,000 kg of steam per hour. The steam is to be generated at 150 bar. The steam leaves the superheater at 143 bar and enters the turbine with 140°C of superheat. The plant is fueled using coal having the following ultimate analysis:

	w/o
Total carbon	55.44
Hydrogen	3.96
Oxygen	6.98
Nitrogen	1.08
Sulfur	4.54
Moisture	13.00
Ash	15.00
Total	100.00

The coal is to be burned with 15% excess air. The radiant section of the boiler is made up of 2-cm-O.D. tubes. The tubes are arranged in two rows on a 3-cm triangular pitch. It may be assumed that all the evaporation takes place within the radiant section. The heat transfer in the economizer and superheater may be assumed to depend primarily on the forced convection heat-transfer coefficient to furnace gas which you may take as 750°C. Radiation is not significant. Assume that the water enters the economizer at 200°C. Calculate:

(a) The flame temperature.

(b) The rate at which coal must be fed at full load.

(c) The number of feet of tubing required and dimensions of the radiant section of the boiler (furnace dimensions).

(d) The number of feet of superheater tubing required (assuring tube to be in 2 cm O.D.).

(e) The number of feet of economizer tubing required if the exit gas temperature is not to exceed 250°C (assuming tubes of 2 cm O.D.).

5.5. A convective superheater is to superheat 3×10^4 kg/h of 120-bar steam from saturation to 250°C above saturation. The hot gases enter the superheater at 650°C and flow across the tube bank at a superficial velocity of 8 m/s. The heat exchanger, which consists of a triangular array of $\frac{3}{4}$-in.-O.D. tubes on a 1.5-in. pitch, has a superficial flow area of 18 m². For simplicity, you may assume that (1) the hot gases have the same physical properties as air, (2) the steam-side heat-transfer coefficient is 1500 W/m² · °C, and (3) the tube walls are 0.1 in. thick. Determine the number of tube rows required.

5.6. A condenser operating at 0.2 bar has a heat-transfer area of 2000 m². The cooling water enters the condenser at 25°C and a flow rate of 4×10^6 kg/h. You may assume that the heat-transfer coefficient inside the tube is 2500 W/m² · °C and that the brass tubes are 2.5 cm O.D. and have a thickness of 0.25 cm. At what rate (kg/h) is steam condensed?

5.7. A boiler is producing steam at 180 bar and 300°C superheat. The restrictions on the turbine operating conditions limit the maximum superheat to 200°C. Determine the quantity of 360°C water per kilogram of steam which would have to be added in a spray desuperheater to limit the superheat to the required value.

5.8. In a simple natural convection boiler system, surface separation of the steam is used. The steam drum is located 15 m above the floor of the furnace. The steam drum feeds a downcomer made of 12-in. (nominal) schedule 80 steel pipe. The downcomer, in turn, feeds a number of risers each of which has an internal diameter corresponding to 1-in. (nominal) schedule 80 pipe. The recirculation ratio (recirculation ratio = inlet water flow to riser/exit steam flow from riser) is not to be lower than 4.0. How many risers are needed to produce 60,000 kg/h of steam at 40 bar? Assume a uniform heat input along each riser.

5.9. In the boiler section of a Benson boiler, the liquid is fully vaporized in 10 m of vertical tubing. The mass velocity of water entering the boiler tubes is 3.5×10^6 kg/m$^2 \cdot$ h. The entering water is at 180 bar and 35°C below saturation. The boiler tubes connect to inlet and outlet plenums which have flow areas three times that of the boiler tubes. What is the pressure drop between the inlet and outlet plenums?

BIBLIOGRAPHY

ASHNER, F., *Planning Fundamentals of Thermal Power Plants.* New York: Wiley, 1978.

Combustion, 3rd ed. Combustion Eng., Inc., 1981.

CULP, A., JR., *Principles of Energy Conversion.* New York: McGraw-Hill, 1979.

FRAAS, A., *Engineering Evaluation of Energy Systems.* New York: McGraw-Hill, 1982.

LYONS, J., "Optimizing Designs of Fossil Fired Generating Units," *Power Engineering*, Feb. 1979.

SKROTZKI, B., and VOPAT, W., *Power Station Engineering and Economy.* New York: McGraw-Hill, 1960.

Steam, Its Generation and Use. Lynchburg, Va.: The Babcock & Wilcox Company, 1978.

WILSON, W., "Procedure of Optimizing the Selection of Power Plant Components," *Transactions of the ASME*, Vol. 79, 1120 (1957).

6

Fossil-Fueled Steam Power Plants: Auxiliary Systems

6.1 AIR CIRCULATING AND HEATING SYSTEM

Air Circulation

The combustion process requires large amounts of air. The air circulation system provides the necessary airflow and remove the gaseous combustion products. To produce the required flow a pressure differential is needed. The term *draft* is used to define the static pressure in the furnace, air passageways, and the stack. *Forced draft* means that the air and gases are circulated at pressures above atmospheric. A forced-draft fan, located near the system inlet, is required for such a system. In an induced-draft system, a fan located near the system exit pulls the air or gases through the system at a pressure below atmospheric.

In a forced-draft system, the fan circulates combustion air at positive pressure to the furnace/boiler. The furnace heat-transfer rate is improved because of the high convection rate through the furnace. However, the benefits of positive furnace pressure are offset by the possibility of combustion gas and soot leakage through any cracks or defects in the furnace.

The induced-draft system solves the leakage problem. The induced-draft fan draws combustion products from the combustion chamber and pumps the gases into the stack for discharge. Since there is a negative pressure in the combustion chamber, air leakage is inward.

Many large power plants employ both a forced-draft and an induced-draft fan system to produce a *balanced draft* system; that is, one in which the furnace operates at approximately atmospheric pressure [generally about -0.1 in. of water (in. H_2O)].

The forced-draft fan is sized for the pressure drops in the fuel handling and air supply ductwork. The induced-draft fan and plant stack are sized to match the pressure drops in the exhaust gas system.

Forced-draft fans are high-speed centrifugal fans with backward-curved blades. Induced-draft fans are low-speed centrifugal fans with forward-curved blades. A typical large power plant (600 MWe) might require two 1750-hp forced-draft fans and two 4000-hp induced-draft fans.

Pressure and flow measurement. The total pressure at a point in the system is the sum of the static and dynamic pressures:

$$p_t = p_s + p_v \tag{6.1}$$

where p_t = total pressure, Pa (N/m²)
p_s = static pressure, Pa (N/m²)
p_v = dynamic pressure, $= \rho_G u^2 / 2 g_c$
ρ_G = gas density, kg/m³
u = velocity, m/s
g_c = gravitational conversion constant (unity in SI units)

Although the SI units of pressure are pascals (newtons per square meter), it is still customary to state the static and dynamic pressures in inches of water since measurements are commonly made using a simple, standard U-tube with water as the manometer fluid.

In terms of inches of water, the dynamic pressure head, p_v, which would be measured by a pitot tube connected to a water-filled manometer, becomes

$$p_h = \frac{\rho'}{5.2} \frac{u^2}{2 g_c} \tag{6.2}$$

and for consistent units,

p_h = average velocity pressure head, in. H₂O

ρ' = gas density, lb/ft³

g_c = 32.2 ft/s² (lb$_m$/lb$_f$)

u = velocity, ft/s

The results obtained from Eq. (6.1) can be used to calculate the quantity of air or gas flowing in a duct or passageway:

$$V_0 = 1100 A \sqrt{\frac{p_h}{\rho'}} \tag{6.3}$$

where V_0 = volumetric flow, ft³/min
A = duct cross-sectional area, ft²

Sec. 6.1   Air Circulating and Heating System   219

The flue gas density depends on the moisture content and the temperature. For 5% moisture and at 550°F, flue gas density is approximately 0.0405 lb/ft³. Combustion air density at 550°F is 0.0397 lb/ft³.

Stack draft. In early boilers, a chimney or stack provided all the draft needed by the furnace. It is generally not feasible to construct a stack tall enough to provide all the draft needed by a modern boiler. However, a stack must be provided to assure adequate dispersal of particulate matter and noxious gases in the plant effluent. The draft created by the stack is used to reduce the head which must be supplied by the fans.

The draft created by a chimney or stack results from the difference in pressures at the inlet of a vertical column of hot gas and ambient air. Since the hot flue gases are of a lower density then ambient air, the pressure at the base of a vertical column of flue gas is lower than that at the base of an air column of the same height. This creates a suction at the base of the stack which is the theoretical stack draft. The stack draft can be computed from the difference in gas density as

$$p_d = 0.1923 L \left(\rho'_a - \rho'_{FG} \right) \tag{6.4}$$

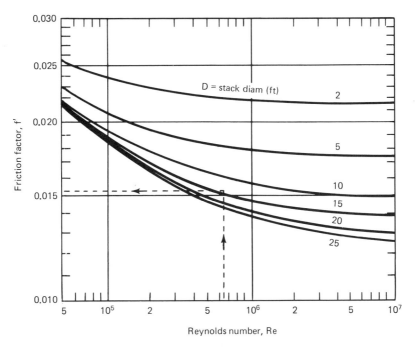

Figure 6.1 Friction factor, f', as related to the Reynolds number and stack diameter. (Reproduced with permission from *Steam, Its Generation and Use*, The Babcock & Wilcox Company, Lynchburg, Va., 1978.)

where p_d = theoretical stack draft, in. H$_2$O
   ρ'_{FG}, ρ'_a = density of the flue gas and ambient air, respectively, lb/ft^3
   L = stack height, above gas entrance, ft

In SI units, the static draft created by a stack height of L_{SI} meters is

$$p_d(\text{mm H}_2\text{O}) = L_{SI}(\rho_a - \rho_{FG}) \tag{6.5}$$

where L_{SI} is the stack height in meters.

To calculate the actual stack draft we must reduce the theoretical draft by the friction and exit losses. The actual draft, p_{ac}, in inches of water, is therefore given

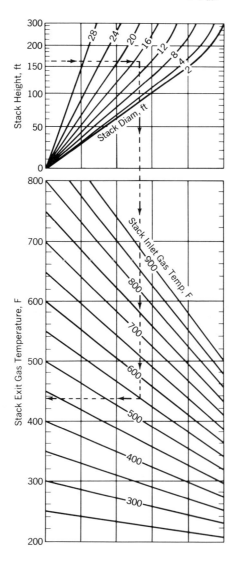

Figure 6.2 Approximate relationships between the stack exit gas temperature and stack dimensions. (Reproduced with permission from *Steam, Its Generation and Use*, The Babcock & Wilcox Company, Lynchburg, Va., 1978.)

by

$$p_{ac} = p_d - \frac{u^2}{2g_c}\left(1 + \frac{f'L}{D}\right)\frac{\rho'_G}{5.2} \tag{6.6}$$

where u = stack exit velocity, ft/s
g_c = gravitational conversion constant, 32.2 ft/s² (lb$_m$/lb$_f$)
D = stack inside diameter, ft
f' = frictional factor, dimensionless
ρ'_G = gas density, lb/ft³

The value of f' is obtained from Fig. 6.1 using a Reynolds number based on the stack diameter. The value of Re may be approximated by Re ≃ $20,000W/(T_gD)$ where W is total flow in lb/h, D is diameter in feet, and T_g is gas temperature in °R. For normal flue gas exit temperatures (~ 500°F), the induced theoretical stack draft would be approximately 0.64 in. H$_2$O per 100 ft of stack height.

The appropriate flue gas temperature to be used in draft calculations is the average of inlet and outlet temperatures. Some temperature losses occur in the stack due to heat transfer and air infiltration. An approximate stack exit temperature based on stack height, diameter, and flue gas inlet temperature may be estimated using the empirical data of Fig. 6.2.

Fans. As previously stated, most large power stations utilize both forced-draft and induced-draft fan systems to achieve a balanced draft system. The forced-draft fan is used to push air through the fuel-handling and combustion air supply systems. The induced-draft fan combines with the stack draft to compensate for the pressure drops in the exhaust gas systems. The volume output of the forced-draft fan must equal the total quantity of combustion air required plus any excess air supplied.

The forced-draft fan shaft power input, P_F, is computed from

$$P_F = \frac{(V_0)_F \Delta p}{\eta_F} = \frac{(V_0)_F \rho_{H_2O} H'_{SI}}{\eta_F} = \frac{(V_0)_F H'_{mm}}{\eta_F}$$

$$P_F(\text{kW}) = \frac{(V_0)_s H'_{mm}}{\eta_F}\left(\frac{\text{kg} \cdot \text{m}}{\text{h}}\right)\frac{\text{h}}{3600 \text{ s}}\left(\frac{1 \text{ kW}}{102 \text{ kg} \cdot \text{m/s}}\right)\frac{T}{293°\text{K}}$$

$$= \frac{0.93 \times 10^{-8}(V_0)_s T H'_{mm}}{\eta_F} \tag{6.7}$$

where $(V_0)_F$ = volumetric flow rate of combustion air at fan conditions, m³/h
$(V_0)_s$ = volumetric flow rate of combustion air at standard conditions, m³/h
T = absolute temperature of gas at fan, °K
H'_{SI} = total fan head, m H$_2$O
H'_{mm} = total fan head, mm H$_2$O
η_F = fan efficiency
ρ_{H_2O} = water density at room temperature (1000 kg/m³)

For example, if an axial forced-draft fan was rated at 500,000 m³/h at STP conditions, with a fan head of 600 mm H$_2$O and had an efficiency of 90%, the fan power would be

$$P_F(\text{kW}) = \frac{0.93 \times 10^{-8}(500{,}000)(273 + 20)(600)}{0.9}$$

$$= 898 \text{ kW}$$

Draft requirements. The major pressure losses in a furnace are due to the flow of the hot gases through the convective heat exchangers. For the most part, the hot gases flow over banks of tubes. The pressure drop for flow of gases across (normal to) a bank of tubes having 10 or more rows is given by the empirical equation

$$\Delta p = \frac{2f'(G'_{max})^2 N_0}{\rho_G}\left(\frac{\mu_w}{\mu_b}\right)^{0.4} \tag{6.8}$$

where Δp = pressure drop, N/m²
μ_b, μ_w = gas viscosity at bulk and wall conditions, respectively, Pa·s
G'_{max} = mass velocity at minimum flow area, kg/m²·s
N_0 = number of rows in the bank

The friction factor, f', is obtained from
Staggered banks:

$$f' = \left\{0.25 + \frac{0.118}{[(S_l - D_o)/D_o]^{1.08}}\right\} \text{Re}_{max}^{-0.16} \tag{6.9a}$$

In-line banks:

$$f' = \left\{0.044 + \frac{0.08 S_l/D_o}{[(S_t - D_o)/D_o]^{0.43 + 1.13 D_o S_l}}\right\} \text{Re}_{max}^{-0.45} \tag{6.9b}$$

where D_o = outside tube diameter
S_t = distance between tube centers in direction normal to flow
S_l = distance between tube centers in direction parallel to flow
Re_{max} = Reynolds number based on G'_{max}, tube O.D., and bulk gas properties = $D_o G'_{max}/\mu_b$

Pressure losses due to flow through the tubes of a recuperative air heater or passages of a regenerative air heater may be calculated in the same manner as conventional pipe friction. The Reynolds number for the passages of a regenerative heat exchanger is based on the equivalent diameter, D_e. Entrance and exit losses, generally amounting to about 1.5 velocity heads, should be added to the tube friction.

Air Heaters

Air heaters use the low-temperature heat from the exiting combustion gases to preheat the incoming combustion air. This also has the effect of lowering the temperature of the combustion gases that are discharged to the atmosphere. The demand for air heaters increased sharply with the use of pulverized coal since the use of preheated combustion air accelerates ignition and promotes rapid and complete combustion of pulverized coal. The heated combustion air is also useful for drying and transportating pulverized coal through the burner.

The use of preheated air results in a net increase in system thermal efficiency as the heat recovered from the combustion gases is effectively recycled. This energy is added to the thermal energy released in the combustion process and used to generate steam in the boiler or superheater. Estimates are that air heaters can effect a fuel

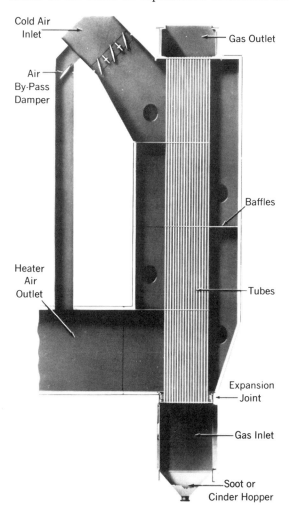

Figure 6.3 Tubular air heater arranged for counterflow of gas and air. An air bypass controls the metal temperature at the air inlet end. (Reproduced with permission from *Steam, Its Generation and Use*, The Babcock & Wilcox Company, Lynchburg, Va., 1978.)

savings of 5 to 10% if the combustion air temperature is increased from 150°C (300°F) to 200°C (390°F).

Air heaters are classified as either recuperative or regenerative. Both types of air heaters are essentially air-to-air heat exchangers. Recuperative air heaters utilize stationary tubes or plates for convective transfer of heat from the exiting hot combustion gases to the incoming combustion air. A tubular recuperative air heater is shown in Fig. 6.3. The incoming combustion air flows over the tube outer surface, while the hot combustion gas flows inside the tube surface.

There are two commonly used types of regenerative air heaters. In one type, the heat-transfer components are moved alternately through the gas and airstreams, thereby undergoing successive heating and cooling cycles. The heat is transferred by means of the thermal storage capacity of the members. The rotor is made up of corrugated elements with many small passages. Rotating slowly, the rotor transfers heat by moving alternately between the combustion gas stream and the incoming airstream. In the other type of regenerative air heater, the heat-transfer elements are fixed in position and the airflow and gas streams are alternated by rotating inlet and outlet connections. Because of the large surface area per unit volume in such a unit, regenerative air heaters are compact. A typical regenerative air heater is illustrated by Fig. 6.4.

Although corrosion is a problem for air heaters, the corrosion problem can be minimized by maintaining combustion gas temperatures above the dew point. This will prevent moisture formation that could combine with sulfur oxides to form acid. The dew-point temperature of the combustion gases can be as high as 150°C because of the combustion products. Since the moisture formation problem would normally occur only in cold weather, preventive measures can be taken to preheat the incoming cold combustion air when the ambient temperature is too low. This can be done by using a steam-heated preheater located ahead of the regular air heater.

Air heaters require regular maintenance and cleaning to prevent fly ash from clogging the air passages. Fixed-air blowers or steam blowers are used.

Air heater analysis. The heat removed from the combustion flue gas is given by

$$q = \dot{m}_{FG}(c_p)_{FG}(T_{FG1} - T_{FG2}) \tag{6.10}$$

where q = rate of heat removal, W
$\dot{m}_{FG}$ = flue gas flow rate, kg/s
$(c_p)_{FG}$ = specific heat of flue gas, J/kg·°C
T_{FG1} = temperature of the flue gas entering heater, °C
T_{FG2} = temperature of the flue gas leaving heater, °C

If we neglect small losses in the regenerative air heater, this heat is added to the incoming combustion air and increase the temperature by

$$T_{a1} - T_{a2} = \frac{q}{\dot{m}_a(c_p)_a} \tag{6.11}$$

Sec. 6.1   Air Circulating and Heating System

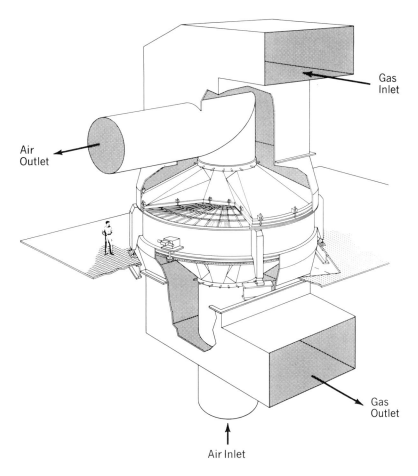

Figure 6.4 Counterflow regenerative air heater with stationary plates. (Reproduced with permission from *Steam, Its Generation and Use*, The Babcock & Wilcox Company, Lynchburg, Va., 1978.)

where T_{a1} = combustion air temperature entering preheater, °C
T_{a2} = combustion air temperature leaving preheater, °C
$\dot{m}_a$ = mass flow of combustion air, kg/s
$(c_p)_a$ = specific heat of air, J/kg·°C

The heat-transfer analysis depends on the design. In a recuperative heat exchanger used as an air heater, the overall rate of heat transfer is determined in the same manner as in the primary system heat exchangers. That is,

$$q = UA_h(\Delta T)_{\mathrm{LM}} F_c \qquad (6.12)$$

where F_c = log mean temperature correction factor (unity for simple counterflow; see any standard heat-transfer text for charts for other configurations)
q = rate of heat transfer required by Eq. (6.10) or (6.11), W
U = overall heat-transfer coefficient, W/m²·°C
A_h = heat-transfer surface area, m²
$(\Delta T)_{LM}$ = log mean temperature difference, °C

As in most heat exchangers, U must include two film coefficients, a tube wall resistance and two fouling factors. The fouling factors account for the buildup of deposits on both the hot flue gas side of the heater and the cool combustion air side of the heater. The general expression for U, based on the internal area, is

$$U_i = \frac{1}{\left(\dfrac{1}{h_i} + \dfrac{1}{(h_f)_i}\right) + r_i \dfrac{\ln(r_o/r_i)}{k} + \dfrac{A_i}{A_o}\left(\dfrac{1}{h_o} + \dfrac{1}{(h_f)_o}\right)} \qquad (6.13)$$

Since the tube wall resistance is generally small in comparison to the other resistances, this reduces to

$$U_i = \frac{1}{\left(\dfrac{1}{h_i} + \dfrac{1}{h_f}\right) + \dfrac{A_i}{A_o}\left(\dfrac{1}{h_o} + \dfrac{1}{h_f}\right)} \qquad (6.14)$$

where h_o = film coefficient on tube outside diameter, W/m²·°C
$(h_f)_o$ = fouling coefficient on tube outside dimater, W/m²·°C
h_i = film coefficient on tube inside diameter, W/m²·°C
$(h_f)_i$ = fouling coefficient on tube inside diameter, W/m²·°C
A_i, A_o = tube inside and outside areas, respectively, m²
r_i, r_o = inside and outside radii, respectively, m
k = tube wall thermal conductivity, W/m·°C

The inner film coefficient, h_i can be found from the Dittus–Boelter equation [see Eq. (5.25)]. The outer film coefficient depends on the flow arrangement through the tube bank. For flow across the tube banks, we may use Eq. (5.26) or (5.27). For air heat exchangers designed for flow along (parallel) to the tube axis, the Dittus–Boelter equation can again be used to evaluate the outside surface film coefficient, h_o. In this case, a unit cell is defined around each tube, as shown in Fig. 6.5. The equivalent diameter, D_e, for the unit cell is given by

$$D_e = \frac{4A_f}{\text{perim.}} \qquad (6.15)$$

where A_f = free-flow area, m²
  $= (S_l S_t) - (\pi/4)D_o^2$
perim. = wetted perimeter = πD_o, m

The value obtained for D_e is used in the Dittus–Boelter equation to calculate h_o.

Sec. 6.1   Air Circulating and Heating System

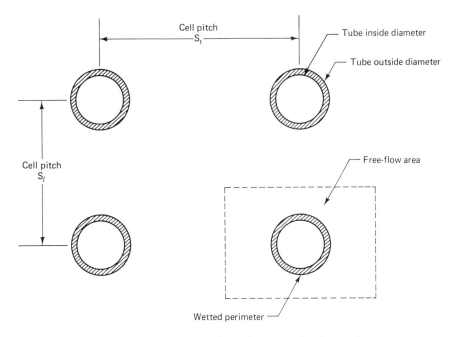

Figure 6.5 Heat-transfer nomenclature for power plant heat exchangers.

The fouling resistance cannot be calculated on the bases of theory, but various tables are available to predict these resistance terms based on practical operating experience. The fouling factor is obtained experimentally from

$$\frac{1}{h_f} = \frac{1}{U_{\text{dirty}}} - \frac{1}{U_{\text{clean}}} \tag{6.16}$$

Table 6.1 provides some typical values.

The full analysis of a rotating regenerative heat exchanger is somewhat complex, but an approximate simplified analysis is readily made. Assume that the rotating unit moves between air at an average temperature, T_a, and flue gas at an average temperature, T_G. Also assume that due to the high conductivity of the metal, the temperature of the metal may be assumed uniform and that at the end of the heating cycle it reaches T_{MH} while at the end of the cooling cycle it is T_{MC}. If the

TABLE 6.1 FOULING FACTORS

	$m^2 \cdot °C/W$
Steam	0.00009
Industrial air	0.0004
Flue gas	0.0008

unit completes one revolution in time t_r, the heat-transfer rate, q, is given by

$$q = (T_{MH} - T_{MC}) \frac{m_r A_h}{t_r} (c_p)_M \qquad (6.17)$$

where m_r = mass of metal per unit heat transfer area
A_h = heat-transfer area
$(c_p)_M$ = specific heat of metal, J/kg·°C

and other quantities have their previous meaning.

We may determine the values of T_{MH} and T_{MC} for a given set of conditions by solving the differential equations describing the change in temperature during the heating and cooling cycle. While a segment is in the heating cycle, the rate at which the temperature of the metal rises is given by

$$h_h(T_G - T_M) = m_r (c_p)_M \frac{dT_M}{dt} \qquad (6.18)$$

where t = time
T_M = metal temperature
h_h = heat-transfer coefficient between gas and metal in heating cycle

If we let $\Theta = T_G - T_M$, our differential equation becomes

$$h_h \Theta = \frac{d\Theta}{dt} m_r (c_p)_M \qquad (6.19)$$

The solution of this differential equation may be written as

$$\Theta = k' \exp\left[\frac{-h_h t}{m_r (c_p)_M}\right] \qquad (6.20)$$

where k' is a constant. This constant is evaluated by noting that when $t = 0$, $\Theta = T_G - T_{MC}$. We then have

$$\frac{T_G - T_M}{T_G - T_{MC}} = \exp\left[\frac{h_h t}{m_r (c_p)_M}\right] \qquad (6.21)$$

Since $T_M = T_{MH}$ when $t = t_r/2$, we have finally

$$\frac{T_G - T_{MH}}{T_G - T_{MC}} = \exp\left[\frac{-h_h t_r}{2 m_r (c_p)_M}\right] = K_2 \qquad (6.22)$$

We proceed similarly for the cooling cycle and write the basic differential equation as

$$h_c(T_M - T_a) = -\frac{dT_m}{dt} m_r (c_p)_M \qquad (6.23)$$

where h_c is the heat-transfer coefficient between air and metal in cooling cycle.

By using the same approach as previously, we finally obtain

$$\frac{T_{MC} - T_a}{T_{MH} - T_a} = \exp\left[\frac{-h_c t_r}{2 m_r (c_p)_M}\right] = K_1 \qquad (6.24)$$

Sec. 6.2   Water Treatment Systems   229

We may solve Eqs. (6.22) and (6.24) simultaneously for T_{MH} to obtain

$$T_{\text{MH}} = \frac{T_G(1 - K_2) + T_a(K_2 - K_2 K_1)}{1 - K_2 K_1} \qquad (6.25)$$

By using this expression for T_{MH} in the solution obtained in the cooling cycle, we can obtain

$$T_{\text{MH}} - T_{\text{MC}} = \frac{T_G(1 - K_2) + T_a(K_2 - K_2 K_1)}{1 - K_2 K_1}(1 - K_1) - T_a(1 - K_1) \qquad (6.26)$$

We can now evaluate the total heat-transfer area required provided that t_r, m_r, h_c, and h_h are known. Values of h_c and h_h are best obtained by empirical measurements for the configuration actually used. If these are unavailable, estimates of the h values may be made using the Dittus–Boelter equation with D_e based on the perimeter and flow area of the average flow channel.

6.2 WATER TREATMENT SYSTEMS

The Need for Water Treatment

Water and steam purity are a major factor ensuring uninterrupted power plant operation. This strong emphasis on water quality is the result of the shift toward higher-pressure and nonrecirculating systems. Feedwater quality or purity becomes extremely important since, with once-through boiler systems, all impurities dissolved in the feedwater are carried by the steam to the turbine. Also, the trend toward high-temperature, high-pressure boiler operation increases the solubility of solids and gases in the water–steam flow.

All natural water (called *raw water*) contains impurities to some degree. These impurities originate from contact with the earth or the air. The impurities can be broadly classified as suspended or dissolved organic and inorganic matter and dissolved gases. The dissolved solids include such minerals as silica, iron, calcium, magnesium, and sodium. When significant quantities of iron, calcium, and magnesium are present in water, the water is said to be *hard*. The dissolved gases are mainly oxygen, nitrogen, and carbon dioxide. The concentration of these impurities is expressed in terms of weight of the impurity per million parts weight of water, or ppm. Gases and trace impurities are often quoted in terms of parts per billion, ppb.

The metallic elements that are present in raw water occur in combination with bicarbonate, carbonate, sulfate, and chloride radicals. In a water solution, these compounds separate into ions. The metal ions, called *cations*, carry a positive charge, while the radicals, called *anions*, carry a negative charge. Scale formation is the result of precipitation of the calcium and magnesium compounds in the water. These precipitates adhere to the boiler and pipe internal surfaces and result in the buildup of a hard layer of deposits (scale) on these surfaces. Scale formation on heat-transfer surfaces retards the flow of heat. This can lead to degradation of power plant performance and overheating of boiler tubes and other metal components.

Sludge, or solid particles suspended in the fluid, can deposit in pipes, pipe bends, and valves. These deposits restrict the water flow and can also insulate in the same manner as the scale deposits, leading to similar problems.

Corrosion, due to acidic conditions or dissolved oxygen, can weaken a boiler due to the loss of metal. Corrosion usually occurs as small holes or pits in localized areas. If allowed to proceed unchecked, these holes deepen and can penetrate through the total metal thickness. Corrosion rates and solid deposits are related. The chemical reactions taking place produce an intergranular attack of the metal that may lead to embrittlement and failure. High concentrations of chlorides and oxygen are known to initiate stress corrosion cracking in many metals, particularly 300 series (18% Cr, 8% Ni) stainless steel.

Water Treatment Terminology

Steam that is condensed in the condenser and returned to the boiler is called *condensate*. *Makeup water* is the name given to the water pumped into the system to replace boiler water lost through blowdown, leakage from the system or discharge.

In a once-through system, there are no steam drums to allow the separation of impurities from the steam. Purification of the water for this type of boiler is accomplished by passing all or most of the condensate through demineralizers. This process is referred to as *condensate polishing*.

Raw Water Treatment

The treatment of raw water for makeup and boiler feedwater involves one or more of the following operations:

1. Removal of suspended solids
2. Chemical treatment for removal of hardness
3. Cation exchange for removal of hardness
4. Demineralization for removal of dissolved solids

Initially, the suspended solids in the raw water are removed by settling, decantation, or filtering. Small particles are removed by coagulation using floc-forming chemicals such as alum or ferrous sulfate to trap the particles in flow. The floc is removed by filtration.

In one approach, the raw water is next treated for removal of hardness. Several exchange processes are available for removal of the scale-forming impurities in the water. Some typical processes used are the lime-soda process, sodium cycle softening, hot lime zeolite softening, and hot lime zeolite–split stream softening.

Sec. 6.2 Water Treatment Systems

The objective of these processes is to replace the iron, calcium, and magnesium ("hard-water") ions, which produce insoluble scale, by sodium ions.

The lime-soda process is used to precipate out calcium and magnesium in a form that can be removed as a sludge. Residual hardness is then removed by *cation exchange*. Here, naturally occurring minerals, such as sodium aluminum silicate, or synthetic resins, such as polystyrenes or phenolic type-materials, exchange sodium ions for calcium and magnesium ions. That is,

$$2[R]Na + Ca^{2+} \rightarrow [R]_2Ca + 2Na^+$$

where [R] represents the polymer structure of the resin. The cation-exchange process removes the calcium and magnesium by passing the water over a bed of granulated zeolite (a natural mineral) particles which retain Ca and Mg. A typical raw water treatment process combines chemical treatment (e.g., hot lime) with cation exchange.

At steam drum pressures over 1000 psi, demineralization or distillation of the makeup water is necessary. In the distillation process, most of the water is evaporated and then condensed. The condensate is nearly pure water and most of the solids remain behind in the still bottoms. Demineralization is the nearly complete removal of dissolved solids by means of ion exchange. There are several types of synthetic organic resins that are capable of selectively removing undesirable cations or anions from water solutions by their exchange for hydrogen or hydroxyl ions. Ion exchange units can be operated separately or in mixed-resin beds. The water leaving these units is almost entirely free of mineral solutes. The ion-exchange reactions may be written

$$[R']H + Na^+ \rightarrow [R']Na + H^+$$
$$[R'']OH + Cl' \rightarrow [R'']Cl + OH^- \quad (6.27)$$
$$H^+ + OH^- \rightarrow H_2O$$

where [R'] and [R''] indicate the polymer structure of the ion-exchange resins.

Condensate Treatment

Protection of turbines operating in high-pressure systems (> 2000 psi) requires water purity levels of 1 ppb or better. In recirculating-boiler-water systems, impurities build up from erosion in the pipes and various other components. Ion exchange of condensate water to remove these impurities is desirable. In once-through boiler systems operating at high pressures, full-flow ion exchange is required (condensate polishing). The condensate polishing system is a bed of resin beds which demineralize the condensate in the same manner as the mixed-resin beds used for raw water demineralization. In addition, the beds also act as filters for removal of suspended material and corrosion products (often called "crud") in the condensate water.

For high-purity demineralization, both cation and anion resin beads are required in a deep-bed demineralizer. Table 6.2 lists the types of resin beads that are

TABLE 6.2 CHARACTERISTICS AND DESIGNATIONS OF RESINS AVAILABLE COMMERCIALLY (TYPICAL)

Resin type	Skeletal structure Chem.	Skeletal structure Phys.	Regeneration Chemical reagent	Regeneration Quantity[a] (lb/ft^3)	Capacity[a] (kg/ft^3)	Manufacturer's designation[b] Amberlite	Manufacturer's designation[b] Dowex	Manufacturer's designation[b] Duolite	Manufacturer's designation[b] Ionac
Cation exchange									
Strongly acidic									
Sodium cycle (water softening)	Sty[c]	Gel	NaCl	6	22	IR-120	50	C-20	C-249
	Sty	Mac[d]	NaCl	6	20	200	CCR-2	C-433	CC
Hydrogen cycle	Sty	Gel	H$_2$SO$_4$[e]	5	12	IR-120	50	C-20	C-249
	Sty	Mac	H$_2$SO$_4$[e]	5	11	200	MSC-1	ES-26	CFP-110
Weakly acidic	Ac[f]	Gel	H$_2$SO$_4$[e]	4–8	30–60	IRC-84	CCR-2	C-433	CC
Anion exchange									
Strongly basic (hydroxide cycle)									
Type I	Sty	Gel	NaOH	4	11	IRA-402	SBR-P	A-101D	ASSB-1-P
	Sty	Mac	NaOH	4	10	IRA-900	MSA-1	A-161	A-641
	Ac	Gel	NaOH	4	12	IRA-458	—	—	—
Type II	Sty	Gel	NaOH	4	21	IRA-410	SAR	A-104	ASB-2
	Sty	Mac	NaOH	4	18	IRA-910	—	—	A-651
Weakly basic	Sty	Mac	NaOH[g]	3.5	20–25	IRA-94	MWA-1	ES-368	AFP-329
	Ac	Gel	NaOH[g]	4	28	IRA-68	—	—	—
	Ep[h]	—	NaOH[g]	4	28	—	WGR	—	—

[a] Average conditions.
[b] Manufacturers are Rohm and Haas (Amberlite), Dow Chemical USA (Dowex), Diamond Shamrock Corp. (Duolite), and Ionac Chemical, Division of Sybron Corp. (Ionac). Note that all provide strongly basic and acidic resins in special grades for condensate-polishing service, designed to minimize pressure drop and physical degradation at high flow rates (50 gpm/ft^2) and to provide the high degree of polishing required for utility use.
[c] Styrene/divinylbenzene.
[d] Macroporous (or macroreticular).
[e] Also HCl, in equivalent quantities (approximately 25% less by weight).
[f] Acrylic.
[g] Also NH$_4$OH, in equivalent quantities (approximately 15% less by weight), and Na$_2$CO$_3$ (33% more).
[h] Epoxy.

Source: Reprinted with permission from *Power,* Sept. 1980; copyright 1980 McGraw-Hill, Inc., New York.

TABLE 6.3 CHEMICAL REACTIONS IN ION-EXCHANGE TECHNOLOGY[a]

(1)	$[R]H + NaCl \rightleftharpoons [R]Na + HCl$
(2)	$2[R]Cl + Na_2SO_4 \rightleftharpoons [R]_2SO_4 + 2NaCl$
(3)	$CaSO_4 + 2[RSO_3]Na \rightleftharpoons [RSO_3]_2Ca + Na_2SO_4$
(4)	$[RSO_3]_2Ca + 2NaCl \rightleftharpoons 2[RSO_3]Na + CaCl_2$
(5)	$Ca(HCO_3)_2 \rightleftharpoons CaCO_3 + CO_2 + H_2O$
(6)	$Ca(HCO_3)_2 + 2[RSO_3]Na \rightleftharpoons [RSO_3]_2Ca + 2NaHCO_3$
(7)	$CaSO_4 + 2[RSO_3]H \rightleftharpoons [RSO_3]_2Ca + H_2SO_4$
(8)	$2NaHCO_3 + H_2SO_4 \rightleftharpoons Na_2SO_4 + 2CO_2 + 2H_2O$
(9)	$[RSO_2]H + NaCl \rightleftharpoons [RSO_3]Na + HCl$
(10)	$[R]OH + HCl \rightleftharpoons [R]Cl + H_2O$

[a] Bracketed terms represent polymer structure of the resin and the functional group (i.e., R or RSO_3), and the associated chemical represents the interchangeable ion.
Source: Reprinted with permission from *Power*, Sept. 1980; copyright 1980 McGraw-Hill, Inc., New York.

frequently used. These beads are housed in the bottom half of a closed cylindrical vessel. Full-flow condensate polishing in mixed resin beds can achieve purity levels down to detectable limits, 0.1 ppb. Condensate polishing actually removes only 1 or 2 ppm from the condensate water, since condensate water already has the purity of distilled water.

The chemical ion-exchange reactions that can be utilized by selecting various types of resin beads are shown in Table 6.3. Recent developments in condensate polishing systems include the use of powdered resins in thin layers, $\frac{1}{8}$ to $\frac{1}{4}$ in. thick, equivalent to 3 to 4 ft of mixed-bed resins. The condensate polishing system must be carefully designed to remove the impurities unique to each power plant system.

Condensate polishing systems require cleaning and regeneration on a regular maintenance cycle. The regeneration of the bed includes backwash to remove solids trapped in the resin bed, regeneration with acid and caustic, and a rinsing operation.

Feedwater Treatment

Boiler feedwater can be condensate, treated makeup water, or a mixture of both. The components in the feedwater train are constructed of steel, metal alloys, and other materials susceptible to corrosion in an acidic solution. Thus the makeup water must be at the correct pH level and free of dissolved gases. A pH level of 8.5 to 9.5 is usually recommended for high-pressure boilers. pH control is generally achieved by the addition of soluble organic chemicals such as neutralizing amines (ammonia, cyclohenylamine, morpholine, or hydrazine) or filming amines (octadecylamine acetate).

Dissolved oxygen is the biggest factor in the corrosion of steel surfaces in contact with water. Hydrazine, which reacts with oxygen to form ammonia and water, is used to effect the removal of oxygen from the makeup water.

To assure that the low oxygen levels required are maintained during operation, it is necessary to remove dissolved gases continuously from the boiler feedwater. As indicated previously, this is done in the deaerating heater. Such heaters are usually large steel shells in which steam and feedwater are intermixed. Feedwater is supplied to the top of the heater and is allowed to spill over a series of trays. Turbine extraction steam rises from the bottom of the heater and heats the feedwater to the temperature corresponding to the saturation pressure of the deaerator. Since the solubility of a permanent gas is very low in water at its boiling point, any dissolved air is released to the vapor. The released gas plus any exiting steam pass through a vent condenser where the steam is condensed and the remaining permanent gases are discharged. The condensed steam is added to feedwater entering the deaerator. Multistage deaerating heaters can reduce oxygen levels to 0.007 ppm or less.

6.3 COOLING TOWERS

General Classifications

A modern fossil-fueled steam turbine power station has a maximum thermal efficiency of approximately 40%. The remainder of the heat generated by the combustion process must be eliminated as waste heat. Although some of this heat is rejected with the gases leaving the stack, the bulk of the waste heat is rejected to the coolant in the condenser.

It was previously noted that many utilities were formerly able to utilize once-through condenser cooling systems in which ambient water, from a river, lake, or ocean, was used for cooling the condenser. The condenser cooling water was returned to the river, lake, or ocean at a temperature a few degrees higher than ambient. Due to new, more stringent environmental regulations, water conservation, or site topology, these once-through systems are being replaced with recirculating systems in which the condenser water is cooled by passage through a cooling tower.

Cooling towers for power plants are classified in accordance with the method of heat transfer from the water to the air and the method utilized to create air movement through the tower. A cooling tower that exchanges heat by the direct contact of air and water is classified as a *wet tower*. The primary mechanism for the transfer of heat is evaporation. In a *dry tower*, the water and air are separated by a conductive surface in the tower and heat is removed by a conduction process. A dry tower is essentially a fluid-to-air heat exchanger. If both types of surfaces are used in a single tower design, it is classified as a *wet/dry tower*.

Air circulation through the tower results, at least in part, from heat transferred to the air, which creates a chimney effect. Towers that rely solely on this bouyancy effect are classified as *natural-draft towers*. When large fans are used to increase the airflow through the tower, the tower is classified as a *mechanical draft tower*. A mechanical draft tower can be either "forced draft" or "induced draft," depending on the fan location.

Since the cooling effect is mainly due to evaporation, part of the water circulating through the system is lost by vaporization, some by drift and some by blowdown. Evaporation loss is approximately 1% for each 10°F of cooling. Drift loss is the loss of fine droplets of water blown out with the air leaving the tower. Drift loss can be held to 0.2% of the circulating water. Finally, blowdown is required to eliminate crud and solids that build up in the system water.

In a typical 700-MWe power plant, a large cooling tower would circulate 350×10^6 gal of water per day through the tower. The makeup water requirement for this plant would be in the order of 6.7×10^6 gal per day.

Flow Patterns and Fill Material

Wet cooling towers are further classified by the relative flow pattern of water and air. In a *cross-flow tower*, air moves across the fill section of the tower. In a *counterflow tower*, the air movement is opposite in direction to the water flow. The various types of cooling towers are depicted in Fig. 6.6. A fill material of baffling is placed within the tower casing to provide large water surface areas for efficient heat transfer. The *fill* is the material that breaks the water up into a fine mist. Two kinds of materials are used: *splash bars*, made of wood, metal transite, or plastic, or *fill pack*, made of thin sections of a material called *cellular fill*. Cellular fill is usually made from polyvinyl chloride (PVC). The splash bar packing breaks the water flow into droplets that fall through the airstream, while film packing turns the droplets into thin films with a larger surface area for more efficient cooling.

Most of the cooling towers now in use are wet cooling towers, with approximately 25% using the natural-draft concept. The latter type of tower is more popular in the eastern United States. All natural-draft towers use the wet design concept. The counterflow design now dominates the natural-draft market.

About 95% of the mechanical draft cooling towers use the cross-flow design concept. This is due to the low-air-side pressure loss through the fill material, resulting in lower fan power consumption.

Cooling Tower Subsystems

The cooling tower subsystems consist of air-moving equipment, a drift elimination system, fill packing or baffling, a water distribution system, and associated structural system.

The air-moving equipment in a mechanical draft cooling tower includes the fan, electric motor, gas engine or steam turbine, and associated belts or gears. The drift elimination system consists of a series of baffles designed to remove entrained water droplets by impingement on the eliminator surfaces. The fill or splash bar system was described previously. The water distribution system is the mechanical method used to pass the hot water over the fill area. Low-pressure spray systems are necessary in counterflow designs, while gravity is used in cross-flow towers. The cold

TYPES OF COOLING TOWERS

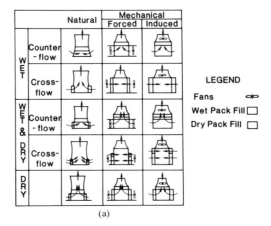

(a)

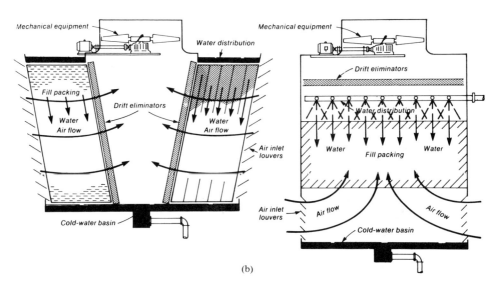

(b)

Figure 6.6 (a) Types of cooling towers; (b) counterflow and crossflow cooling towers. [(a) reprinted with the permission of Sargent & Lundy, Chicago; (b) reprinted with permission from *Power*; copyright 1979 McGraw-Hill, Inc., New York.]

water reaching the bottom of the tower is returned to the condenser system with a sump pump.

The structural members support all the tower live and dead loads. They consist of columns, horizontal ties, diagonals, joists, and beams. Cooling towers are often divided into *cells*. A cell would consist of one complete unit with its own distribution system, and mechanical and structural partitions.

Wet Cooling Tower Performance

The performance of a wet cooling tower is dependent on four major factors:

1. The amount of moisture the air can retain
2. The amount of heat to be dissipated
3. The contact time between water and air
4. The water-to-air weight ratio

Evaporative cooling accounts for 85% of the heat dissipated from a wet cooling tower. The remainder of the heat loss is through conduction and convective losses. The heat dissipation required depends on the generating capacity of the power plant. Tower performance can be varied by increasing or decreasing the water flow across the tower fill sections. As the water flow across a fill section increases, the weight ratio of water to air increases. This results in a higher exit water temperature from the tower. This same result can be achieved by holding the tower water flow constant and decreasing the airflow.

Because there are many variables to contend with in evaluating the thermal performance of a cooling tower, testing is an important part of cooling tower acceptance. The most accepted test is the one devised by the Cooling Tower Institute. The test procedure is outlined in detail in the Cooling Tower Institute Acceptance Test Code (Bulletin ATC 105 or ASME PTC 1958). This test code defines a uniform method of evaluating the tower's water cooling capability.

In northern climates, cooling towers, regardless of type, can have freezing problems. In most towers there is a cold water basin which serves as a collecting point for all the cold water leaving the tower. This basin is subject to freezing, especially if exposed to the wind. Warm-water recirculation can be used to prevent basin freezing. This warm water is obtained by diverting some of the warm inlet water flow from the tower to the basin.

Another freezing problem occurs with exterior ice forming on the air inlets of the tower. On a counterflow tower, this icing occurs in the tower's structural supports and is called *perimeter ice*. In a cross-flow tower, ice forms on inlet louvers. These icing conditions can be controlled by the use of perimeter de-icing lines. These lines divert some of the tower's hot water to the air inlets.

Ice buildup inside the tower can also occur. This interior ice increases the static loading on the structure and can cause damage. Ice control can be achieved by diverting additional hot-water flow to certain areas of the tower. This increased hot-water flow through the active sections raises the temperature in these sections and reduces ice formation.

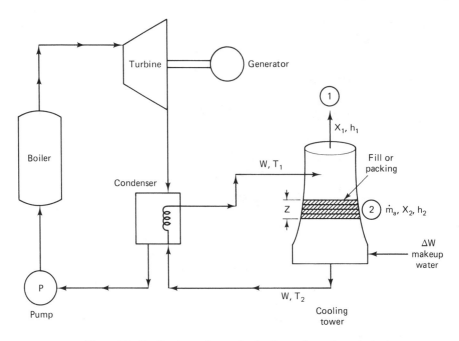

Figure 6.7 Cooling tower flow paths for thermodynamic analysis.

Cooling Tower Performance Criteria

Since most of the cooling towers in operation today are wet cooling towers, we shall confine our consideration to this design. We have observed previously that most of the temperature reduction is due to energy given up through evaporation of a small portion of the water. The cooling effect is therefore dependent on the initial relative humidity of the cooling air.

A typical heat balance for a 1000-MWe fossil-fueled steam power station is shown in Fig. 6.7. Since the cooling tower must dissipate the heat removed from the condenser unit, the quantity of heat removed per megawatt generated is dependent on the thermal efficiency of the particular plant. In modern fossil plants, where an overall thermal efficiency of approximately 40% is achieved, about 45% of the heat input to the furnace/boiler is discharged to the condenser cooling water. The remaining 15% is lost in other plant systems (e.g., discharge of hot gases to stack).

Cooling towers usually cool the water to within 3 to 12°C of the wet-bulb temperature.† We define the *wet-bulb approach*, ΔT_{WB}, as

$$\Delta T_{WB} = T_{CW} - T_{WB} \qquad (6.28)$$

---

†The wet-bulb temperature may be approximately defined as the equilibrium temperature an evaporating film of water will reach in a high-velocity stream of air at a given humidity and bulk temperature.

where ΔT_{WB} = wet-bulb approach, °C
T_{CW} = cooling-water temperature, °C
T_{WB} = wet-bulb temperature, °C

The wet-bulb approach is one of the terms used by power plant engineers to define the performance of a cooling tower. In practice, most towers are rated for wet-bulb approach temperature differences (at the water exit) between 6°C (11°F) and 8°C (14.5°F). The design value is seldom below 3°C.

A second parameter used to define cooling tower performance is the cooling range (ΔT_{CR}). The *cooling range* is defined as the temperature difference between the hot water entering (at T_1) and the cold water leaving (at T_2). That is,

$$\Delta T_{CR} = T_1 - T_2 \qquad (6.29)$$

Values of 6 to 10°C are common for ΔT_{CR} in fossil steam power plants. The relationships among T_1, T_2, ΔT_{CR}, and ΔT_{WB} are shown in Fig. 6.8. Note that the exit water temperature may actually be below the inlet water temperature. This is possible since vaporization, and hence cooling, can take place as long as the partial pressure of the water vapor at the cooling tower surface exceeds the partial pressure of water vapor in the bulk airstream.

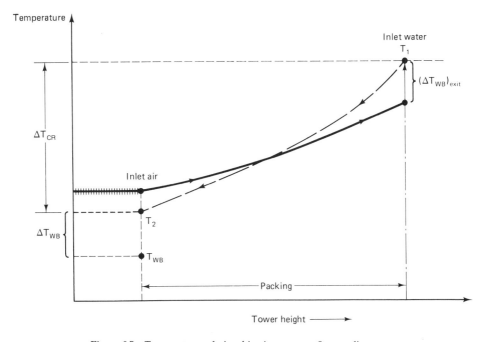

Figure 6.8 Temperature relationships in a counterflow cooling tower.

The term *cooling efficiency*, η_c, is used to define the ratio of actual cooling to the maximum possible cooling. This can be evaluated as

$$\eta_c = \frac{T_1 - T_2}{T_1 - T_{WB}} \tag{6.30}$$

Cooling-water losses can be a problem in some regions of the country. The makeup water requirement, shown as Δw in Fig. 6.7 is the sum of these three components of the losses:

$$\Delta w = \Delta w_e + \Delta w_d + \Delta w_b \tag{6.31}$$

where Δw_e = evaporation loss
Δw_d = drift loss
Δw_b = blowdown loss

The evaporation loss and the mass flow of air required for a cooling tower can be calculated from a heat and mass balance taken over the cooling tower. By following Fig. 6.7, we obtain for water the mass balance through the tower:

$$\Delta w_e = (X_1 - X_2)\dot{m}_a \tag{6.32}$$

where $\dot{m}_a$ = flow of air through the tower, kg/h
X_1 = water content of the air at point 1, kg H_2O/kg dry air
X_2 = water content of the air at point 2, kg H_2O/kg dry air

Alternatively, Eq. (6.32) can be written in terms of airflow requirements based on evaporation loss:

$$\dot{m}_a = \frac{\Delta w_e}{X_1 - X_2} \tag{6.33}$$

A heat balance around the tower can be used to obtain

$$\dot{m}_a(h_1 - h_2) = wc_p T_1 - (w - \Delta w_e)c_p T_2 \tag{6.34}$$

where h_1 = specific enthalpy of the air at 1, kJ/kg
h_2 = specific enthalpy of the air at 2, kJ/kg
c_p = specific heat of the water, kJ/kg·°C
w = water flow rate, kg/h

When tower measurements of air conditions and water temperature and inlet flow are available, we may combine Eqs. (6.33) and (6.34). After rearrangement we get an expression that can be utilized to calculate the evaporative water loss:

$$\Delta w_e = \frac{w(T_1 - T_2)}{\dfrac{h_1 - h_2}{c_p(X_1 - X_2)} - T_2} \tag{6.35}$$

Sec. 6.3   Cooling Towers

Under standard atmospheric conditions, the evaporation loss is approximately 1% for every 6°C cooling range. For a condenser with a cooling range of 8°C, the evaporation loss would be around 1.33% of the water cooled by the tower.

Based on the results of the evaporative loss estimates, the required airflow can be calculated from Eq. (6.33). Practical air-to-water flow ratios are 0.6 to 0.8 kg of air per kilogram of cooling water.

The drift loss Δw_d is a function of tower design and local weather parameters (such as wind). For natural-draft towers, $\Delta w_d \approx 0.3$ to 0.4%. Mechanical-draft towers generally experience a lower drift loss, 2 to 3%. Drift eliminators can reduce this loss to as low as 0.05% in either natural-draft or mechanical-draft towers.

The blowdown loss, Δw_b, is dependent on the concentration limit for dissolved salts in the cooling tower water. Blowdown requirements are given in terms of CL, the multiple of the concentration of the original content of dissolved solids which is allowed. A concentration level of CL = 2 would imply that the tower water contains double the concentration of dissolved solids in the makeup water added to tower coolant. The rate of blowdown necessary to maintain a specified concentration level, CL, is found by writing a solids mass balance:

solids added in makeup = solids lost in blowdown

or

$$\Delta w_e(\%) + \Delta w_d(\%) + \Delta w_b(\%) = (\text{CL})(\Delta w_d + \Delta w_b) \qquad (6.36)$$

Solution of this equation for the blowdown loss gives

$$\Delta w_b = \frac{\Delta w_e}{\text{CL} - 1} - \Delta w_d \qquad (6.37)$$

The blowdown loss is strongly dependent on the allowable concentration level in the system. This, in turn, depends on the quality of the water available for makeup and on the materials and design of the tower.

The total makeup water, ΔW, required for a cooling tower generally ranges from 1.25 to 4.5% of the cooling-water flow. For a large fossil-fired steam plant (800 MWe), the cooling-water flow is on the order of 112,000 m³/h. The cooling range is 7 to 8°C. If we assume a typical evaporation loss of 1.25%, a drift loss of 0.05%, and a blowdown loss of 0.80%, the total quantity of makeup water required for the cooling tower would be 2352 m³/h.

Analyses of Cooling Tower Behavior

The design of a cooling tower begins with a knowledge of the total heat dissipation (cooling load) and cooling-water flow rate. The cooling water flow rate is determined by an energy balance between the pressure loss across the condenser and the temperature drop the cooling tower is to provide. With the cooling load, Q_r, and water flow rate, W, known, we obtain the cooling range, ΔT_{CR}, as

$$\Delta T_{\mathrm{CR}} = \frac{Q_r}{w c_p} \tag{6.38}$$

where c_p is the specific heat of the condenser cooling water.

The temperature of the cooling water entering the condenser is chosen so that the wet-bulb approach under adverse summer conditions is within the acceptable range. If a utility has its peak load in the winter, the design wet-bulb temperature is usually 4 to 5°C below the highest recorded wet-bulb temperature. If the utility is faced with a summer peak, the highest observed wet-bulb temperature is likely to be used as the design basis.

In most modern cooling towers the air and water are in counterflow and once the wet-bulb temperature of the entering water is established and ΔT_{WB} is selected, the temperature of the water leaving the cooling tower, T_2, is obtained from

$$T_2 = T_{\mathrm{WB}} + \Delta T_{\mathrm{WB}} \tag{6.39}$$

and further,

$$T_1 = T_2 + \Delta T_{\mathrm{CR}} \tag{6.40}$$

From metereological data, we may determine the bulk air temperature associated with the design wet-bulb temperature of the inlet air. With the wet- and dry-bulb temperatures known, the percent humidity is obtained from the intersection of these values on a psychrometric chart (see Fig. 6.9). We may then calculate h_2, the enthalpy of the entering air, from

$$h_2 = h_{\mathrm{dry}} + \left(h_{\mathrm{sat}} - h_{\mathrm{dry}} \right) \times \frac{\% \text{ humidity}}{100} \tag{6.41}$$

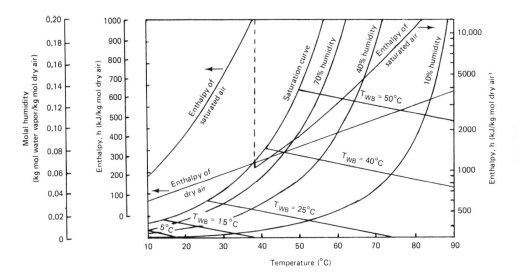

Figure 6.9 Simplified psychrometric chart.

Sec. 6.3 Cooling Towers

where h_{dry} = specific enthalpy of dry air at inlet temperature
h_{sat} = specific enthalpy of saturated air at inlet temperature

By selecting a reasonable wet-bulb approach, $(\Delta T_{WB})_{exit\,air}$ for the exiting air, we calculate T_{WB1}, the wet-bulb temperature of the exiting air:

$$T_{WB1} = T_1 - (\Delta T_{WB})_{exit\,air} \tag{6.42}$$

We then obtain h_1, the enthalpy of the exit air, by noting that the enthalpy remains essentially constant along a line of constant wet-bulb temperature. We therefore take h_1 as the enthalpy of saturated air at T_{WB1} and read this value from the psychometric chart.† With this we find the needed airflow rate from Eq. (6.34).

We are now in a position to determine the needed packing height. In a very simplified picture of the transfer processes, we may imagine that the water surfaces are surrounded by a very thin film of air which is saturated with water vapor (100% humidity) and is at the same temperature as the liquid water. The primary resistance to energy transfer may be pictured as residing in this film. The driving force for energy transfer from the film to the bulk gas stream may be expressed in terms of the enthalpy difference between the film and the bulk gas stream. From an energy balance the rate of energy transfer, dq, in a differential element of length dz, is

$$dq = dh\,\dot{m}_a \tag{6.43}$$

where $\dot{m}_a$ = flow of air through tower, kg/h
h = specific enthalpy of air at bulk gas temperature, kJ/kg

The energy transfer may also be described in terms of a mass-transfer coefficient (K) and driving force ($h^* - h$).

$$dq = (h^* - h)Ka\,dz\,A_t$$

where K = overall mass-transfer coefficient, kg water/m²·h (kg H$_2$O/kg air)
a = effective transfer area per unit tower volume, m²/m³
A_t = total cross-sectional area of tower, m²
h^* = specific enthalpy of saturated air at liquid temperature, kJ/kg

After rearrangement, we have

$$\frac{A_t}{\dot{m}_a} Ka\,dz = \frac{dh}{h^* - h} \tag{6.44}$$

which becomes in integration

$$z = \frac{\dot{m}_a}{KaA_t} \int_{h_2}^{h_1} \frac{dh}{h^* - h} \tag{6.45}$$

By defining an appropriate mean driving force, we may rewrite the foregoing as

$$z = \frac{\dot{m}_a}{KaA_t} \frac{h_1 - h_2}{(h^* - h)_{mean}} \tag{6.46}$$

---

†Note that Fig. 6.9 provides the molal enthalpy (kJ/kg mol dry air). To obtain h in (kJ/kg dry air) it is necessary to divide by 29.0, the molecular weight of air.

The mean driving force may be evaluated by applying a correction factor to the driving force determined for the mean water temperature, T_m, where

$$T_m = \frac{T_1 + T_2}{2} \qquad (6.47)$$

The value of h_m^* corresponding to T_m is readily obtained from the psychrometric chart. To obtain the driving force $(h_m^* - h_m)$ for T_m, we get h_m, the mean air

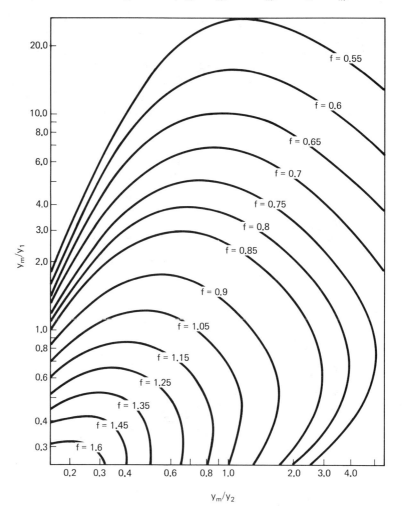

KEY:

$y_1 = (h_1^* - h_1)$; top of tower
$y_m = (h_m^* - h_m)$; mean position
$y_2 = (h_2^* - h_2)$; bottom of tower

Figure 6.10 Graphical solution for the mean driving force in a cooling tower.

Sec. 6.3   Cooling Towers

enthalpy, from the analog of Eq. (6.34):

$$\dot{m}_a(h_m - h_2) = wc_p T_1 - \left(w - \frac{\Delta w_e}{2}\right) c_p T_m \quad (6.48)$$

We then use the f correction factors of Fig. 6.10 to write

$$(h^* - h)_{\text{mean}} = (h_m^* - h_m) f \quad (6.49)$$

The packing height may now be evaluated if values of K and a are known. It is very difficult to separate the values of K and a and hence they are usually combined and the product, Ka, determined from empirical data. The value of Ka depends on the superficial liquid and gas mass flow rates (G_L', G_G') and the packing type. Table 6.4 provides typical equations for evaluation of this parameter.

With the packing height and airflow evaluated, the tower height (chimney height) above the packing may be determined if a natural-draft tower is used. The tower above the packing acts as a stack with the density difference between the warm air in the tower and the ambient air providing the theoretical draft, p_t. The

TABLE 6.4 EMPIRICAL EQUATIONS FOR MASS-TRANSFER COEFFICIENT FOR USE IN COOLING TOWER DESIGN[a]

Tower and packing	Flow-rate range (kg/m² · h)	Equation	Reference
6-ft² tower; 11 ft 3 in. packed height; wood slats, $\frac{3}{8}$ by 2 in., spaced parallel, 15 in. between tiers	$G_L' = 1550–13,550$ $G_G' = 3000–7600$	$Ka = 0.472 G_L'^{0.4} G_G'^{0.5}$	J. Lichtenstein, Trans. ASME, Vol. 65, 779 (1943)
Carbon slats—$1 \times \frac{1}{8}$ in. cross section; bottom edge serrated; $\frac{3}{4}$-in. horizontal centers; $1\frac{1}{4}$-in. vertical centers; alternate ties at right angles	$G_L' = 4200–9100$ $G_G' = 4500–13,500$	$Ka = (1.18 \times 10^{-4} G_L' + 1.45) G_G'^{0.8}$	W. S. Norman, Trans. Inst. Chem. Lond., Vol. 29, 226 (1951)
$41\frac{5}{8} \times 23\frac{7}{8}$ in. tower; $41\frac{3}{8}$-in. packed height; wood slats, $\frac{1}{4} \times 2 \times$ 23.5 in., bottom edge serrated; $\frac{5}{8}$-in. horizontal center; $3\frac{5}{8}$- to $2\frac{5}{8}$-in. vertical centers	$G_L' = 3900–6700$ $G_G' = 3100–6700$	$Ka = 1.2 \times 10^{-4} G_L' G_G'$ $- 0.233 G_G'$ $- 0.27 G_L' + 3100$	W. M. Simpson and T. K. Sherwood, Refrig. Eng., Vol. 52, 53 (1946)

[a] Ka = kg/[h (kg H$_2$O/kg dry air) m³], G_G', G_L' = kg/(m² · h).

theoretical draft, p_t, is equated to the pressure drop through the packing, friction loss through the tower, and exit loss from the tower. Since a natural-draft tower has a hyperbolic shape, the gas velocity is not constant, and we have

$$p_d = \left(\frac{k_p u_p^2}{2g_c} \Delta z + \frac{u_e^2}{2g_c} + \frac{u_{av}^2 f' L}{2g_c D} \right) \frac{\rho_G}{5.2} \tag{6.50}$$

where D = average tower diameter, ft
k_p = empirical packing pressure loss coefficient per unit depth[†]
u_p = superficial gas velocity through packing, ft/s
g_c = gravitational conversion constant, 32.2 ft/s² (lb$_m$/lb$_f$)
L = height of chimney, ft
p_d = theoretical draft, in. H$_2$O
f' = friction factor for stack (see Fig. 6.1)
u_{av} = average gas velocity in tower, ft/s
u_e = gas velocity at tower exit, ft/s
ρ_G = tower gas density, lb$_m$/ft³
Δz = height of packing, ft

The value of L, the tower height above the packing, is then estimated from Eq. (6.4) with ρ_{FG} replaced by the density of the moist air in the cooling tower. Only a minimum of iteration is required, as the frictional loss in the tower itself is small.

Note that the hyperbolic shape used for natural-draft cooling towers allows the simple theoretical draft equation to be used despite the very large diameter of such towers. The hyperbolic shape eliminates internal recirculation within the tower as well as eliminating high-velocity exit jets, which would increase exit losses.

6.4 EMISSION CONTROL SYSTEMS

Introduction to Federal Regulations

The role of the power plant engineer in emission control is twofold: (1) a system must be designed to satisfy all local, state, and federal emission levels, while simultaneously finding a method to dispose of the waste; and (2) the capital cost and energy requirements of this nonproductive process must be kept to a minimum.

Emissions of air pollutants came under federal regulation with the passage of the Clean Air Act of 1963. Since then the law has been amended twice to establish national primary (health related) and secondary (air quality) standards. Also, regulations have been introduced limiting discharge of specific air pollutants from new power plants.

In 1970, amendments to the Clean Air Act gave greater authority to state pollution control agencies. The Environmental Protection Agency (EPA) was also

---

[†] If experimental values for k_p are lacking, a very approximate estimate of wood-slat packing Δp may be obtained by treating the packing as a tube bank with a flow of air across it.

Sec. 6.4 Emission Control Systems

chartered by Congress in the same year. Legislation was soon passed that established emission control limits on all new steam generating stations with a thermal input of more than 250×10^6 Btu/h.

The EPA adopted new source performance standards in 1971 and promulgated these in Section 40 of the Code of Federal Regulations Part 60 (40 CFR 60), in subpart D. These standards covered emissions of SO_2, nitrogen oxides (NO_x), and particulate matter. Sulfur oxides were limited to 0.80 lb per million Btu of input thermal energy for an oil-fired system and to 1.20 lb per million Btu of input thermal energy for a coal-fired system. Nitrogen oxides were limited to 0.30 lb for an oil system and to 0.30 lb for a coal-fired system per million Btu of input energy. Particulate matter was limited to 0.1 lb per million Btu of input energy. Visual emissions from the stack were not to exceed 20% opacity.

Further amendments affecting utility power boilers were made to the Clean Air Act in 1977. The ambient air quality standards and emission release limits were not changed, but were made subject to review every five years. States were required to prepare a *State Implementation Plan* (SIP) to comply with the Clean Air Act. These amendments also required the use of the *best available control technology* (BACT) that will give the maximum reduction in emissions. States were encouraged to establish state EPA agencies and enforce regulations. States may impose stricter emission control limits in order to meet the air quality standards deemed desirable by that state.

Revised *performance standards for new sources* (NSPS) were issued by the EPA in 1979. These standards were the most restrictive issued thus far. In summary, the new emission standards require that particulate matter not exceed 0.03 lb per 10^6 Btu of heat input. The sulfur dioxide emission level for coal remains at 1.2 lb per 10^6 Btu of heat input, but the regulations now require up to a 90% reduction in potential emissions. A reduction of 90% is required unless uncontrolled discharges to the atmosphere are less than 0.6 lb per 10^6 Btu of heat input. In that case, a 70% reduction in emissions is required. For oil-fired systems, the maximum SO_2 emission rate was established at 0.8 lb per 10^6 Btu plus a 90% reduction in potential emissions unless uncontrolled discharges to the atmosphere are less than 0.2 lb per 10^6 Btu.

The new 1979 regulations effectively require a flue-gas-desulfurization system on every new utility boiler, even when burning low-sulfur coal. The regulations would permit the use of a less expensive dry scrubbing system with a 70% removal efficiency if the fuel sulfur content is less than 0.8%.

The maximum NO_x emission levels were reduced to 0.5 lb per 10^6 Btu for subbituminous coal and 0.6 lb per 10^6 Btu for bituminous coal. For oil-fired power stations, the NO_x limit is 0.3 lb per 10^6 Btu. Future EPA regulations are expected to further reduce this NO_x limit to 0.2 to 0.3 lb per 10^6 Btu for coal-fired boilers. Even more stringent local regulations may be applied. In one locality, an NO_x emission level of 0.1 lb per 10^6 Btu has been proposed.

Figure 6.11 shows the emission control decisions that must be made by today's power plant engineer in order to meet the Clean Air Act requirements. Particulate emission standards are now rigid enough to require the use of filters and/or

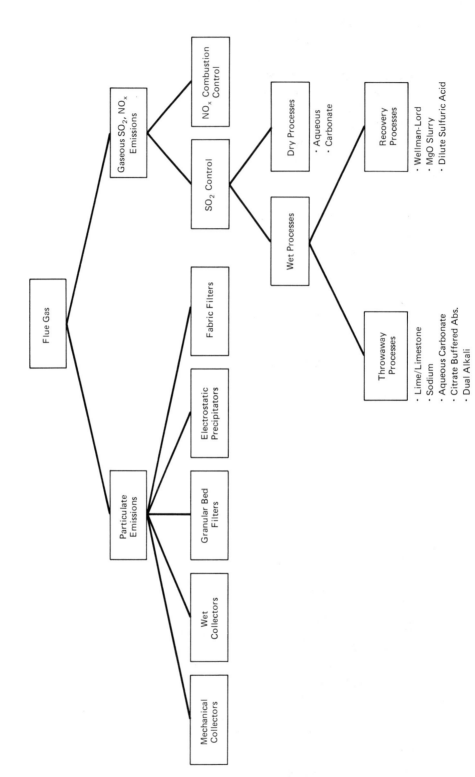

Figure 6.11 Sulfur dioxide control techniques at a coal-fired power station.

Sec. 6.4   Emission Control Systems

electrostatic precipators. Gaseous emission of NO_x can still be controlled through the use of combustion control techniques, while most utilities are currently using a lime/limestone process for SO_2 removal. This section contains a brief description of the systems and processes under development to meet clean air standards.

Sulfur Dioxide Removal Techniques

Sulfur dioxide removal can be accomplished by a number of processes (see Fig. 6.11). The wet processes are the most efficient, but the less efficient dry process is the most economical. The wet processes, usually referred to as *scrubbing processes*, can be further divided into throwaway or recovery processes. As implied by the name, the *throwaway processes* yield a wet sludge that must be disposed of, while the *recovery processes* yield a salable by-product. The most practical recovery processes are the Wellman–Lord, magnesium oxide slurry, and the dilute sulfuric acid/gypsum process. The better developed throwaway processes include the lime/limestone, sodium, aqueous carbonate, and dual alkali processes.

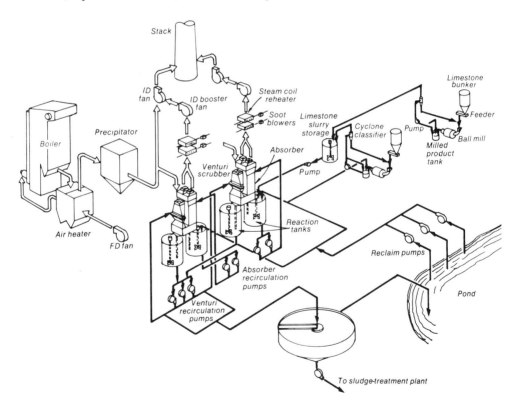

Figure 6.12 Flue gas desulfurization by the lime/limestone scrubbing process. (Reprinted with permission from *Power*; copyright 1978 McGraw-Hill, Inc., New York.)

The most common method of flue gas desulfurization is the *lime/limestone scrubbing process* (Fig. 6.12), used in about 90% of the utility power plants that have SO_2 removal systems. In this process, SO_2 is absorbed in an alkaline water solution. Calcium in the form of slaked lime, $Ca(OH)_2$, or milled limestone, $Ca(CO_3)$, is reacted with SO_2 in a slurry of about 10% solids to form a mixture of insoluble $CaSO_3$ and $CaSO_4$. The principal ionic reactions are

$$SO_2 + H_2O \rightleftharpoons SO_3^{2-} + 2H^+$$
$$2OH^- + SO_3^{2-} \rightleftharpoons H_2O + SO_4^{2-}$$
$$Ca^{2+} + SO_4^{2-} \rightleftharpoons CaSO_4(s) \tag{6.51}$$

The solids are precipitated in a thickener for disposal. Auxiliary power requirements for this process are high and the process can consume between 7 and 10% of the station power output.

Sodium sulfite is also used as a scrubber agent. In the *sodium/calcium dual-alkali system*, soluble Na_2SO_3 reacts with SO_2 in the scrubber to produce $NaHSO_3$. Calcium is added in a regeneration loop to form $CaSO_3$ and regenerate Na_2SO_3. Removal efficiencies as high as 90 to 95% can be achieved with this system.

The most popular SO_2 recovery system appears to be the *Wellman–Lord system*. In this system, $NaSO_3$ reacts with SO_2, forming $NaHSO_3$. A bleed stream is sent to a reactor, where SO_2 is driven off as a gas. The gas can be converted to SO_2, sulfuric acid, or elemental sulfur. Another popular recovery process, the magnesium oxide (MgO) *wet recovery slurry process*, uses a conventional contactor/absorber to react the MgO with the SO_2 to form $MgSO_3$. The blowdown from the scrubber is centrifuged, dried in a rotary drier, and transported to a calcining facility for MgO regeneration. Removal efficiencies of up to 90% are achieved with this system.

Dry scrubbing systems, utilizing either sodium or calcium, are a recent development. A typical system is the aqueous carbonate process. A sodium carbonate solution is sprayed into the hot flue gas in a drying chamber. The drying is controlled at saturation temperature to produce a sodium sulfate powder. If the system is regenerative, the powdered product goes through a reduction with coke in a reactor at 982°C (1800°F) to form sodium sulfide. This is quenched and carbonated to form sodium carbonate and hydrogen sulfide gas. The sodium carbonate is recycled back to the absorbing unit. This process has a 90 to 95% removal efficiency and consumes less energy than do wet removal systems. System complexity has posed some operational difficulties.

Controlling NO_x Emissions

Oxides of nitrogen, NO_x, are produced in all combustion processes occurring in air. They are formed initially as nitric oxide, NO. The nitric oxide gradually combines with oxygen to form nitrogen dioxide, NO_2. The NO_2, in turn, reacts with hydrocarbons and ozone, with sunlight as the catalyst; to produce smog and compounds that irritate eyes, may aggravate certain respiratory diseases, and damage vegetation.

NO_x emissions are now suspected to be a contributor to the acid rain problem, which is of concern to many eastern states and Canada.

The oxides of nitrogen are derived from nitrogen in the combustion air (thermal NO_x) and the nitrogen in the fuel itself (fuel NO_x). Some fuels, such as natural gas and light distillate oil, contain very little nitrogen. Unfortunately, coal and the heavy fuel oil burned in power plants contain high quantities of nitrogen. In a typical modern power plant, about 70% of the NO_x generated is due to fuel nitrogen. Considerable research is therefore being directed toward ways to reduce the nitrogen content of fuels. For the present, modifying combustion process conditions is the only economically feasible method to reduce NO_x emissions.

Combustion control can be utilized to effectively reduce NO_x emissions by up to 60%. Thermal NO_x originates in the hottest regions of the combustion process. The production of NO_x is affected by such parameters as flame temperature, residence time, and the fuel-to-air ratio. Research indicates that solutions to the thermal NO_x problem involve primarily a reduction of the flame temperature. The flame temperature, in turn, is affected by such parameters as load, furnace heat release rates, burner settings, and flue gas recirculation.

The most effective combustion control technique for reducing NO_x emissions is staged combustion. The objective of staged combustion is to create the formation of localized fuel-rich conditions, where both thermal and fuel NO_x are minimized. Substoichiometric conditions are created in the primary combustion zone, with complete combustion occurring downstream of the burners. NO_x reductions of up to 50% can be achieved by staged combustion.

Figure 6.13 shows a two-stage venturi furnace for reducing NO_x emissions. Coal is burned in two separate chambers. In the initial chamber, coal is burned at substoichiometric conditions, resulting in incomplete combustion. This partial burning also allows the desired N_2 producing reactions to occur. Secondary air is added at the entrance to the second chamber in order to achieve oxidizing conditions during transport through the second combustion chamber. Other processes are under development based on similar principles.

NO_x emissions can also be reduced by *flue gas recirculation* (FGR). NO_x formation is reduced due to a lowering of the bulk flame temperature and by slightly reducing the flame oxygen concentration. The flue gas is mixed with the combustion air in the windbox or burner throat. Flue gas recirculation can reduce NO_x levels between 20 and 50%, depending on the type of fuel. Since FGR inhibits only thermal NO_x, it is more useful for boilers using gas or light distillate oil. Initial installation costs are high.

Another technique for reducing oxygen availability and furnace temperatures is firing with low excess air (LEA). This reduces the production of NO_x from both fuel and thermal sources. Utility boiler manufacturers incorporate methods for LEA control to comply with NO_x emission limits required for new boiler construction. Firing with low excess air is standard practice on new oil-fired utility boilers. Reduction of NO_x levels by 15 to 20% have been achieved by this technique alone. For new coal-fired power plants, LEA techniques have reduced NO_x emissions by as

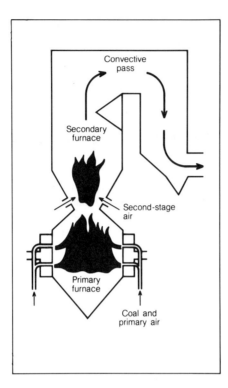

Figure 6.13 Babcock & Wilcox two-stage venturi furnace. (From the *EPRI Journal*, Dec. 1980.)

much as 20%. LEA levels can cause operational problems where the excess air levels are below manufacturers-specified levels.

Controlling Particulate Emissions

Particulate emissions are the easiest of the power plant pollutants to control. The particulate matter is usually classified by particle size and source. For example, particulate matter from coal ash is called *fly ash* if the particle diameter is less than 1×10^{-4} m. Smoke particles have a diameter of 1×10^{-5} m or less and behave as a suspension of particles. Particles with sizes less than 1×10^{-6} m are classified as dust particles. The five basic methods for reducing particulate emissions are indicated in Fig. 6.11. Only fabric filters and electrostatic precipators are feasible systems for large utility plant applications. Mechanical collectors, wet collection systems, and granular bed filters are used primarily for particulate removal for industrial boilers and small utility boiler applications. The choice remaining for large power plant applications is between electrostatic precipitators and fabric filters. In some cases, fabric filtration to remove the larger particles may be followed by electrostatic precipitation.

Sec. 6.4 Emission Control Systems

Electrostatic precipitators. If an electrostatic precipitator, flue gas containing particulate matter flows between rows of discharge electrodes and grounded collecting plates as shown in Fig. 6.14(a). A high-voltage direct current is applied across the collecting electrodes, accelerating free electrons present in the gas. The fast-moving electrons collide with the gas molecules, stripping them of an electron. The positive ions migrate to the negative electrode, where they are collected. The electrons are collected by the grounded collecting electrode.

As the electrons leave the corona region near the discharge electrode, their velocity decreases permitting capture by other gas molecules. These molecules become negative ions. As these ions move to the collecting electrode, they give a negative charge to fly ash particles in the flue gas stream. The fly ash particles are driven to the collecting plate by the force which is proportional to the product of this

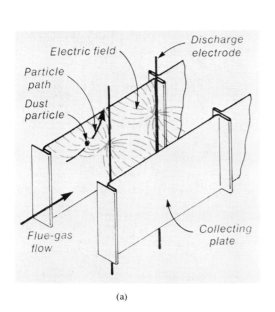
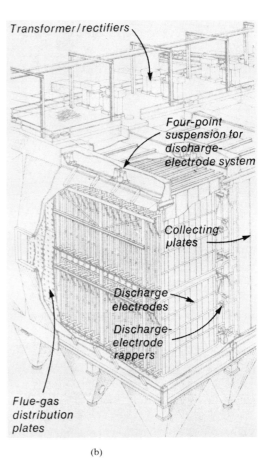

Figure 6.14 (a) Electrostatic precipitator plates: (b) electrostatic precipitator unit. (Reprinted with permission from *Power*; copyright 1980 McGraw-Hill, Inc., New York.)

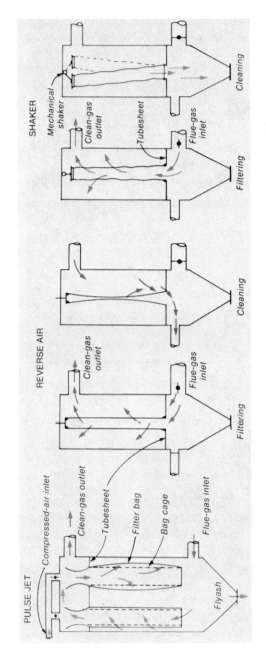

Figure 6.15 Baghouse design concepts. (Reprinted with permission from *Power*; copyright 1980 McGraw-Hill, Inc., New York.)

charge and the strength of the electric field. The particles are collected on the collecting electrode. Some positive ions, although moving more slowly, impart a positive charge to some of the flue gas particulate matter and this material is also collected on the appropriate collecting electrodes.

Collected particulate matter must be removed from the collecting plates on a regular schedule to ensure efficient collector operation. Removal is usually accomplished by a mechanical hammer or rapping system. The vibration knocks the particulate matter off the collecting plates and into a hopper at the bottom of the precipitator [Fig. 6.14(b)].

Electrostatic precipitators can remove up to 99.8% of the particulate matter in the flue gas stream. Precipitators can remove submicron particles from the effluent stream over a broad temperature range, from just above the dew point to metal temperature design limits. These precipitators are highly reliable and have a low-pressure drop. However, electrostatic precipitators have difficulty removing high-resistivity fly ash, such as that which originates from the burning of some low-sulfur coal.

Fabric filters. The use of fabric filters is an old technique for removing particulate matter from flue gas streams. There is now renewed interest in fabric filters due to the fact that sulfur content of the fuel does not influence collection efficiency. Fabric filter systems, called *baghouses*, can be designed for maintenance while the system is in operation. Baghouses are characterized according to the method used to remove fly ash from the filter system. The basic concepts are illustrated by Fig. 6.15.

A variety of filter materials are available. A typical filter bag, one of perhaps 100 required for a baghouse, would utilize woven fiberglass coated with a chemical to protect the fibers and improve abrasion resistance. Manufacturers claim particulate removal efficiency of up to 99.8%. Maximum temperature of operation is around 500°F for most materials. Consideration must be given to maintenance problems and filter bag lifetime for the application being considered.

6.5 WASTE DISPOSAL

Ash Handling and Disposal

Boilers burning pulverized coal have dry bottom furnaces. The large ash particles are collected under the furnace in a water-filled ash hopper. Fly ash is collected, as described in Section 6.4, by either an electrostatic filter system or a baghouse. A pulverized coal boiler generates approximately 80% fly ash and only 20% bottom ash. Three major factors should be considered for ash disposal systems: (1) plant siting, (2) fuel source, and (3) environmental regulations.

Plant siting considerations are due to the need for water and land for many ash handling systems. The fuel source is a major consideration due to the vastly different quantities of ash produced from regional coal. Table 6.5 illustrates this large difference for three regional coals. Thus a power plant burning Illinois coal

TABLE 6.5 ASH GENERATION FROM COAL (600-MWe POWER PLANT)

	West Virginia	Illinois	Wyoming
High heating value (Btu/lb)	13,000	10,000	8,000
Ash in coal (%)	7	16	7
Ash produced (tons/h)	15	44	24

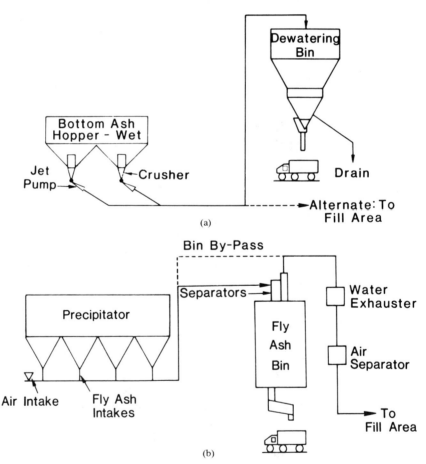

Figure 6.16 (a) Bottom ash-sluice conveyor: (b) fly ash-hydraulic vacuum conveyor. (Reprinted with the permission of Sargent & Lundy, Chicago.)

would produce almost three times the ash as the same size unit burning West Virginia coal. Environmental regulations often make it difficult to obtain approval for ash storage and disposal sites. Local legislation must also be considered.

Ash handling systems are available in a wide variety of design configurations:

Bottom ash systems:

Sluice conveyor

Recirculating sluice conveyor

Fly ash systems:

Hydraulic vacuum conveyor

Pneumatic vacuum conveyor

Pneumatic pressure conveyor

Vacuum and pressure conveyor

All of these systems are being used in present power plants. The sluice conveyor system [Fig. 6.16(a)] is the most widely used for bottom ash handling while the hydraulic vacuum conveyor [Fig. 6.16(b)] is the most frequently used system for fly ash systems.

Disposal of ash from power plants is a difficult task due to the quantity of ash involved. A 600-MWe power plant, over a 40-year lifetime, could produce as much as 5 million cubic yards of ash. Landfill is one obvious solution. If landfill sites are not available, specially designed sites must be created to hold the ash. One such site consists of a compacted earth toe embankment at the base of a sloping disposal site. The embankment face consists of compacted fly ash, which is intended to assure a stable storage site. The finished site is covered with topsoil and seeded to resist erosion.

Disposal of Flue Gas Wastes

The disposal of solid wastes, called *sludge*, from flue gas cleaning systems has become very important due to the stringent air pollution regulations, which require a higher percentage of SO_2 removal from the flue gas, and the Resource Conservation and Recovery Act of 1976, which regulates the disposal of the scrubber wastes. In the past, these wastes have been disposed of in large storage ponds. Usually, there were no provisions to control overflow or seepage into groundwater. The new regulations cover the design, handling, and disposal of flue gas cleaning wastes. Criteria have also been established which define the location, design, and construction of disposal facilities.

The chemical constituents and the crystalline structure of the sludge contribute to handling and disposal difficulties. Flue gas scrubber sludge is composed of sulfur salts of calcium together with varying amounts of calcium carbonate ($CaCO_3$) and unreacted lime (CaO). The conventional lime scrubber described in Section 6.4

produces sludge composed of calcium sulfite hemihydrate ($CaSO_3 \cdot \frac{1}{2}H_2O$), calcium sulfate dihydrate ($CaSO_4 \cdot 2H_2O$), or gypsum, and some free lime. Most of the unreacted lime forms $CaCO_3$ with the carbon dioxide from the flue gas.

The crystalline structure of the calcium-based sludge affects both its chemical and physical properties. When limestone is the reagent in the scrubbing process, sulfite crystals predominate, whereas when lime is used in the scrubbing process, sulfate crystals predominate in the sludge. Because of their microstructure, the sulfate crystals provide greater mechanical stability and remain in place better after final sludge disposal. In contrast, if the sludge contains too many sulfite crystals, it tends to creep or flow at the disposal site if loaded statically or dynamically.

The slurry wastes originally contain only 5 to 15% solid sludge. Dewatering is required to improve the physical properties of the sludge and reduce the total volume for disposal. A variety of dewatering operations are used, some as simple as pond settling. A few power plants utilize thickening and centrifugation to produce a sludge with minimum water content and minimum volume for disposal. One of the by-products of the dewatering operation, gypsum, has commercial value.

Selection of a containment site for the final disposal of flue gas sludge is of major environmental importance. Several factors must be considered, such as groundwater intrusion, surface-water runoff, trace metal emissions or fugitive particle emissions, and physical instability. Economic factors no longer dictate disposal site selection and design.

Wet ponding is still the most common method of disposal. The sludge is not dewatered. Some of the ponds are lined with clay to prevent groundwater intrusion. Wet ponding is the most economic solution, but requires large land areas. It is estimated that sludge wet ponding land requirements are on the order of 0.5 acre-ft[†]/MW · year. Reclamation is not possible when the pond is full. Lining of wet ponds with clay, cement, or plastic materials helps to prevent seepage into the ground.

Dry disposal methods are receiving additional attention because of many new local environmental regulations. The sludge is dewatered and often stabilized further by the addition of a binder or matrix material. Chemical fixation techniques are also being investigated. These techniques include the interblending of portland cement, fly ash, and flue gas desulfurization sludge. Other commercial fixation processes are available. The dry, stabilized final product can be used in a landfill, followed by land reclamation.

SYMBOLS

a effective transfer area per unit tower volume, m^2/m^3

A area, m^2 or ft^2

[†] The quantity 0.5 acre-ft can be interpreted as an area of 1 acre filled to a depth of 0.5 ft.

Chap. 6 Symbols

A_f	free-flow area, m²
A_h	heat-transfer area, m²
A_i	tube inside area, m²
A_o	tube inside area, m²
A_t	total cross-sectional area, m²
c_p	specific heat, J/kg · °K
$(c_p)_M$	specific heat of metal in regenerative heat exchanger, J/kg · °K
CL	concentration level
D	stack or tower inside diameter, ft
D_e	equivalent diameter, m
D_o	tube outside diameter, m
f	correction factor used to obtain mean enthalpy driving force, dimensionless
f'	friction factor (for stack, see Fig. 6.1)
F_c	log mean temperature correction factor
g_c	gravitational conversion constant (1.0 in SI units)
G'_G, G'_L	superficial mass velocity of liquid and gas, respectively, based on total cross-sectional area of tower, kg/hr · m²
$G'_{\max}$	maximum mass velocity, kg/m² · s
h	specific enthalpy of fluid at bulk conditions, J/kg or kJ/kg
h_{dry}	specific enthalpy of dry air, kJ/kg
h_{sat}	specific enthalpy of saturated air, kJ/kg
h^*	specific enthalpy of saturated air at liquid temperature, kJ/kg
h	film or heat-transfer coefficient, W/m² · °C
h_c	heat-transfer coefficient between gas and metal in regenerative heat exchanger cooling cycle, W/m² · °K
h_f	fouling coefficient, W/m² · °C
h_h	heat-transfer coefficient between gas and metal in regenerative heat exchanger heating cycle, W/m² · °K
h_i	film coefficient, tube inside diameter, W/m² · °C
h_o	film coefficient, tube outside diameter, W/m² · °C
H'_{mm}	total fan head, mm H₂O
H'_{SI}	total fan head, m H₂O
k	thermal conductivity, W/m · °C
k_p	empirical packing pressure loss coefficient per unit depth
K	overall mass-transfer coefficient, kg (H₂O)/m² · h
L	height, ft

L_{SI}	height, m
m_r	mass of metal in regenerative heat exchanger, per unit of heat-transfer area, kg/m²
$\dot{m}$	flow rate, kg/h or kg/s
N_0	number of rows in tube bank
p_{ac}	actual stack draft, in. H₂O
p_d	theoretical stack draft, in. H₂O
p_h	average velocity pressure head, in. H₂O
p_s	static pressure, Pa (N/m²)
p_t	total pressure, Pa (N/m²)
p_v	dynamic pressure, Pa ($u^2/2g_c$)
Δp	pressure drop, Pa (N/m²)
P_F	forced-draft fan shaft power input, kW
q	rate of heat removal or transfer, W
Q_r	cooling heat load, W
r_i	inside tube radius, m
r_o	outside tube radius, m
Re	Reynolds number
S_t, S_l	distance between tube centers, normal and parallel to flow, respectively, m
t	time
t_r	time to complete one revolution
T	temperature, °C or °K
T_1	temperature of water entering cooling tower, °C
T_2	temperature of water leaving cooling tower, °C
T_a	average air temperature, °C
T_{a1}	combustion air temperature entering preheater, °C
T_{a2}	combustion air temperature leaving preheater, °C
T_{CW}	cooling water temperature, °C
T_{FG1}	temperature of flue gas entering air heater, °C
T_{FG2}	temperature of flue gas leaving air heater, °C
T_g	gas temperature, °K
T_G	average flue gas temperature, °C
T_M	mean temperature or temperature of the regenerative heat exchanger metal, °C
T_{MC}	temperature of regenerative heat exchanger at end of cooling cycle, °C

T_{MH}	temperature of regenerative heat exchanger at end of cooling cycle, °C
T_{WB}	wet-bulb temperature, °C
T_{WB1}	web-bulb approach temperature for exit cooling tower air, °C
ΔT_{CR}	cooling range of cooling tower, °C
$(\Delta T)_{LM}$	log mean temperature difference, °C
ΔT_{WB}	wet-bulb approach, °C
u	velocity, m/s or ft/s
u_{av}	average velocity in tower, ft/s
u_e	gas velocity at tower exit, ft/s
u_p	superficial gas velocity through packing, ft/s
U	overall heat-transfer coefficient, W/m² · °C
V_0	volumetric flow rate, volume/time
w	water flow rate, kg/h
Δw_b	blowdown loss, kg H₂O/h
Δw_d	drift loss, kg H₂O/h
Δw_e	evaporation low, kg H₂O/h
Δw	total makeup loss, kg H₂O/h
X	water content of the air, kg H₂O/kg air
z	height, in. or ft
Δz	average packing height in tower, ft

Greek Symbols

η	efficiency
η_c	cooling tower efficiency
η_F	fan efficiency
Θ	$T_G - T_M$
μ_b	gas viscosity at bulk fluid conditions, Pa · s
μ_w	gas viscosity at wall fluid conditions, Pa · s
ρ	density, kg/m³
ρ'	gas density, lb/ft³

Subscripts

1, 2, 3, ...	locations 1, 2, 3, ...
a	air

F	fan
FG	flue gas
G	gas
H_2O	water
i	at inner diameter
o	at outer diameter
m	mean or average conditions
s	standard conditions

PROBLEMS

6.1. The pressure drop across a furnace may be approximated by $\Delta p_b = 0.35\ (W/10^5)^2$, where Δp_b is in in. H_2O and W is in lb/h. If (1) the temperature of the gases entering the stack is 600°F, (2) the stack is 5 ft in diameter and 150 ft high, and (3) no fans are used, what is the total gas flow through the furnace?

6.2. A furnace having a 4 m × 4 m cross section contains a convective heat exchanger having the following characteristics:

Number of rows: 180
Design: staggered
$S_t = 4$ cm, $S_l = 4$ cm
Tube diameter = 2 cm

The furnace gases leaving the convective heat exchanger at 400°C and enter a 6-ft-diameter stack which is 30 m high. If the furnace burns 300 gal of fuel oil (specific quantity = 0.85) per hour with 20% excess air, estimate the horsepower of the fan needed to provide the required airflow.

6.3. Assume that the steel tubes in the convective heat exchanger described in Problem 6.2 are 0.16 cm thick and contain flowing steam displaying a heat-transfer coefficient of 850 W/m² · °C. If the steam enters the heat exchanger at 350°C and leaves at 800°C, how much heat will be transferred per hour if the exit gas temperature is maintained at 400°C? (*Hint*: Use the gas flows computed for Problem 6.2 to get the needed heat balance.)

6.4. A furnace burning 2 tons per hour of coke in 10% excess air heats the incoming air in a recuperative heat exchanger to 200°C. The furnace gases entering the recuperative heat exchanger are at 350°C. Assume that the heat exchanger is to consist of an in-line tube bank in which $S_t/S_l = 1$ and the tube pitch is 2.5 times the tube diameter. Design a heat exchanger that will transfer the desired amount of heat, fit within a duct having a 5 m × 5 m cross section, and have a pressure drop through the tubes of less than 1 in. of water. Assume that the incoming air flows inside the tubes.

6.5. A regenerative heat exchanger is being designed as a replacement for the recuperative exchanger described in Problem 6.4. The exchanger is to consist of a series of concentric cylinders made of 1-cm-thick rolled steel sheet. The outermost cylinder has a diameter of 4 m, the innermost cylinder a diameter of 0.1 m, and intermediate cylinders are spaced every 0.1 m (centerline of one steel sheet to centerline of next steel sheet). Assume that the unit rotates at a speed of 5 rpm.

(a) Estimate the length of the unit needed to transfer heat at the desired rate.

(b) With the assumption that the air flows through half of the open area and the hot gases through the other half, compare the air-side pressure drop to the air-side pressure drop in Problem 6.4.

6.6. A boiler contains 40,000 ft of 1-in. 12 BWG (Birmingham Wire Gage) tubing and steam drums which hold a water volume equal to 30% of the tubing volume. It is estimated that the tubing steel corrodes at a rate of 100 mg/m$^2 \cdot$ yr. Estimate the rate at which a bypass stream should be drawn off for purification if the corrosion product level in the boiler water is not to exceed 5 ppm. You may assume that the purified water contains a negligible quantity of corrosion products.

6.7. (a) Condenser cooling water at 100°F flows to a wet cooling tower at the rate of 150,000 kg/hr and is to be cooled to 80°F by upward-flowing air. You may assume that ambient air is at 70°F and has a relative humidity of 50%. Assume that the cooling section of the tower is packed with carbon slats having the characteristics given in Table 6.4. If the air-mass flow rate is 5000 kg/m$^2 \cdot$ h, determine the cross-sectional area and packing height needed.

(b) If the design of the tower is fixed at that determined in part (a), to what temperature could condensed cooling water be cooled if the ambient air went to 80°F and a relative humidity of 90%?

(c) Estimate the chimney height needed to provide the draft required for the condenser of part (a) if it is assumed that $k_p = 0.05$, that the entering tower area and exit area are each equal to the superficial flow area of the packing, and that no circulation fans are used.

6.8. A steam plant produces 500 MW of electrical power with an overall thermal efficiency of 35%. The coal used as fuel has a lower heating value of 25,000 kJ/kg and a sulfur content of 2.3%. The stack gas scrubber removes 95% of the SO$_2$ produced and discharges a slurry of CaSO$_4$ which is 45% w/o CaSO$_4$. What volume of slurry is produced per year if the plant operates at a 75% capacity factor?

BIBLIOGRAPHY

ASCHNER, F., *Planning Fundamentals of Thermal Power Plants*. New York: Wiley, 1978.

CHEREMISINOFF, N. P., and P. N. CHEREMISINOFF, *Cooling Towers: Selection, Design and Practice*. Ann Arbor, Mich.: Ann Arbor Science Publishers, 1981.

FRAAS, A., *Engineering Evaluation of Energy Systems*. New York: McGraw-Hill, 1982.

GAFFERT, G., *Steam Power Stations*. New York: McGraw-Hill, 1952.

SKROTZKI, B., and W. VOPAT, *Power Station Engineering and Economy*. New York: McGraw-Hill, 1960.

Steam, Its Generation and Use, Lynchburg, Va.: The Babcock & Wilcox Company, 1978.

WOODRUFF, E., and H. LAMMERS, *Steam Plant Operation*. New York: McGraw-Hill, 1977.

7

Nuclear Power Stations

7.1 PRINCIPLES OF NUCLEAR FISSION

Structure of the Atom

We are all aware that all matter is made up of unit particles called *atoms*. The central part of the atom is the *nucleus*, which contains the relatively heavy positively charged and neutral particles (*protons* and *neutrons*). The nucleus is surrounded by relatively light negatively charged particles which orbit the nuclei (*orbital electrons*). These electrons are often depicted as an *electron cloud*. An atom is electrically neutral; thus an atom contains exactly the same number of electrons and protons. The number of orbital electrons determine the chemical family and chemical behavior of the atom, while the number of protons and neutrons in the nucleus establish the nuclear characteristics of the atom.

Since the nucleus contains the heavy protons and neutrons (often called *nucleons*), most of the weight of the atom is concentrated in the nucleus. The radius of the nucleus is in the order of 10^{-16} m, considerably smaller than the radius of the atom, which is of the order of 10^{-11} m. Thus the bulk of the atomic mass is concentrated in a small portion of the atom. Figure 7.1 illustrates the atomic structure of the simplest atoms.

Since all of the positive charges (protons) are confined to a small central region, very strong short-range nuclear forces exist that overcome the Coulomb repulsive forces. After establishing some basic nomenclature, we will calculate the energy that holds the nucleus together.

Sec. 7.1   Principles of Nuclear Fission

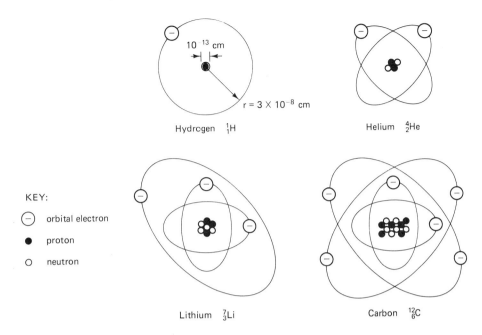

KEY:
⊖ orbital electron
● proton
○ neutron

Figure 7.1 Structure of the atom.

A specific *nuclide* is generally represented symbolically by the notation, $^{A_m}_{Z_e}Y$. In this notation, Y is the symbol for the chemical element, Z_e is called the atomic number and A_m represents the atomic mass number. The atomic number, Z_e, defines the number of protons in the nucleus, which is, as already indicated, the same as the number of electrons in orbit about the atom. Each atom has an atomic number that defines the chemical element and is synonymous with the chemical symbol.

The mass number, A_m, represents the total number of nucleons in the nucleus, protons plus neutrons. Since the mass of a subatomic particle is very small, their mass is often represented in atomic mass units or amu (1 amu = 1.6605×10^{-27} kg). Table 7.1 lists the weight in kilograms and amu of the fundamental particles. As

TABLE 7.1 DATA ON FUNDAMENTAL PARTICLES

Particle	Symbol	Charge, e	Mass (kg)	Mass (amu)
Electron	e^-	-1	9.109×10^{-31}	5.486×10^{-4}
Positron	e^+	$+1$	9.109×10^{-31}	5.486×10^{-4}
Proton	e^+	$+1$	1.673×10^{-27}	1.007277
Neutron	n	0	1.674×10^{-27}	1.008665
Hydrogen atom	^1_1H	$+1$	1.673×10^{-27}	1.007825
Alpha particle	α	$+2$	6.644×10^{-27}	4.002603

shown in the table, the mass of a proton or neutron is approximately 1 amu. Thus the nucleus has a mass approximately equal to the mass number, A_m, the sum of the number of neutrons and protons in the nucleus. The number of neutrons, n_n, is easily found from the difference between the mass number, A_m, and the atomic number, Z_e, or $n_n = A_m - Z_e$.

As noted previously, the number of electrons in an atom establishes the chemical family. When atoms of an element have the same number of protons but a different number of neutrons, they are called *isotopes* of the element. They are indistinguishable chemically but exhibit different nuclear characteristics. For example, the simplest three isotopes are members of the hydrogen family: hydrogen, ^1_1H (see Fig. 7.1); deuterium, ^2_1H; and tritium, ^3_1H. Each atom contains one electron and one proton. In addition, deuterium has one neutron in the nucleus and tritium has two neutrons in the nucleus. Each shows the same chemical behavior but exhibit markedly different nuclear characteristics. All forms of water contain some tritium, T, in HTO and some deuterium, D, in D_2O.

Mass Defect and Binding Energy

The mass of an atomic nucleus is less than the sum of the masses of the protons and neutrons that comprise the nucleus. Thus difference in mass is called the *mass defect*. The energy equivalent of this mass defect is apparently the "glue" that holds the nucleus together. The mass defect is found by adding up all the individual particle weights and subtracting the actual mass of the atom:

$$\Delta = n_n m_n + Z_c(m_p + m_e) - {}^A_Z m \qquad (7.1)$$

where Δ = mass defect, kg
 m_n = mass of neutron, kg
 m_p = mass of a proton, kg
 m_e = mass of an electron, kg
 ${}^A_Z m$ = mass of atom with mass number A_m and atomic number Z_e, kg

This mass defect can be converted to energy by the *Einstein equation*:

$$\mathscr{E} = mc^2 \qquad (7.2)$$

where $\mathscr{E}$ = energy, J
 c = velocity of light, m/s
  $\sim 3.0 \times 10^8$ m/s
 m = mass, kg

The energy associated with the mass defect is known as the *binding energy* (BE) of the nucleus. As an example of the application of Eq. (7.2), the energy equivalent of 1 g of mass can be calculated. It is

$$\mathscr{E} = 1 \times 10^{-3} \text{ kg} \times (3.0 \times 10^8 \text{ m/s})^2 = 9.0 \times 10^{-13} \text{ J}$$

The energy released in nuclear reactions is usually expressed in units of electron

Sec. 7.1 Principles of Nuclear Fission

volts (eV) or millions of electron volts (MeV). An *electron volt* is defined as the amount of energy acquired by an electron when it is accelerated in an electric field which is produced by a potential difference of 1 V. It is a small quantity of energy.

$$1 \text{ eV} = 1.6 \times 10^{-19} \text{ J}$$
$$1 \text{ MeV} = 1.6 \times 10^{-13} \text{ J}$$

The energy equivalent of 1 amu of mass is often used to facilitate calculations of energy released in nuclear reactions. By using Eq. (7.2) this energy equivalent can be shown to be

$$1 \text{ amu} = 931 \text{ MeV}$$

Therefore, if 1 amu of mass could be completely converted to energy, 931 MeV would result.

The binding energy of the nucleus is usually expressed in binding energy per nucleon. Consider the binding energy per nucleon for deuterium as a simple example. Deuterium was shown previously to consist of one electron, one proton, and one neutron. Using the data from Table 7.1, we would expect the weight (in amu) of the deuterium atom ($^2_1$H) to be

$$m(^2_1\text{H}) = m_p + m_e + m_n$$
$$= 1.007277 + 0.000549 + 1.008665$$
$$= 2.01649 \text{ amu} \qquad (7.3)$$

The actual mass of deuterium is 2.01410 amu, leaving a mass defect of

$$\Delta(\text{amu}) = 2.01649 - 2.01410 = 0.00239 \text{ amu}$$

This mass defect is converted to binding energy by Eq. (7.2) or approximated by

$$\text{BE(MeV)} = \Delta(\text{amu}) \times 931$$
$$= 0.00239 \text{ amu} \times 931 \text{ MeV/amu}$$
$$= 2.23 \text{ MeV}$$

The binding energy per nucleon is simply

$$\text{BE/nucleon} = 2.23 \text{ MeV}/2 \text{ nucleons}$$
$$= 1.11 \text{ MeV/nucleon}$$

By a similar calculation, the binding energy per nucleon can be calculated for all the isotopes. The result of this calculation can be presented in a graph which shows the binding energy per nucleon versus mass number (Fig. 7.2). The higher the binding energy per nucleon, the more stable is the isotope.

The binding-energy curve that shows the most stable isotopes are in the intermediate mass number range. If low-mass-number isotopes were to fuse together, there would be an increase in stability. If the higher-mass-number isotopes were to undergo fission, this would lead to more stable elements. Thus light isotopes such as hydrogen, deuterium, and tritium are candidate isotopes for fusion reactions, while the heavier isotopes, such as uranium, are likely prospects for the fission reaction.

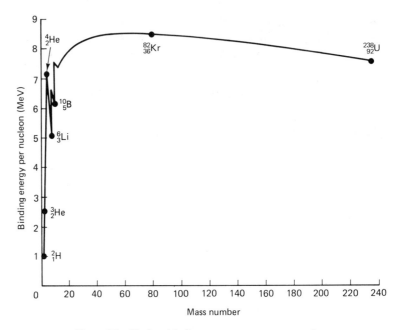

Figure 7.2 Nuclear binding energy versus mass number.

The average binding energy per nucleon is theoretically the minimum energy that must be supplied to a nucleus to remove a proton or neutron from the nucleus. Thus energy can be supplied in the form of excitation energy that is gained by a nucleus after it absorbs a neutron or proton. The excitation energy needed is the binding energy of the particle plus the kinetic energy of the particle. Subsequent analysis will show how particles which are this energetic can cause nuclear fission of various isotopes.

Radioactive Decay

Most isotopes that occur in nature are stable chemical elements. This stability is a result of a tightly bound nucleus due to a favorable combination of neutrons and protons. Some natural isotopes and many human-made isotopes consist of nuclei which are loosely bound and can attain a more stable nuclear configuration by emitting radiation. In such cases, a spontaneous disintegration process, called *radioactive decay*, occurs. The resulting nucleus is called the *daughter* and the original nucleus is called the *parent*. The daughter product may be stable or radioactive. If the daughter product is radioactive, the series of decays that result are called *decay chains*.

A typical radioactive decay process is shown in Eq. (7.4). In this process, radioactive cobalt 60 decays to nickel 60 with the release of two gamma particles, a

Sec. 7.1 Principles of Nuclear Fission

beta particle, and an antineutrino ($\bar{\nu}$).

$$^{60}_{27}\text{Co} \rightarrow ^{60}_{28}\text{Ni} + ^{0}_{1}\beta + 2\gamma + \bar{\nu} \qquad (7.4)$$

Experiments show that the probability of radioactive decay is a constant for each specific radioactive species (radionuclide). The decay probability is a fundamental physical property of the radioisotope and cannot be changed by temperature, pressure, or chemical reaction. The probability that a given nucleus will decay in a unit time is called the *decay constant* and it is designated by λ, with units of time^{-1}, usually s^{-1}.

In a sample containing n_a atoms of a radionuclide, the expected value of the decay rate is λn_a decays per second. Since n_a is very large, on a statistical basis, the expected value is the frequency at which disintegration will be observed. The decay rate, λn_a, is called the *activity*, ACT:

$$\text{ACT} = \lambda n_a = -\frac{dn_a}{dt} \qquad (7.5)$$

The traditional unit for activity has been the curie, Ci, which is defined as 1 curie = 3.7×10^{10} disintegrations per second. The newer SI unit is the becquerel, defined as 1 disintegration per second.

We may integrate Eq. (7.5) to obtain a simple exponential relationship:

$$n_a(t) = n_a(0) e^{-\lambda t} \qquad (7.6)$$

where $n_a(0)$ = radioactive atoms present at time 0
$n_a(t)$ = radioactive atoms present at time t
t = time, s
λ = decay constant, s^{-1}

Because of the exponential nature of the radioactive decay process, it is useful to define a new term called the *half-life*. The half-life, $t_{1/2}$, is the time required for one-half of the radioactive atoms in a sample to disintegrate. It is related to the decay constant by

$$t_{1/2} = \frac{0.693}{\lambda} \qquad (7.7)$$

Extensive data on the physical properties and decay schemes of radioisotopes can be found in General Electric Co.'s *Chart of the Nuclides* included in the Bibliography for this chapter.

Frequently, the daughter product of a radioactive decay is itself radioactive. A succession of radioactive decays, called *chain decay*, results. A chain can be represented by

$$Y_1 \xrightarrow{\lambda_1} Y_2 \xrightarrow{\lambda_2} Y_3 \xrightarrow{\lambda_3} \cdots Y_n \qquad (7.8)$$

where Y_1 is the original radioactive member of the chain.

Mechanics of Nuclear Reactions

In most nuclear reactions, two particles or nuclei react to form two different nuclei. Such reactions are written as

$$a + x \rightarrow b + y \qquad (7.9)$$

Equation (7.9) can be written in a shorter form as

$$x(a,b)y$$

A practical example of this type of reaction occurs in a nuclear reactor when boron 10 (an isotope of natural boron used to help control a reactor) absorbs a neutron in a $^{10}\text{B}(n, \alpha)^7\text{Li}$ reaction. Since an α particle is the nucleus of an He atom, the reaction is written as

$$^1_0\text{n} + ^{10}_5\text{B} \rightarrow ^4_2\text{He} + ^7_3\text{Li} \qquad (7.10)$$

Energy Released in Nuclear Reactions (Q_{NR} Values)

There is always a net difference in mass between the reacting particles [LHS of Eq. (7.9)] and the product particles [RHS of Eq. (7.9)]. Conservation of total energy and the equivalence of mass and energy must be applied to nuclear reactions. One concludes that in an exothermic reaction, some mass must be converted to energy, whereas in an endothermic reaction, the reverse occurs. The Q_{NR} value of the reaction defined by Eq. (7.9) would be the energy equivalent of the difference in masses between the reactants and products:

$$Q_{NR} = \left[(m_a + m_x) - (m_b + m_y)\right]c^2 \qquad (7.11)$$

Example 7.1

Find the Q_{NR} value for the reaction described by Eq. (7.10), $^{10}\text{B}(n, \alpha)^7\text{Li}$.

Solution The masses in amu are:

$^{10}\text{B} = 10.1294 \qquad ^7\text{Li} = 7.01601$

$^1_0\text{n} = 1.008665 \qquad ^4_2\text{He} = 4.0026$

$$Q = \left(\sum m_r - \sum m_p\right)c^2 = \frac{\sum m_r - \sum m_p}{m_{\text{amu}}}\left(931 \frac{\text{MeV}}{\text{amu}}\right)$$

$$= \left[(10.01294 + 1.008665) - (7.01601 + 4.0026)\right] \text{amu}$$

$$\times 931 \frac{\text{MeV}}{\text{amu}} = 2.790 \text{ MeV}$$

Mechanics of the Fission Process

A nuclear reaction of special importance is the fission process. In this process, a neutron collides with a heavy fissionable isotope, is absorbed, and for a very brief period of time, 10^{-19} s, forms a compound nucleus in an excited energy state. The

Sec. 7.1 Principles of Nuclear Fission

compound nucleus may *fission*, that is, break into two unequal fragments with the release of energy, or it may release the energy in the form of gamma radiation. The latter behavior is known as *radiative capture*. The reaction process for uranium 235 is shown in Fig. 7.3 and Eqs. 7.12(a) and 7.12(b).

radiative capture: $\quad {}^{235}_{92}U + {}^{1}_{0}n \rightarrow {}^{236}_{92}U + {}^{0}_{0}\gamma \quad$ (7.12a)

fission: $\quad {}^{235}_{92}U + {}^{1}_{0}n \rightarrow K^{A_1}_{Z_1}Y + {}^{A_2}_{Z_2}Y + 2.4\,{}^{1}_{0}n + \text{energy} \quad$ (7.12b)

where ${}^{A_1}_{Z_1}Y$ and ${}^{A_2}_{Z_2}Y$ are the radioactive fission fragments.

The fission process results in several hundred different radionuclides being formed. Some radionuclides are formed directly in the fission process, as described by Eq. (7.12b). The direct yield of *fission products* is described in terms of the percent yield per fission for each radionuclide. The mass distribution of fission products from the fission of uranium 235 is shown by Fig. 7.4.

Nearly all of these fission products are radioactive and decay by beta or gamma emission to daughter-product radionuclides. Often, a series of radioactive

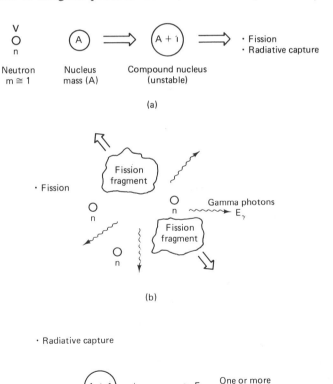

Figure 7.3 Neutron absorption by fissile material.

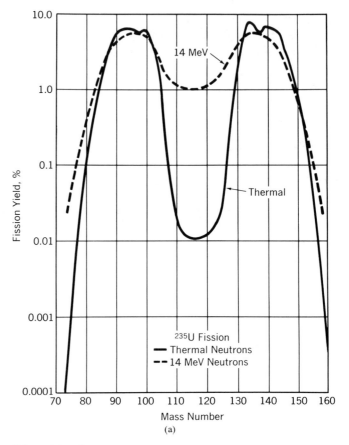

Figure 7.4 (a) Mass distribution of fission products from thermal and fast fission of uranium 235. (Reproduced with permission from *Steam, Its Generation and Use*, The Babcock & Wilcox Company, Lynchburg, Va., 1978.)

decays occur by the chain decay process described by Eq. (7.8). Decay continues until a stable element is reached. A typical decay process for the mass 90 chain would be as follows:

$$^{90}_{34}\text{Se} \rightarrow {}^{90}_{35}\text{Br} \rightarrow {}^{90}_{36}\text{Kr} \rightarrow {}^{90}_{37}\text{Rb} \rightarrow {}^{90}_{38}\text{Sr} \rightarrow {}^{90}_{39}\text{Y} + {}^{90}_{40}\text{Zr} \qquad (7.13)$$
$$\text{stable}$$

Some mass is converted into energy during the fission process. Approximately 200 MeV of energy is released during the fission process and is distributed as shown in Table 7.2. This energy release corresponds to approximately 3.2×10^{-11} J (W·s) per fission or the more useful factor, 3.1×10^{10} fissions per second equal 1 watt (thermal).

Most of this useful fission energy is deposited in the reactor fuel material. The fission fragments are relatively massive particles that travel only a short distance in

Sec. 7.1   Principles of Nuclear Fission

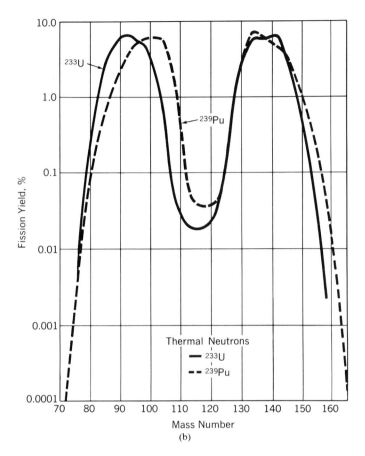

Figure 7.4 (b) Mass distribution of fission products from thermal fission of uranium and plutonium. (Reproduced with permission from *Steam, Its Generation and Use*, The Babcock & Wilcox Company, Lynchburg, Va., 1978.)

TABLE 7.2   ENERGY PRODUCED BY THE FISSION PROCESS

	MeV
Instantaneous energy	
Kinetic energy of the fission fragments	165
Kinetic energy of the fission neutrons	5
Electromagnetic energy of the fission gammas	7
Neutrino energy (not usable)	(10)
Delayed energy	
Beta decay energy from fission products	8
Gamma decay energy from fission products	7
Total usable energy	~ 192

the fuel material and deposit their kinetic energy in the form of heat. Similarly, the β energy is deposited in the fuel material, but the prompt gamma energy may be deposited in the water coolant, structural material, or shielding is well as the reactor fuel.

7.2 NUCLEAR CORE ANALYSIS

The power output of a nuclear power station is directly proportional to the rate at which fissions occur in the nuclear fuel. The fission reaction rate is dependent on the number of fissionable atoms present, a neutron flux which characterizes the neutron population, and a proportionality constant, unique to each isotope. This section will define the terminology used and develop some basic calculational techniques for nuclear reactor core analysis.

Reaction Rates, Cross Section, and Neutron Flux

The extent to which neutrons interact with nuclei can be described in terms of a proportionality constant called the *microscopic cross section*. The microscopic cross section can be understood most easily by considering the interaction of a collimated beam of neutrons, I_n, with a target material of area A and thickness l_t, as shown in Fig. 7.5. The interaction rate has been experimentally shown to be

$$R_a(\text{reactions per second}) = \sigma_t I_n N V_m \qquad (7.14)$$

where R_a = reaction rate, interactions per second
I_n = intensity of the neutron beam, n/cm$^2\cdot$s
N = atom density in the target material, atoms/cm^3
V_m = volume of the target material, cm^3 = Al_t
σ_t = proportionality constant called the total microscopic cross section, cm^2

The microscopic cross section is equivalent to the number of collisions per second per target nucleus for a unit intensity neutron beam (single neutron). The probability per unit path length that a neutron in the beam will collide with a nucleus in the target material is $\sigma_t N$. Alternatively, we may view the cross section as the *effective target area* presented by the nucleus to an approaching neutron.

Microscopic cross sections are expressed in units of area, square centimeters (cm^2) or barns. One barn is defined as 1 barn = 1×10^{-24} cm^2.

Cross sections may be further subscripted to indicate the relative probability of various types of reactions that may occur. For example:

σ_a = microscopic cross section for neutron absorption

σ_f = microscopic cross section for fission

σ_c = microscopic cross section for neutron capture

σ_t = total microscopic cross section for all reactions = $\sigma_a + \sigma_f + \sigma_c + \cdots$

Sec. 7.2   Nuclear Core Analysis

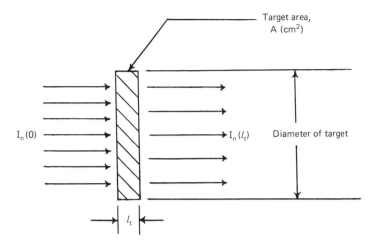

Figure 7.5 Interaction of a neutron beam with a target.

Often, as in Eq. (7.14), the product of the atom density, N, and the microscopic cross section occur together. This product is called the total macroscopic cross section, Σ_t (cm^{-1}).

$$\Sigma_t = N\sigma_t \tag{7.15}$$

The macroscopic cross section may also be used to designate the particular reactions of interest:

Σ_a = macroscopic absorption cross section, cm^{-1} ($\Sigma_a = N\sigma_a$)

Σ_f = macroscopic fission cross section, cm^{-1} ($\Sigma_f = N\sigma_f$)

Σ_t = macroscopic total cross section, cm^{-1} = $\Sigma_a + \Sigma_f$

We have already noted that $\sigma_t N$, the macroscopic cross section, can be viewed as the relative probability per unit path length that a neutron will interact with a nucleus.

Using these concepts and Fig. 7.5, it can be shown that the number of neutrons in the collimated beam which pass through the target of thickness without any interaction, $I_n(l_t)$, is

$$I_n(l_t) = I_n(0)e^{-\Sigma_t l_t} \tag{7.16}$$

The previous examples dealt with a monoenergetic beam of neutrons moving in a specified direction. This is a special laboratory situation. Neutrons in a power reactor, like molecules in a gas, have a completely random motion. In Fig. 7.5, the neutron beam intensity, I_n, is the product of the neutron density, n, in the beam, times the velocity of the neutrons, u_n, or $I_n = nu_n$. For the general case of random neutron motion in a nuclear reactor, this product is given the special name of *neutron flux,* ϕ.

$$\phi = nu_n \quad \text{neutrons/cm}^2 \cdot \text{s} \qquad (7.17)$$

The neutron flux is the total path length traveled by all the neutrons in 1 cm³ in 1 s.

The neutron flux can be used to calculate the reaction rate of neutrons with material i:

$$\text{reaction rate} = R_a = N_i \sigma_i \phi = \Sigma_i \phi \qquad \begin{array}{l}(7.18a)\\(7.18b)\end{array}$$

where R_a = reactions/cm³·s
N_i = atom density of material, atoms/cm³
σ_i = microscopic cross section for the reaction of interest, cm²
Σ_i = macroscopic cross section for the reaction of interest, cm⁻¹

The number of fission reactions, for example, would be

$$R_f = N_f \sigma_f \phi$$

where N_f designates the number density of fissionable atoms present. For a reactor system of core volume, V_m, the total fission rate, R'_f (fissions/s), would be

$$R'_f = \Sigma_f V_m \phi \qquad (7.19)$$

Since there are 3.1×10^{16} fissions/MW$_{th}$·s, the fission rate in a reactor producing $P_T \text{MW}_{th}$ is

$$R'_f = 3.1 \times 10^{16} P_T \qquad (7.20)$$

$$= \bar{\sigma}_f N_f \bar{\phi}_m = \bar{\Sigma}_f V_m \bar{\phi} \qquad (7.21)$$

In this equation $\bar{\sigma}_f$, $\bar{\Sigma}_f$, and $\bar{\phi}$ represent average values of these neutron energy-dependent parameters.

Cross-Section Energy Dependence

The reaction-rate equation, Eq. (7.14), is deceptively simple. The neutron cross section is a function of both the type of reaction and the energy or velocity of the incoming neutron. Cross sections are determined experimentally and are normally reported in units of barns as a function of neutron energy. Neutrons are born during the fission process in the MeV range (fission neutrons are thus at high velocities) and slow down in a reactor due to elastic collisions with water and other materials. Reactions occur between nuclei and neutrons at all neutron energy levels in the reactor. Although the reaction rate can still be predicted by a simple equation such as (7.18), the situation is complicated because the neutron flux and reaction cross section both vary with energy over a wide spectrum of energy levels. Figure 7.6 illustrates this variation in cross section for uranium 238. Energy-dependent cross sections have been determined and cataloged for essentially all materials.

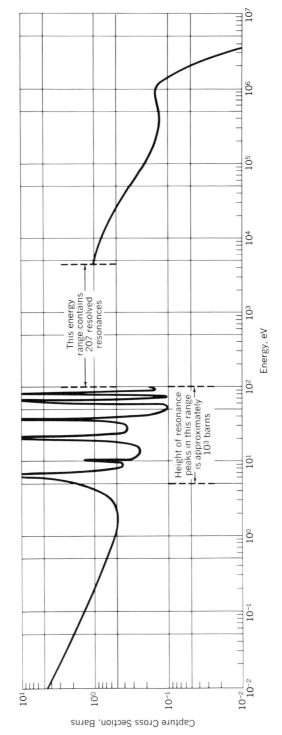

Figure 7.6 Capture cross section of uranium 238. (Reproduced with permission from *Steam, Its Generation and Use*, The Babcock & Wilcox Company, Lynchburg, Va., 1978.)

In order to develop concepts and perform simple hand calculations, this complication due to energy dependence is avoided by using one or two average neutron energy groups. The most important group, the thermal energy group, represents neutrons with a low kinetic energy that are in thermal equilibrium with the surrounding media.† In the usual power reactor, the neutrons are in thermal equilibrium with the water cooling the core. Neutrons in thermal equilibrium with the surrounding media would have a most probable velocity and corresponding kinetic energy expressed by

$$\text{kinetic energy} = \text{KE} = k'_B T \tag{7.22}$$

$$\text{neutron speed, } u_n = \sqrt{\frac{2\text{KE}}{m_n}} \tag{7.23}$$

where k'_B = Boltzmann constant, 1.38×10^{-23} J/°K
m_n = neutron mass, kg
T = temperature of surrounding medium, °K

At 20°C (293°K), the energy and velocity would be

$$\text{KE} = 1.38 \times 10^{-23} \text{ J/°K} \times 293°\text{K} = 4.023 \times 10^{-21} \text{ J} = 0.025 \text{ eV}$$

$$u_n = \left[\frac{2(4.023 \times 10^{-21})}{1.68 \times 10^{-27}}\right]^{1/2} = 2197 \text{ m/s}$$

Thermal neutrons actually exhibit a distribution of energies around the kinetic energy expressed by Eq. (7.22). The distribution of neutron energy is approximately described by the Maxwellian distribution with a mean value of $\frac{3}{2} k'_B T$. The thermal energy region is usually defined to include all neutrons up to the low-keV energy level.

Cross-section data are usually listed at 2200 m/s, which corresponds to a thermal neutron at 20°C. In the low-energy region, cross sections often vary inversely with the neutron velocity, called $1/v$ *behavior*. For these situations, the cross section at some slightly elevated temperature T (with kinetic energy, KE) can be obtained from the 2200-m/s cross section by

$$\sigma(\text{KE}) = \sigma(\text{th}) \sqrt{\frac{\text{KE(th)}}{\text{KE}(T)}} \tag{7.24}$$

where $\sigma(\text{KE})$ = cross section at temperature, T, barns
$\sigma(\text{th})$ = 2200-m/s cross section, barns
$\text{KE}(T)$ = neutron energy at temperature, T, eV
KE(th) = thermal neutron energy, 0.025 eV

As the neutron energy increases, the cross sections begin to develop sharp peaks, called *resonances* (see Fig. 7.6). Because the cross section changes so rapidly

---

†Neutrons that have been slowed down by elastic collisions to the point where on the average they neither gain nor lose energy by further collision.

Sec. 7.2 Nuclear Core Analysis

with neutron energy, an integral cross-section value is reported, called the *resonance integral* (RI):

$$\text{RI} = \int_{\text{KE}_1}^{\text{KE}_2} \sigma(\text{KE}) \frac{d\text{KE}}{\text{KE}} \qquad (7.25)$$

The resonance region extends from neutron energy level KE_1 to neutron energy level KE_2.

This integral is generally experimentally evaluated over the energy range $\text{KE}_1 \simeq 0.5$ eV to $\text{KE}_2 \simeq 0.1$ MeV. The reaction rate at any energy level follows the form of Eq. (7.18a) or (7.18b) but is complicated by the variation of neutron population with energy and the energy-dependent nature of the cross sections.

Nuclear Reactor Processes

Although we have indicated that uranium undergoes fission, it would, perhaps, have been more proper to indicate that only one of the naturally occurring isotopes of uranium fissions readily. Natural uranium consists primarily of ^{238}U and about 0.7% ^{235}U. It is only the ^{235}U isotope that fissions readily, and further, its fission cross section, σ_f, is high only at low neutron energies (thermal energies).

Since the neutrons produced by fissioning ^{235}U are at high energy levels (~ 2 MeV), we must slow down the fission neutrons if we wish to use a uranium fuel containing only a small fraction of ^{235}U. We do this placing a *moderator* material in the reactor. The moderator is low-atomic-mass material having a relatively low absorption cross section (hydrogen and carbon are two common moderators) which slow down the neutrons to thermal energies by elastic collisions. The thermalized neutrons are then absorbed by the ^{235}U to produce fission. The fission reaction not only produces energy but about 2.4 neutrons per fission. If one of these neutrons escapes radioactive capture prior to being absorbed by a ^{235}U atom, we have a self-sustaining *chain reaction*.

Because of the large number of ^{238}U atoms present, a number of the fission neutrons will be absorbed by ^{238}U. The sequence of reactions is

$$^{238}\text{U}(n, \gamma) \rightarrow {}^{239}\text{U} \xrightarrow{\beta} {}^{239}\text{Np} \xrightarrow{\beta} {}^{239}\text{Pu} \qquad (7.26)$$

Although plutonium (Pu) undergoes α decay, it has a very long half-life and can be considered stable. Plutonium has a significant fission cross section and it can be recovered from fuel discharged from a reactor and used to fuel other reactors. At the end of the irradiation period for a reactor fuel element, the fissions occurring in plutonium cannot be ignored.

Although nearly all the fissions in a reactor at the beginning of life occur in ^{235}U, a few percent occur in ^{238}U. Both ^{235}U and ^{238}U have a small fission cross section in the fast energy range. Although the ^{235}U fast fission cross section in larger, there is a large amount of ^{238}U present. Hence most fissions due to fast neutrons originate from fast neutrons colliding with ^{238}U atoms. Although this is helpful to the designer, this effect is overshadowed by the removal of neutrons by ^{238}U before they are thermalized (resonance absorption). In most thermal reactors, it is necessary to improve the probability that a neutron will be absorbed in ^{235}U by increasing its ^{235}U concentration over that found naturally. Such fuel is said to be *enriched*. Reactors moderated by ordinary water (H_2O in contrast to D_2O) generally use uranium enriched to a ^{235}U concentration four to five times that occurring naturally.

Neutron Balance in an Operating Reactor

In a power reactor, the neutron population is a function of the production rate, absorption rate, and leakage rate. In general, the time rate of change of the neutron population is

change in neutron population = production rate from fission
— absorption rate — leakage rate

The neutron leakage from a reactor core is difficult to calculate, so the idealized concept of an infinite system with homogeneous material properties is used to simplify initial calculations. For this case, an infinite multiplication factor is defined as

$$k_\infty = \frac{n^1}{n} = \frac{\text{number of neutrons in new generation}}{\text{number of neutrons in the previous generation}} \quad (7.27)$$

When the reactor is at constant power, the neutron population is constant. For an infinite system at constant power, the number of neutrons produced must equal the number absorbed, or

$$k_\infty = \frac{\nu \Sigma_f \phi_{th}}{\Sigma_a \phi_{th}} = \frac{\nu \Sigma_f}{\Sigma_a} \quad (7.28)$$

where ν is the average number of neutrons produced per fission event (2.418 for ^{235}U).

Even though many large power reactors can be approximated by infinite systems, some leakage of neutrons does occur from the outer surfaces of the reactor core. This leakage can be calculated by the rigorous mathematical methods developed in reactor theory. When the leakage term is considered, the infinite multiplication factor is converted to an effective multiplication factor, k_{eff}.

$$k_{eff} = k_\infty \mathscr{P}_{NL} \quad (7.29)$$

where $\mathscr{P}_{NL}$ is the nonleakage probability evaluated for a finite reactor system.

In order for a power reactor to operate, the neutron population must reach a constant or steady-state value consistent with the reactor power level [Eq. (7.27)]. For such a condition, $k_{eff} = 1.0$ and $k_\infty > 1$ to allow for neutron leakage.

Four-Factor Formula

A simple method that can be used to evaluate thermal reactor systems is called the *four-factor formula*. Although it is less complex than the methods currently used for design, it does provide a reasonable approximation and it provides a good physical understanding of some basic reactor physics parameters. For an infinite system, the four-factor formula is written

$$k_\infty = \eta \varepsilon' p_{re} f_{th} \qquad (7.30)$$

where η = thermal fission factor
ε' = fast fission factor
p_{re} = resonance escape probability
f_{th} = thermal utilization factor

The fast fission factor, ε', is defined as the ratio

$$\varepsilon' = \frac{\text{total number of fast neutrons produced in the core}}{\text{number of fast neutrons produced by thermal fission}}$$

The ε' factor accounts primarily for the fast neutron fission in uranium 238. If there were no uranium 238 present, we would have $\varepsilon' = 1.0$. In light-water reactors, there is a large amount of uranium 238 and $\varepsilon' \simeq 1.03$. Thus the neutrons born from thermal fission are augmented by those born directly in fast fission. All of these fast neutrons begin to slow down, but some are absorbed in the high resonance absorption peaks of uranium 238 (see Fig. 7.6). This absorption is accounted for by the resonance escape probability term.

The resonance escape probability, p_{re}, is defined as

$$p_{re} = \frac{\text{number of neutrons reaching thermal energy}}{\text{total number of fast neutrons starting to slow down}}$$

The resonance escape probability is the fraction of neutrons that escape capture in uranium 238 resonances while slowing down. We have seen that the neutrons captured in the uranium 238 produce plutonium 239, a useful nuclear fuel. However, enough neutrons must escape capture and reach the thermal energy region to continue the neutron chain reaction.

Once reaching the thermal region, neutrons may be captured in the fuel material, or any other material present in the reactor core. The thermal utilization factor, f_{th}, is the ratio of

$$f_{th} = \frac{\text{neutrons absorbed in the fuel material}}{\text{neutrons absorbed in all core materials}}$$

$$f_{th} = \frac{(\Sigma_a)_{fuel}}{(\Sigma_a)_{fuel} + (\Sigma_a)_{\text{all other material}}} \qquad (7.31)$$

For each neutron absorbed in the fuel, there are η fast neutrons produced. Thus η is defined as

$$\eta = \frac{\text{number of fast neutrons produced by thermal fission}}{\text{number of thermal neutrons absorbed in the fuel}} \qquad (7.32)$$

We may compute η from

$$\eta = \frac{\nu\Sigma_f}{(\Sigma_a)_{\text{fuel}}}$$

For light-water reactor systems using slightly enriched fuel, η ranges from 2.05 to 2.10. A typical value would be 2.09.

Thus, for an infinite system, the ratio of next-generation fission neutrons to initial-generation fission neutrons is given by the four-factor formula, $k_\infty = \eta\varepsilon p_{\text{re}} f_{\text{th}}$. It has been pointed out that for a finite system we must obtain k_{eff} by multiplying our result by $\mathscr{P}_{\text{NL}}$, the nonleakage probability. It is generally desirable to express $\mathscr{P}_{\text{NL}}$ by two leakage terms which account for fast and thermal neutron leakage separately or $\mathscr{P}_{\text{NL}} = \mathscr{P}_{\text{th}} \times \mathscr{P}_f$. Thus Eq. (7.29) becomes

$$k_{\text{eff}} = k_\infty \times \mathscr{P}_{\text{th}} \times \mathscr{P}_f \tag{7.33}$$

where $\mathscr{P}_{\text{th}}$ = thermal nonleakage factor
$\mathscr{P}_f$ = fast nonleakage factor

The thermal nonleakage factor is the fraction of neutrons that do not escape from the reactor core after reaching thermal energy. These neutrons are available for absorption in the fuel to continue the chain reaction. The thermal nonleakage factor can be expressed as

$$\mathscr{P}_{\text{th}} = \frac{1}{1 + L_{\text{th}}^2 B_g^2} \tag{7.34}$$

Two new terms have been introduced in this equation, the thermal neutron diffusion length, L_{th}, and the geometric buckling, B_g^2. The thermal diffusion length is a measure of the distance that a neutron travels while slowing down. It is one-sixth of the mean square of the distance that a neutron travels from the point where it becomes a thermal neutron to the point where it is absorbed. Values of the thermal diffusion length are well established for most materials used in reactors.

The geometric buckling is a measure of the actual physical size of the fueled region of the reactor core. Most power reactors have a fueled core region that is cylindrical in shape. The geometric buckling for a cylindrical-shaped core is

$$B_g^2 = \frac{(2.405)^2}{R_c^2} + \frac{\pi^2}{z_c^2} \tag{7.35}$$

where R_c = radius of the active core, m
z_c = height of the active core, m

Notice that the buckling becomes smaller as the physical size of the core increases. This is consistent with the actual neutron balance. As the size of the fueled core region increases, the neutron leakage decreases.

The second leakage term is the fast nonleakage probability, $\mathscr{P}_f$. This term expresses the probability that a fast neutron will not escape from the system during the slowing-down process, that is, from its birth as a fast neutron until it reaches thermal energy. This term may be approximated by the equation

$$\mathscr{P}_f = \frac{1}{e^{\tau B_g^2}} \quad (7.36)$$

The new variable, τ, is called the *Fermi age*. It is a measure of the distance that a fast neutron travels while slowing down. It is one-sixth of the mean square of the distance that a neutron travels from the point where it is born as a fast neutron to the point where it reaches thermal energy. It is the fast neutron analog of the diffusion length. Values of the Fermi age are available for the materials used in power reactors. Using these new nonleakage terms, the effective multiplication factor becomes

$$k_{\text{eff}} = \eta \epsilon p_{\text{re}} f_{\text{th}} \frac{e^{-B_g^2 \tau}}{1 + L_{\text{th}}^2 B_g^2} \quad (7.37)$$

Equation (7.37) should be viewed as a neutron balance equation for a thermal reactor of finite size. Each term is a measure of a portion of the neutron economy of the system. Figure 7.7 illustrates the neutron balance through the use of a neutron cycle diagram. Most of the terms in this diagram coincide with those of the four-factor formula.

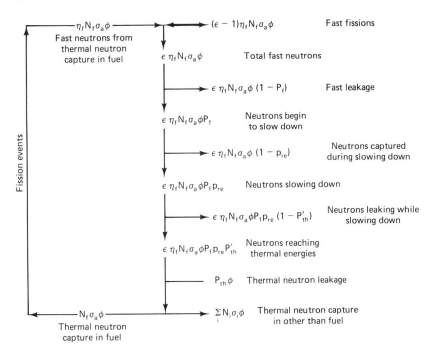

Figure 7.7 Neutron life cycle in a thermal reactor.

Equation (7.37) is a highly simplified treatment of reactor core physics. It is valid for use as an initial approximation of reactor core materials and geometry required to achieve a critical condition. More rigorous treatment can be found in the texts included in the Bibliography for this chapter.

Example 7.2

A cylindrical reactor has a radius of 120 cm and a height of 120 cm. The fuel rods are clad with 0.025-cm-thick aluminum and have an outer radius of 0.55 cm. The fuel rods are arranged in a square lattice with a pitch (center-to-center distance) of 0.72 cm. The fuel rods contain UO_2 at 95% of the theoretical density and enriched to 1.6 atom % ^{235}U. You may assume $\varepsilon' = 1.02$, $p_{re} = 0.86$, $\tau = 36.1$ cm², and $L_{th}^2 = 2.22$ cm². Determine k_{eff} at the beginning of life at room temperature.

Solution From the dimensions of a unit cell, we calculate the following volume fractions:

Material	Volume fraction
Water	0.6485
Fuel	0.3215
Cladding (aluminum)	0.030

We then calculate the number densities, $(N)_j$, of each component from the volume fractions by using

$$(N)_j = \sum_j \frac{V'_j \rho_j i_j r (Av)}{MW}$$

where V'_j = fraction of material j
  ρ_j = density of material j, g/cm³
  i_j = number of atoms of component i per molecule of j
  r = ratio of atoms of given isotope to total number of atoms of component (unity except for ^{235}U)
  Av = Avogadro's number = 6.03×10^{23} molecules/g mol
  MW = molecular weight of material

The values of $(N)_j$ are then used with the values of σ_a and σ_f obtained from standard tables to obtain Σ_a and Σ_f. We then have the following:

Component	$N \times 10^{-24}$	σ_a (barns)	σ_f (barns)	Σ_a (cm^{-1})
^{235}U	1.18×10^{-4}	683	577	0.08
^{238}U	7.09×10^{-3}	2.71	—	0.0192
Hydrogen	4.338×10^{-2}	0.332	—	0.0144
Oxygen	3.58×10^{-2}	—	—	—
Aluminum	1.81×10^{-3}	0.241	—	0.00044

We now compute the parameters of the four-factor formula:

$$f_{\text{th}} = \frac{\Sigma_{a_U}}{\Sigma_{a_U} + \Sigma_{a_{H_2}} + \Sigma_{a_{Al}}} = \frac{0.08 + 0.0192}{0.0992 + 0.0144 + 0.00044} = 0.87$$

$$\eta = \frac{\nu \Sigma_f^{235}}{\Sigma_a^{235} + \Sigma_a^{238}} = 2.418 \left(\frac{0.000118 \times 577}{0.0992} \right) = 1.66$$

$$k_\infty = p_{\text{re}} \varepsilon' \eta f_{\text{th}} = (0.86)(1.02)(1.66)(0.87) = 1.27$$

To obtain k_{eff}, we first calculate B_g^2:

$$B_g^2 = \left(\frac{2.405}{R_c} \right)^2 + \left(\frac{\pi}{z_c} \right)^2 = \left(\frac{2.405}{120} \right)^2 + \left(\frac{3.1417}{120} \right)^2 = 1.09 \times 10^{-3} \text{ cm}^{-2}$$

We may now compute k_{eff}:

$$k_{\text{eff}} = \frac{k_\infty e^{-B_g^2 \tau}}{1 + L_{\text{th}}^2 B_g^2} = 1.27 \left(\frac{\exp\{-[1.09 \times 10^{-3}(36.1)]\}}{1 + (2.22)(1.09 \times 10^{-3})} \right) = 1.20$$

From the foregoing, it is seen that at the beginning of life, additional absorbing material will have to be introduced into the core so that $k_{\text{eff}} = 1.0$.

Physical Significance of the Neutron Balance

When k_{eff} is equal to 1.0, the reactor is just critical and for each neutron causing fission, there is one neutron available for fission in the next generation. The number of neutrons in each successive generation remains constant. This implies that the neutron flux is constant and the reactor power is constant. When $k_{\text{eff}} = 1.0$, the reactor is operating at a steady-state condition.

When $k_{\text{eff}} < 1.0$, the reactor power level and flux are decreasing and the reactor is said to be *subcritical*. The number of neutrons in each successive generation is decreasing.

For conditions where $k_{\text{eff}} > 1.0$, the reactor is called *supercritical*. The power, flux level, and the number of neutrons in each successive generation are increasing. We cannot allow this except during reactor startup. We therefore introduce absorber materials, in the form of control rods, to maintain $k_{\text{eff}} = 1.0$. We gradually remove the control rods from the core during the fuel life to compensate for the absorbing material (poisons) introduced by the fission products and the decrease in ^{235}U concentration.

Thermal and Fast Reactors

Power reactors are categorized as thermal reactors or fast reactors, depending on the average energy level of the neutron flux spectra. A thermal reactor is one in which the largest percentage of the reactor power is produced through fissions that occur

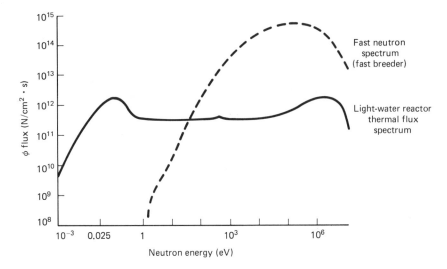

Figure 7.8 Neutron flux spectra for thermal and fast reactors.

with neutrons of thermal energy 10^{-3} to 1 eV. Conversely, a fast reactor is one in which the power is generated primarily by neutrons of high energy, 1 to 10^7 eV. Figure 7.8 shows the neutron flux spectra for these two types of reactors. The neutron balance equations developed previously apply only to thermal reactors.

All of the large light-water power reactors are thermal reactors. Collisions with the hydrogen atoms in the water molecules slows down the neutrons to thermal energy where fission is more likely to occur. As described in Section 7.3, fast reactors utilize either a liquid metal or gaseous coolant that does not slow the neutrons down. Fission occurs at fast- or high-neutron-energy levels. We must use a fuel containing a large percentage (e.g., 20 atom %) of fissionable isotopes in such a reactor.

Reactor Dynamic Behavior

We have noted that the neutron multiplication expressed by Eq. (7.33) must be adjusted. We may do this by changing production, absorption, or leakage. The usual method of control in power reactors in absorption. Three techniques are employed to adjust neutron absorption during power operation: (1) the use of solid movable absorbers (control rods), (2) the use of solid fixed absorbers in the fuel rod itself (burnable poisons) and (3) the use of absorbing material in the primary coolant (soluble poison).

Since k_{eff} represents the multiplication factor for a particular system, then $(k_{\text{eff}} - 1)$ is the excess in the multiplication factor over that needed for criticality. The term $(k_{\text{eff}} - 1)/k_{\text{eff}}$ is the fractional excess multiplication, which has been given the name *reactivity*, ρ_r:

$$\rho_r = \frac{k_{\text{eff}} - 1}{k_{\text{eff}}} = \frac{\delta k}{k_{\text{eff}}} \qquad (7.38)$$

where ρ_r = reactivity
k_{eff} = effective multiplication factor
$\delta_k = k_{\text{eff}} - 1$

A parameter used to evaluate dynamic changes in the reactor system is the mean neutron generation time, t^*. It is the time required for neutrons in one generation to produce the fissions that yield the next generation of neutrons. The mean neutron generation time is

$$t^* = \frac{1}{u_n \nu \Sigma_a} \qquad (7.39)$$

The mean generation time is divided into two components, the time spent as a fast neutron and the time spent as a thermal neutron. The fast-neutron slowing-down time is in the order of 10^{-7} s. Thermal neutrons tend to diffuse around for 10^{-4} s before being absorbed. Based on these crude estimates, fast reactors have $t^* = 10^{-7}$ s, while thermal reactors have $t^* = 10^{-4}$ s.

During startup, the neutron flux level increases exponentially with time and in a simplified way can be expressed as

$$\phi(t) = \begin{cases} \phi(0)\exp\left(\frac{\delta k}{t_a}t\right) & 0 < \delta k \leq \beta^* \\ \phi(0)\exp\left[\frac{t(\delta k - \beta^*)}{t^*}\right] & \delta k > \beta^* \end{cases} \qquad (7.40)$$

where t_a is the average lifetime of prompt and delayed neutrons $\simeq 0.085$ s. Recall that δk depends on the value of k_{eff} and can be adjusted by control rod movement. The new term introduced, β^*, represents the fraction of neutrons that are not released promptly at the fission process. These so-called *delayed neutrons* appear anywhere from a few tenths of a second to minutes after fission. Their effective lifetime is much longer than $1/u_n\nu\Sigma_f$. The net result is that the time constant for the reactor power level to increase by a factor of e is much longer than that from prompt neutrons alone. This enables the reactor control system operating at reasonable speeds to effect a positive control of reactor power level as long as we keep $\delta k < \beta^*$.

7.3 POWER REACTOR SYSTEMS

Power reactors are classified according to the type of fuel utilized, the neutron flux spectrum, the coolant utilized, and in thermal reactors, the type of moderator (see Table 7.3).

TABLE 7.3 REACTOR CLASSIFICATION

Flux spectrum	Moderator	Coolant	Fuel material
Thermal	Light water	Light water	Enriched uranium
	Heavy water	Heavy water	Natural uranium
	Graphite	Gas	Natural or enriched uranium
Fast		Liquid-metal	Plutonium
		Helium	

TABLE 7.4 TYPICAL POWER REACTOR CHARACTERISTICS

	PWR	BWR	LMFBR	HTGR
Electric power (MWe)	1300	1050	1000	330
Thermal power (MW$_{th}$)	3800	3000	2750	842
Specific power (kW$_{th}$/kg heavy metal)	33	26	575	50
Power density (MW$_{th}$/m^3)	100	60	300	10
Core height (m)	4.25	3.75	1.50	5.0
Core diameter (m)	3.50	4.90	3.25	5.9
Coolant	H$_2$O	H$_2$O	Liq. Na	He
Pressure (MPa)	15.5	7.2	0.8	4.8
Inlet temperature (°C)	280	275	330	400
Outlet temperature (°C)	310	285	500	770
Coolant flow rate (Mg/s)	20	12	11	0.45
Average linear heat rate (kW/m)	22.5	20	30	

Light-water cooled and moderated reactors (LWRs) using slightly enriched uranium fuel are the type most commonly used for power production. These reactors are further divided into boiling-water reactors (BWRs) or pressurized-water reactors (PWRs). High-temperature gas-cooled thermal reactors (HTGRs) have been used on a limited basis.

Fast reactors, using a liquid-metal coolant, with either plutonium or a plutonium–uranium mixture for fuel, have been operated in this country and elsewhere around the world. Many countries, France most notably, are looking toward liquid-metal fast breeder reactors (LMFBRs)[†] as the source of electrical power for the future. These reactor types, plus several other power reactor concepts, are reviewed in this section. The characteristics of several different reactor systems are listed in Table 7.4.

Although our initial discussion was devoted to nuclear analysis, the actual system design is not dominated by nuclear considerations. A nuclear power reactor produces very large amounts of heat and it is the provision of adequate heat removal that usually dominates the system design.

[†]A *breeder reactor* is a nuclear reactor in which the rate of fissionable isotope production (e.g., production of Pu) exceeds the rate at which fissionable isotopes are consumed.

Pressurized-Water Reactor

The pressurized-water reactor is moderated and cooled by light water. The system pressure is maintained at about 15.5 MPa (2260 psi), which is high enough to prevent bulk boiling of the coolant in the exit line from the reactor core. The major U.S. suppliers of the nuclear steam supply systems for PWRs are the Westinghouse Electric Company, the Babcock & Wilcox Company, and Combustion Eng., Inc. Overseas manufacturers of PWR systems are located in Germany, France, Japan, and the USSR.

Figure 7.9 shows the PWR reactor system and internal components. The primary coolant water exchanges heat in the steam generators and returns to the primary core vessel through the inlet nozzles. The water flows down through the annular region between the outside of the active core and the inside of the steel reactor vessel. The water cools these components as it flows downward to the bottom of the vessel. The water than returns and flows upward through the active core region. After removing heat from the fuel rods, the water enters a common upper plenum and exits through the outlet nozzles. From the outlet nozzles, the heated water flows to the steam generators and the process is repeated.

A typical PWR core contains about 200 fuel assemblies, as shown in Fig. 7.10. Each assembly is an array of rods. In a typical fuel assembly, such as shown in Fig. 7.10, there are 264 fuel rods and 24 guide tubes or thimbles for control rods. A central tube is available for instrumentation. Although all fuel assemblies are designed to accept control rods, only approximately one-third of the assemblies in the core actually contain control rods.

The grid assemblies, as shown in Fig. 7.10, are spaced at regular intervals along the fuel bundle. These grid spacers are designed to maintain a separation between the fuel rods, prevent excessive vibration, and allow some axial thermal expansion of individual fuel rods.

The top and bottom nozzles somewhat control the flow of coolant water through the fuel assembly, but since there are no cans around the assemblies, there is significant lateral coolant flow in the core itself.

The control rods described above provide for short-term reactivity changes, such as increasing or decreasing the power level of the reactor. Intermediate to long-term reactivity control is accomplished through the use of a soluble poison added to the coolant water. This minimizes the need for routine control rod insertion, leading to a more uniform power distribution.

PWR steam generating systems. Most PWR systems consist of three or four independent "loops." Each loop has its own steam generator and usually a single pump. The pumps are commonly a single-stage vertical centrifugal type driven by a 6-MW (8000-hp) motor. Such pumps can deliver about 6 m^3/s (95,000 gpm) at a differential pressure of 0.70 MPa (100 psi).

Two different types of steam generators are used in PWR systems, although they are both shell-and-tube designs. The primary coolant from the reactor flows

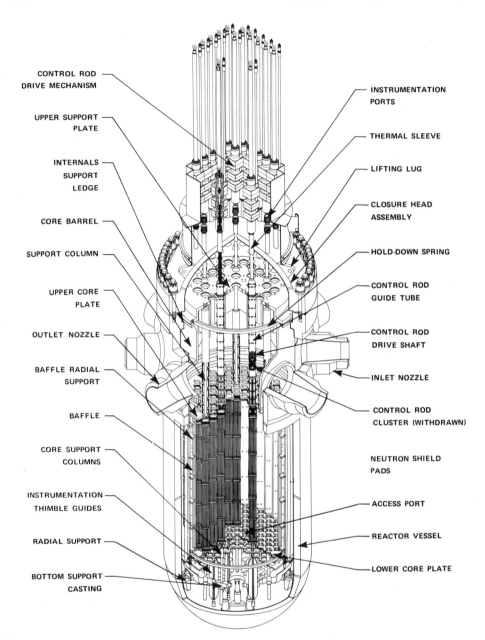

Figure 7.9 PWR reactor vessel internals. (Courtesy of Westinghouse Electric Co., Nuclear Energy Systems, Pittsburgh, Pa.)

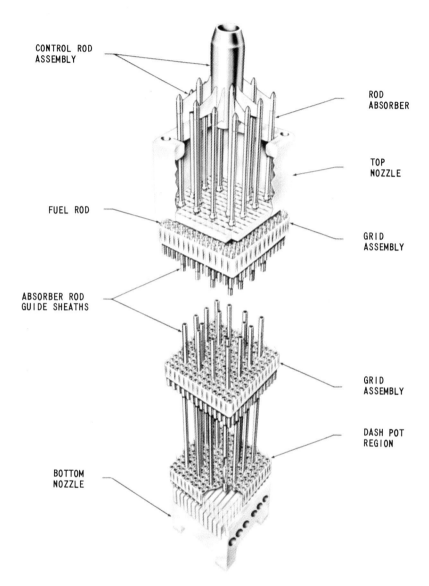

Figure 7.10 Fuel element assembly consisting of a 17×17 array of fuel rods and a rod cluster poison control assembly. (Courtesy of Westinghouse Electric Co., Nuclear Energy Systems, Pittsburgh, Pa.)

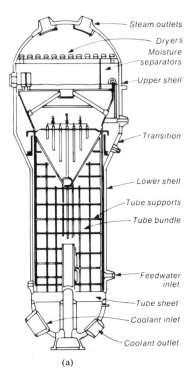

Figure 7.11 (a) U-tube recirculating-type steam generator. (Reprinted with permission from *Power*; copyright 1979 McGraw-Hill, Inc., New York.)

inside the tubes, while the secondary loop water fills the region between the outside of the tubes and the inside of the shell. Most vendors utilize a U-tube steam generator shown in Fig. 7.11(a), while one vendor utilizes a once-through type of steam generator [Fig. 7.11(b)].

In the U-tube design, primary coolant from the reactor core enters the bottom primary inlet, flows through the U-shaped path shown, and leaves through the primary outlet on the opposite side. In the once-through steam generator design, the primary coolant enters at the top of the steam generator and flows downward through straight tubes. The major difference in the designs is that only a fraction of the secondary water flowing upward is vaporized in the U-tube design. The steam is separated at the top of the unit and the remaining water recirculated. In a once-through design all the entering secondary water is evaporated.

The system pressurizer is another vital component of the steam generation system. It is a large cylindrical tank, connected to one of the primary coolant loops. During normal operation, the pressurizer contains 50 to 60% water, with steam occupying the remaining fraction of the tank volume. Electric immersion heaters can heat the water to increase pressure and a cold-water spray can be used to reduce steam pressure and temperature. The pressurizer utilizes these heating and cooling features to control primary system pressure. If the pressure exceeds a predetermined value, power-operated valves vent the system. If the pressure continues to rise, spring-loaded relief valves are actuated. Steam from the steam generators flows to

Sec. 7.3 Power Reactor Systems   293

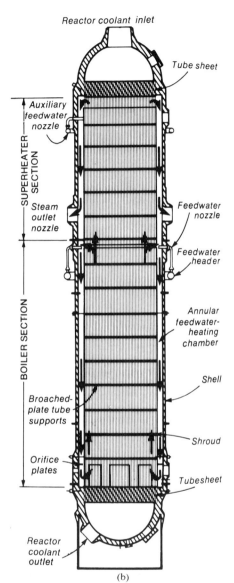

Figure 7.11 (b) Once-through straight-tube steam generator. (Reprinted with permission from *Power*; copyright 1979, McGraw-Hill, Inc., New York.)

the high-pressure turbine. The remainder of the steam plant operation has been described previously.

Boiling-Water Reactors

The current version of the boiling-water reactor (BWR) was commercialized by the General Electric Company. The first commercial prototype was Dresden I, which achieved full power in 1960. The BWR differs primarily from the PWR in that the

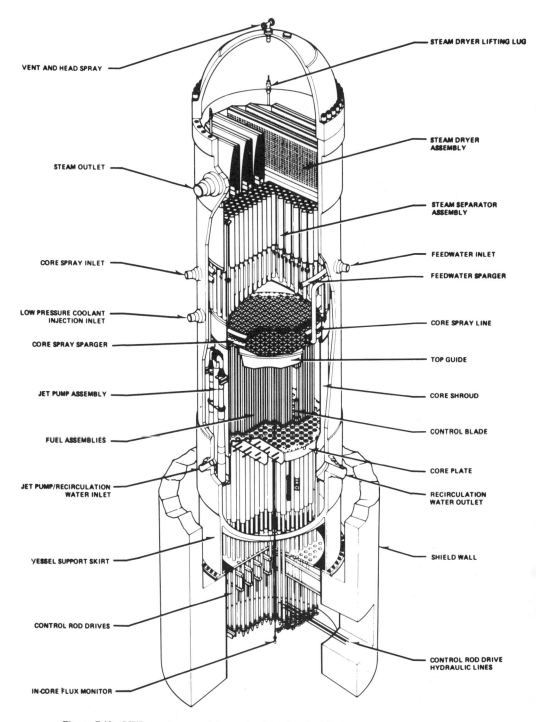

Figure 7.12 BWR reactor vessel intervals. (Reprinted with the permission of the General Electric Company, Nuclear Energy Operations, San Jose, Calif.)

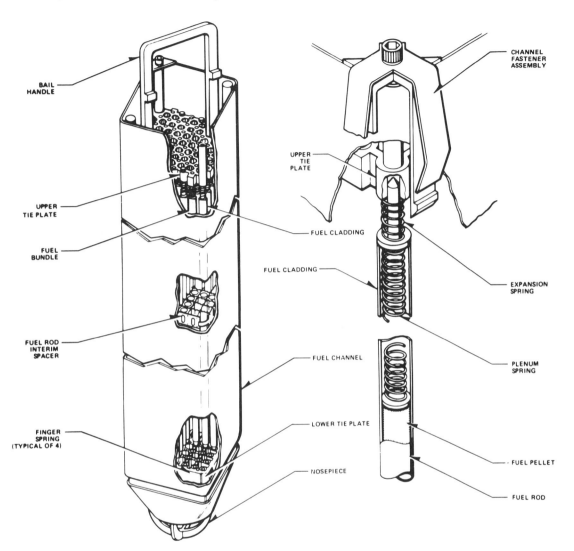

Figure 7.13 BWR fuel assembly. (Reprinted with the permission of the General Electric Company, Nuclear Energy Operations, San Jose, Calif.)

steam flowing to the turbine is produced directly in the reactor core. The steam is separated and dried by mechanical devices located in the upper part of the pressure vessel assembly shown in Fig. 7.12. The dried steam is sent directly to the high-pressure turbine, thereby eliminating the need for steam generators.

The reactor power is appreciably affected by the quantity of steam in the reactor core. An increase in the volume of steam (or the boiling rate) displaces water (moderator) in the core and reduces the ability of the moderator to thermalize neutrons and hence reduces the reactor power level. At power levels above 60% of nominal, the fraction of steam in the core can be kept nearly constant by varying the coolant circulation rate. To attain lower power levels, and to compensate for reactivity changes with lifetime, reactivity is controlled by a large number (about 160 in large plants) of cruciform-shaped control rods. Because of the steam drying equipment in the top of the reactor vessel, the control rods for a BWR enter from the bottom of the core. The absorber rods are used for reactor startup and shutdown and also for reactor power flattening in addition to control of operating power at low power levels.

The coolant recirculation system is a unique design feature of the BWR. Approximately 15% of the total weight of reactor coolant water is converted to steam. The remaining 85% of the flow is recirculated. About 30% of the core inlet flow is bled off to two or more recirculation loops. Centrifugal pumps circulate the water at high pressure and velocity and deliver this water to the 10 or 12 jet pumps located circumferentially around the vessel. High-velocity flow through the jet nozzles causes suction, which draws additional water into the jet pumps. The net effect is that the required core coolant flow rate can be supplied by circulating only a fraction of the flow through external pumps. The size of the external loop and circulating pump are therefore reduced.

Core and fuel elements. The active or fueled core region consists of approximately 800 fuel assemblies. Each typically contains an 8 by 8 array of fuel rods and associated hardware as shown in Fig. 7.13. Each assembly contains 62 fuel rods and two water rods. The water rods help keep the moderator to fuel ratio in the desired range. The Zircaloy channel around the fuel rods prevents cross-flow in the reactor core. The control rod elements are contained within a stainless steel cruciform-shaped blade that moves between a set of four fuel element assemblies.

Most of the fuel element assemblies are supported from the top of the control rod guide tubes, which obtain support from the control rod penetration nozzles in the bottom of the reactor vessel. The core plate, shown in Fig. 7.12, is used for lateral support and vertical support for peripheral assemblies. Fuel assemblies can be provided with orifices to adjust and regulate the coolant flow for more uniform power distribution.

BWR fuel rods, shown in Fig. 7.13, are slightly larger than PWR fuel rods. A typical pellet diameter is 10.6 mm (0.41 in.) with an outside cladding diameter of 12.5 mm (0.49 in.). The average fuel enrichment varies from 1.9 to 2.6%. Fuel design is discussed in more detail in Section 7.5.

Liquid-Metal Fast Breeder Reactor

General concepts. The liquid-metal fast breeder reactor is not yet (1984) considered to be in commercial operation, but it is being studied extensively. The first electricity from nuclear power was generated by the experimental breeder reactor (EBR) in 1952. Several small units were built and operated in the United States: the EBR I, the EBR II, and the Enrico Fermi reactor.

At present (1984), the United States has built and is operating the Fast Flux Test Facility. Other nations, especially the French, are aggressively pursuing the large LMFBR concept. The French Super Phenix, a 1200-MWe LMFBR, is under construction and the Super Phenix II (1500 MWe) is being designed. Demonstration plants are also being operated by the USSR.

Fast breeder reactors are designed to create or breed new fissile material while producing useful electric power. Most produce fissile plutonium from fertile uranium 238, as described previously. The fuel rods in the core region thus contain a mixture of fissile plutonium and uranium 238. The active core region is surrounded by a so-called "blanket" of fertile uranium 238. This blanket region captures neutrons that would otherwise be lost to the system through leakage, thus producing additional fissile material. There is both an axial breeding blanket and a radial breeding blanket.

Since the fission neutrons are not thermalized, the microscopic fission cross section is low. As noted previously, fuel with a substantial fraction of Pu (about 20 atom %) is therefore required to achieve criticality.

Steam generation cycles. All LMFBR designs use liquid sodium as the coolant. Liquid sodium has excellent heat-removal capability and low-neutron-absorption properties. However, some absorption does occur and the liquid sodium becomes radioactive. It is also highly reactive in contact with water. For these reasons, an intermediate or secondary, clean sodium loop is required.

Two conceptually different arrangements have been devised to provide for these intermediate loops. The first is a pool-type system, illustrated by Fig. 7.14(a), in which the reactor core, primary pumps, and intermediate heat exchangers are all placed in a large pool of liquid sodium contained within the reactor vessel. The liquid sodium is discharged from the intermediate heat exchanger to the pool. It is then pumped upward through the core and reenters the heat exchanger. Alternatively, a loop-type system, such as shown in Fig. 7.14(b), may be used. In a loop system, the intermediate heat exchanger is located outside the reactor vessel. The primary sodium coolant enters near the bottom of the vessel, flows upward through the core, and exchanges heat with the clean sodium loop outside the core vessel. The pool type of configuration is widely used in Europe, while the loop system is used in the United States for the FFTF and Clinch River reactors.

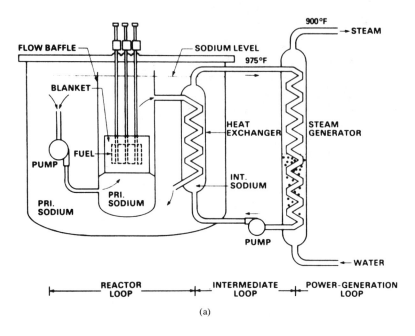

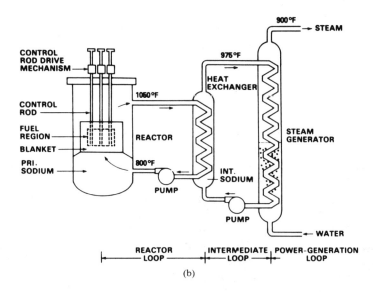

Figure 7.14 (a) Pool-type liquid-metal fast breeder reactor; (b) loop-type liquid-metal fast breeder reactor. (Reproduced with the permission of the Breeder Reactor Corporation, Oak Ridge, Tenn.)

Sec. 7.3 Power Reactor Systems

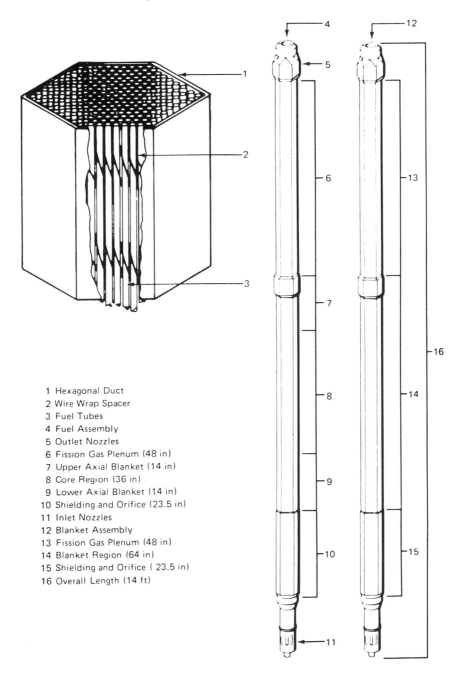

1 Hexagonal Duct
2 Wire Wrap Spacer
3 Fuel Tubes
4 Fuel Assembly
5 Outlet Nozzles
6 Fission Gas Plenum (48 in)
7 Upper Axial Blanket (14 in)
8 Core Region (36 in)
9 Lower Axial Blanket (14 in)
10 Shielding and Orifice (23.5 in)
11 Inlet Nozzles
12 Blanket Assembly
13 Fission Gas Plenum (48 in)
14 Blanket Region (64 in)
15 Shielding and Orifice (23.5 in)
16 Overall Length (14 ft)

Figure 7.15 Fast reactor fuel assembly and fuel rod configuration. (Reproduced with the permission of the Breeder Reactor Corporation, Oak Ridge, Tenn.)

Figure 7.15 shows the fuel assembly design for the Clinch River breeder reactor (funding for which was cut off by Congress in 1983). The fuel is mixed oxide (UO_2–PuO_2), clad with stainless steel. These small-diameter fuel pins, approximately 7.9 mm (0.31 in.) O.D., consist of a central fueled region equal to the length of the active core plus depleted uranium "fertile" pellets on each end to form the axial blanket. In the radial blanket region, the pins are fabricated entirely of depleted uranium.

Each fuel assembly is orificed to control the flow of coolant through the assembly. Although fuel assemblies operate at various power levels due to differences in neutron flux and fuel depletion, good overall power plant performance dictates that all fuels operate at approximately the same maximum cladding temperature. The primary heat transport system in a typical LMFBR includes three independent loops, each with its own pump and intermediate heat exchanger (IHX). Each of the three IHX systems has an independent steam generator in the secondary loop.

Pressure loss through the close-packed array of fuel pins shown in Fig. 7.15 is quite high, about 8×10^5 Pa (115 psi). The coolant increases in temperature as it passes through the core, reaching a maximum temperature of about 540°C (1000°F). The temperature is limited to well below the boiling point of liquid sodium, which is 880°C (1615°F).

Gas-Cooled Reactors

The first gas-cooled reactors, known as the Magnox type, were developed in England during the late 1950s. Carbon dioxide (CO_2) at a pressure of 1.6 MPa (230 psi) was used as the coolant. The fuel was a natural uranium, clad with an alloy of magnesium called Magnox. The moderator material was a high-purity graphite.

Several types of gas-cooled reactors have been designed and built, with England developing an advanced gas-cooled reactor system (AGR) and West Germany and the United States developing helium-cooled, graphite-moderated systems (HTGRs). The AGR still uses CO_2 as the coolant but changed the fuel design to use stainless-steel-clad UO_2 pellets to improve reactor performance.

HTGR reactors. The basic HTGR design in the United States was commercially developed by the General Atomic Company. Peach Bottom, a 40-MWe prototype unit, began initial operation in 1967. A second and larger prototype unit, (350 MWe) at Fort St. Vrain, was built for the Public Service Company of Colorado. Several larger commercial units were sold to utilities but later canceled due to adverse economic conditions.

The HGTR is a graphite-moderated, helium-cooled reactor system. It is designed to use uranium 233 as the fissile material and thorium as the fertile material. Initially, the system would have to be fueled with uranium 235, until sufficient uranium 233 is available for makeup fuel. (These fuel cycles are described in Section 7.4.) The entire core is constructed of nonmetallic materials and the reactor is

Sec. 7.3 Power Reactor Systems   301

capable of very high temperature operation. Additionally, the graphite has a high heat capacity that can absorb much of the decay heat for long periods of time after emergency shutdown without the problem of core meltdown.

The HTGR system is completely contained within a prestressed concrete pressure vessel (PCPV) together with the helium coolant and steam generator. The system is capable of producing steam conditions of 510°C (960°F) and 17 MPa (2500 psi), resulting in an overall system thermal efficiency of about 38%. The PCPV can house up to six sets of helium circulators and associated steam generators. As shown in Fig. 7.16, helium flows from the top of the reactor core downward through cooling holes in the fuel blocks, then upward through the steam generators and back to the top of the core.

The HTGR fuel and moderator assembly is shown in Fig. 7.17. Each assembly is a hexagonal-shaped graphite block that contains the fuel sticks and helium coolant holes. The entire core is made up of a series of these fuel/moderator assemblies, stacked to the predetermined core height and diameter. A reflector of pure graphite blocks surround the active core region.

The basic fuel forms are small beads or spheres of fissile and fertile material as carbides, UC_2 or ThC_2. The fissile beads are 0.35 to 0.50 mm in diameter, the fertile beads are 0.6 to 0.7 mm in diameter. Each bead is coated with either two or three layers of carbon and silicon carbide. These coatings are designed to prevent fission products from escaping from the particles and are a vital part of the reprocessing plan.

A carbonaceous matrix binder is used to form these small fuel particles with a fuel stick. The fuel sticks are inserted into selected holes in the fuel/moderator assembly block. The helium coolant passes downward through other holes in the fuel block.

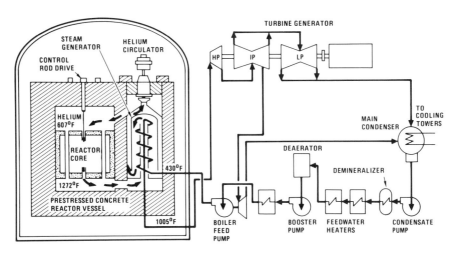

Figure 7.16 Simplified flow diagram for a high-temperature gas-cooled reactor (HTGR). (Courtesy of GA Technologies, Inc., San Diego, Calif.)

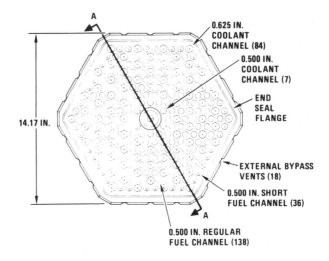
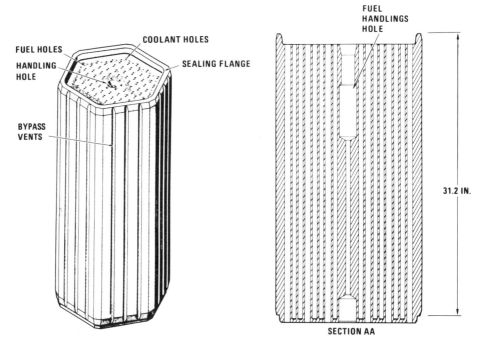

Figure 7.17 Standard graphite fuel element for the HTGR reactor. (Courtesy of GA Technologies, Inc., San Diego, Calif.)

Sec. 7.3 Power Reactor Systems

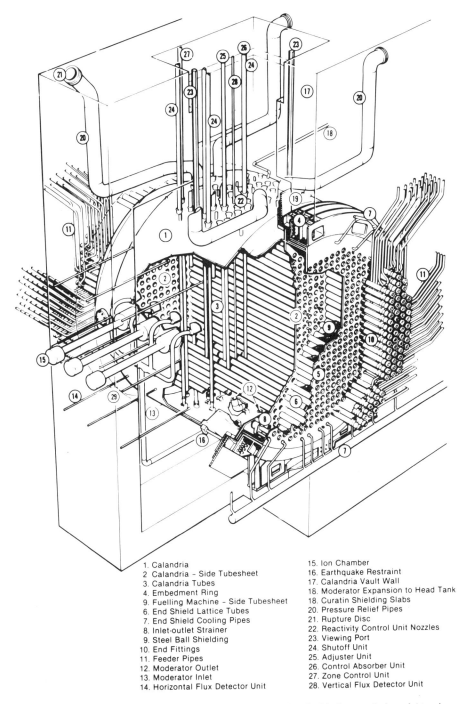

1. Calandria
2. Calandria - Side Tubesheet
3. Calandria Tubes
4. Embedment Ring
5. Fuelling Machine - Side Tubesheet
6. End Shield Lattice Tubes
7. End Shield Cooling Pipes
8. Inlet-outlet Strainer
9. Steel Ball Shielding
10. End Fittings
11. Feeder Pipes
12. Moderator Outlet
13. Moderator Inlet
14. Horizontal Flux Detector Unit
15. Ion Chamber
16. Earthquake Restraint
17. Calandria Vault Wall
18. Moderator Expansion to Head Tank
18. Curatin Shielding Slabs
20. Pressure Relief Pipes
21. Rupture Disc
22. Reactivity Control Unit Nozzles
23. Viewing Port
24. Shutoff Unit
25. Adjuster Unit
26. Control Absorber Unit
27. Zone Control Unit
28. Vertical Flux Detector Unit

Figure 7.18 CANDU reactor assembly intervals. (Reproduced with the permission of Atomic Energy of Canada Limited, Missisauga, Ontario, Canada.)

The HTGR concept can be modified for use in a gas turbine cycle for electric power production, or it can be used to produce high-temperature process heat.

Other Power Reactor Concepts

The Canadians have developed a power reactor concept called CANDU-PHW (Canadian Deuterium Uranium–Pressurized Heavy Water), shown as Fig. 7.18. This system utilizes natural uranium as the fuel, and heavy water serves as both the moderator and primary coolant. The two circuits are completely separated and independent of each other. The heavy water in the pressure tube containing the fuel is pressurized to prevent boiling, but the separate moderator circuit which is kept cool does not require high pressure.

The primary coolant enters the regular array of pressure tubes at 260°C (500°F) and 11 MPa (1600 psi), flows through the fuel element, and leaves the pressure tube at 320°C (600°F). Overall power plant efficiency is somewhat below that of an LWR system at about 28 to 29%.

The active core region is approximately 6 m high, with a diameter of 7 to 8 m. The reactor vessel, called the *calandria*, contains up to 380 horizontal pressure tubes, called *calandria tubes*. The calandria tubes are welded to the tube sheets at each end of the vessel. The moderator temperature is held to 70°C (150°F) and low pressure to reduce heavy water losses.

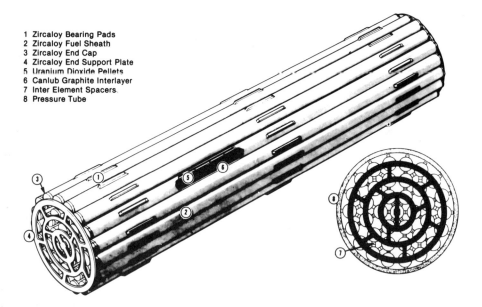

1 Zircaloy Bearing Pads
2 Zircaloy Fuel Sheath
3 Zircaloy End Cap
4 Zircaloy End Support Plate
5 Uranium Dioxide Pellets
6 Canlub Graphite Interlayer
7 Inter Element Spacers
8 Pressure Tube

Figure 7.19 CANDU fuel element assembly consisting of 37 fuel rods. (Reproduced with the permission of Atomic Energy of Canada Limited, Missisauga, Ontario, Canada.)

Sec. 7.4   Fuel Design and Analysis   305

Each calandria tube houses one pressure tube. Twelve fuel elements of the type shown in Fig. 7.19 are placed in series in each pressure tube. The heavy-water coolant flows through the tubes and fuel bundles. The system is designed to permit on-line refueling with the use of two refueling machines, operating in tandem. One machine inserts a fuel bundle in a pressure tube while the other machine removes a spent fuel assembly from the opposite end.

The fuel assembly, shown in Fig. 7.19, contains 37 fuel rods. Each rod contains natural uranium dioxide fuel pellets with 0.38 mm (0.015 in.) zircaloy cladding. Each rod bundle is approximately 0.1 m (4.0 in.) in diameter and 0.5 m (20 in.) long.

The CANDU design is unique because of the use of natural uranium, heavy water, and the capability for on-line refueling.

7.4 FUEL DESIGN AND ANALYSIS

The primary objective of fuel design and engineering is to provide for a geometric arrangement of fissile, fertile, and other materials that will lead to efficient power generation and good thermal hydraulic design. In addition, it is necessary to design a fuel rod that will be easy to mass produce and will have a long, failure-free life while in the reactor core.

The first nuclear reactors, which were used for production of plutonium rather than generation of power, used cylindrical rods of uranium metal clad in aluminum as fuel elements. Although uranium metal is subject to severe radiation damage, the fuel remained in these early reactors such a short time that the radiation damage was within acceptable limits. Further, these early reactors were cooled by room temperature water so that the poor corrosion resistance of uranium metal and aluminum in hot water was not a problem.

To generate useful power from a water-cooled reactor, it was necessary to find a fuel material that had much better behavior under irradiation and a much higher corrosion resistance. Improved behavior can be obtained by alloying small amounts of such materials as molybdenun or zirconium with the uranium metal. However, alloys with only moderate amounts of metal such as molybdenum were marginally acceptable at best. A fully satisfactory alloy can be obtained from uranium and zirconium if uranium makes up only a few percent of the alloy. This requires the use of highly enriched uranium which makes the fuel uneconomic for civilian power plants. However, such uranium–zirconium alloys have had military applications.

Ceramics were found to show much greater dimensional stability under irradiation than metallic fuels. The ceramic of greatest interest is uranium dioxide, UO_2. Not only does the fuel have good dimension stability, but its corrosion rate in hot water is low. Acceptable fuel elements can be produced by cladding the UO_2 with stainless steel or an alloy of zirconium (Zircaloy). Both of these cladding materials have very good corrosion resistance and are compatible with UO_2.

First-generation commercial reactors utilized stainless steel cladding. Because of the high rate of neutron absorption in stainless steel, the stainless steel cladding was replaced by Zircaloy in second-generation reactor designs. Today, virtually all light-water reactors use a fuel rod design that consists of ceramic UO_2 pellets with Zircaloy cladding. Stainless steel cladding is, however, used for fast reactor fuel elements which operate at temperatures above the design range for Zircaloy. We shall confine our fuel element discussion to the design utilizing UO_2 pellets in steel or Zircaloy tubes.

Fuel Design for Light-Water Reactors

Although all LWR fuel consists of UO_2 fuel pellets clad with Zircaloy, the size of the pellets, clad thickness, the outside clad diameter, and overall rod length vary from manufacturer to manufacturer. A typical rod configuration for a boiling-water reactor is shown in Fig. 7.20.

The UO_2 in the ceramic fuel pellets is enriched in the isotope ^{235}U and usually contains between 1 and 4% ^{235}U. The pellets, which are produced by pressing and sintering UO_2 powder, contain some voids. Because of the voids, density is generally about 95% of the theoretical UO_2 density. Pellet sizes vary from about 0.32 to 0.42 in. in diameter, depending on the reactor system. The pellets are between 0.375 and 0.55 in. long and "dished" in the center to minimize axial thermal expansion of the pellet stack during operation. In general, the PWR fuel pellets and rods have a smaller outside diameter than BWR fuel pellets. Current designs use smaller rod diameters than first used in order to provide lower pellet and cladding temperatures during loss-of-coolant accident situations.

The fuel cladding is a long, thin-walled, cylindrical Zircaloy tube. The cladding thickness ranges from 0.57 mm (0.0225 in.) to 0.86 mm (0.034 in.). Zircaloy, which is an alloy of zirconium plus about 1.5% tin and traces of other metals, is used as a cladding material because of its low absorption of thermal neutrons and good mechanical strength. Addition of tin to zirconium appreciably improves the mechanical properties. The use of Zircaloy cladding in place of stainless steel reduced enrichment requirements by about 1%.

The pellets are stacked in the cladding tube to a height equal to the desired active core height of the reactor. A typical pellet stack height would be 3.7 m (145 in.). The top part of the fuel rod, less than 0.3 m (1 ft) in length, contains a void or plenum for fission gas collection and retention. A spring located in the plenum region (Fig. 7.20) exerts a downward force on the pellets to maintain their positions in the fuel rod.

The fuel pellets and cladding tube are designed such that there is a small gap between the fuel pellet outside diameter and the clad tube inside diameter. Dimensionally, this radial gap is designed to be 0.10 to 0.15 mm (0.004 to 0.006 in.) at room temperature. This gap is required to allow easy assembly of the fuel rod and to minimize contact between the fuel pellets and cladding. When the fuel rods are

Sec. 7.4   Fuel Design and Analysis

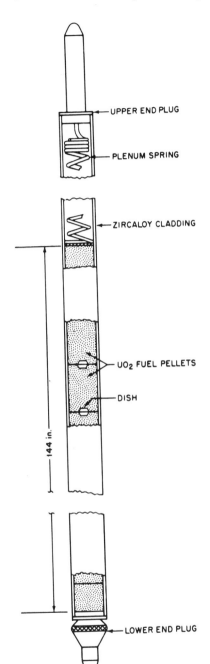

Figure 7.20 BWR fuel rod. (From *Reactor Technology*, Vol. 14, No. 1, 1972; reprinted with the permission of the American Nuclear Society, LaGrange Park, Ill.)

actually assembled, the rod is evacuated and then back-filled with helium to improve the heat transfer across this gap.

We have previously indicated that the individual fuel rods are placed into a regular square array called a fuel element assembly. Although the BWR fuel assembly shown previously in Fig. 7.13 contains a 7×7 array, most current assemblies contain an 8×8 array of fuel rods. These rods are spaced and supported by upper and lower tie plates and intermediate spacer rods. The top plate has a 304 stainless steel handle which is used for fuel assembly handling. BWR fuel assemblies often contain fuel rods of four different enrichments. The enrichment variation is designed to achieve more uniform heat generation in the fuel assembly. PWR fuel assemblies such as the one shown in Fig. 7.10 consist of a regular array of between 15×15 and 17×17 fuel rods. Fuel enrichment is generally uniform throughout the assembly.

Fuel Design for Liquid-Metal Fast Breeder Reactors

As presently envisioned, the fuel design for the LMFBR would consist of 15 to 20% PuO_2 mixed with depleted UO_2. These small-diameter (0.6-cm) mixed-oxide fuel rods would utilize 316 stainless steel fuel cladding. A wire of 316 stainless steel wrapped around the fuel pins ensures the required spacing between fuel rods in the fuel bundle. A typical fuel assembly contains about 217 fuel pins.

Fuel and Cladding Changes during Power Operation

Fuel behavior. During the fission process, several changes occur in the UO_2 fuel. The microstructure is altered (thus changing the physical and mechanical properties of the UO_2), some volumetric swelling occurs, the pellet cracks due to thermal temperature gradients, and fission gas is released to the void regions and fuel rod plenum. The Zircaloy cladding also undergoes property changes due to neutron irradiation. The most significant change is a decrease in ductility due to neutron irradiation and hydriding of the Zircaloy.

Due to the changes noted above, plus the differential thermal expansion between the fuel and cladding, physical contact can occur between the fuel pellet and cladding during reactor irradiation. If accompanied by high stresses in the cladding, this pellet–cladding interaction (PCI) can lead to undesirably high fuel failure rates.

The volumetric fuel swelling that occurs is estimated most accurately by computer programs which simulate the physical processes involved. Alternatively, a number of empirical correlations are available. One simple correlation that has been used is

$$\frac{\Delta V}{V_0} = d_e \alpha \left[\exp\left(-F_g \frac{\Delta V}{V_0}\right) - 1 \right] + \beta_f (1 - \alpha) B \qquad (7.41)$$

Sec. 7.4 Fuel Design and Analysis

where $\Delta V/V_0$ = relative volume expansion by swelling
  B = burnup level, MWd/MTU [MW-days/(thousand tons of uranium)]
  α = void fraction in the fuel
  β_f = swelling rate of pure, 100% dense UO_2 ($\sim 1.6 \times 10^{-6}$ MWd/MTU at LWR conditions)
  d_e = density correction factor = $0.149\alpha^{-0.746}$
  F_g = geometry factor = 100

The fuel radial thermal expansion may be determined from a polynomial fit of UO_2 thermal expansion data:

$$\frac{\Delta L}{L_0} = -4.972 \times 10^{-4} + 7.107 \times 10^{-6}T + 2.58 \times 10^{-8}T^2 + 1.14 \times 10^{-13}T^3$$

(7.42)

where $\Delta L/L_0$ = fractional change in length or diameter
  T = temperature (°C)

Since UO_2 is a ceramic material, there is very little plasticity below 1000°C. The stress–strain curve is essentially linear and Hooke's law can be applied. However, the change in dimensions with stress are so small that to a reasonable approximation, the UO_2 may be considered incompressible.

Under neutron irradiation during power reactor operation, uranium dioxide creeps at temperatures below the normal creep range. This fission-enhanced creep is linearly proportional to the fission rate and can be predicted from the empirical equations. However, fuel creep is normally not important in LWR designs and needs to be considered only at the higher temperatures and exposures used in fast reactor design.

Cladding behavior. The cladding thermal expansion is estimated from the expression

$$\frac{\Delta L}{L_0} = 8.2 \times 10^{-4} - 7.8 \times 10^{-6}T + 1.2 \times 10^{-8}T^2 - 6.10 \times 10^{-12}T^3 \quad (7.43)$$

with T being in units of degrees Kelvin. Within the elastic limit of the Zircaloy, *Young's modulus* is given by the expression

$$E = 6.0 \times 10^7 T \quad (7.44)$$

with T in units of degrees Kelvin and E in units of N/m^2.

During initial reactor operations, the cladding is subjected to the system coolant pressure. This results in an inward creep of the cladding toward the fuel pellets. Subsequently, the fuel may have swelled sufficiently so that the UO_2 contacts the cladding. The cladding then creeps outward at a rate equal to the swelling rate. The Zircaloy cladding creep rate has been measured as a function of stress and temperature and the results cast into empirical equations. One such equation is

$$\dot{\varepsilon} = K\phi\left[\sigma'_{ts} + B\exp(C\sigma'_{ts})\right]\exp\left(-\frac{10{,}000}{RT}\right)t^{-1/2} \quad (7.45)$$

where $\dot{\varepsilon}$ = transverse creep rate, m/m·s
$K = 5.129 \times 10^{-29}$
$C = 4.967 \times 10^{-8}$
T = temperature, °K
ϕ = fast neutron flux, n/m²·s (KE > 1.0 MeV)
σ'_{ts} = transverse stress, N/m²
$B = 7.252 \times 10^2$
R = ideal gas constant = 1.987 cal/mol·°K
t = time, s

Equation (7.45) enables us to compute the initial inward creep and then subsequently to compute the stress needed to sustain the outward creep rate.

Fuel Rod Behavior during Reactor Operation

During reactor operation, the fuel rod experiences a high internal heat generation leading to high temperatures, a high radial thermal gradient, and a high external pressure. In addition, the rod is subjected to a high neutron fluence and to an internal pressure buildup due to fission product gas release. The design and analysis of the fuel rod must consider each of these effects.

The temperature gradient across the fuel pellet and pellet-to-clad gap is a function of the thermal conductivity through the pellet and gap, as well as the internal heat generation rate in the local region. If the local volumetric heat generation rate is q''' (W/m³), the heat flux at the surface of the pellet q'' (W/m²) is given by

$$q'' = \frac{q''' r_{fp}}{2} \quad (7.46)$$

The temperature drop across the pellet-to-clad gap is then

$$\Delta T_{gap} = \frac{q''}{h_g} \quad (7.47)$$

where h_g is the gap conductance. During initial power operation, h_g can be represented by the equation

$$h_g = \frac{g^* k_g}{l_0} + (1 - g^*)\frac{k_g}{l_{fc} + l_0} \quad (7.48)$$

where h_g = gap conductance at operating radial gap thickness, W/m²·°C
l_{fc} = gap thickness (neglects pellet cracking or eccentricity), m
l_0 = root mean square of fuel and cladding surface projections, m

k_g = thermal conductivity of the gas in the gap, W/m·°C
g^* = empirical function describing the fraction of the pellet surface in contact with the cladding at zero contact pressure
$= C_1 + (1 - C_1)C_2^{l_{fc}C_3/D_p}$

where C_1 = fraction of the cracked fuel pellet assumed to be in contact with cladding which is independent of gap size
C_2, C_3 = empirical constants
D_p = hot pellet diameter (expanded diameter), m

Based on experimental data, reasonable values for these constants appear to be

$$C_1 = 0.3 \quad C_2 = 0.2 \quad C_3 = 50$$

The gas conductivity, k_g, required by equation (7.48) can be calculated assuming the gap contains a mixture of fission gas (xenon and krypton) and helium.

$$k_g = (k_{He})^{M'_{He}} (k_{FG})^{1 - M'_{He}} \tag{7.49}$$

where k_g = thermal conductivity of gas mixture in the gap (W/m·°C)
k_{He} = thermal conductivity of helium (W/m·°C)
k_{FG} = thermal conductivity of fission gas (W/m·°C)
M'_{He} = mole fraction of helium

and the fission gas thermal conductivity is

$$k_{FG} = 147 \left(\frac{T}{302} \right)^{0.86} \tag{7.50}$$

where T is the gas temperature (°K).

The temperature rise across the UO_2 fuel pellet itself may be computed by the usual expression of the temperature rise across an internally heated cylinder with negligible axial conduction. However, consideration must be given to the substantional variation of the thermal conductivity of UO_2 with temperature. The expression for the total temperature rise is therefore often written as

$$\int_{T_s}^{T_c} \bar{k} \, dT = \int_0^{T_c} \bar{k} \, dT - \int_0^{T_s} \bar{k} \, dT = \frac{q''' r_{fp}^2}{2} \tag{7.51}$$

where $\bar{k}$ = thermal conductivity of UO_2
T_c, T_s = center and surface temperature of pellet, respectively, °C
q''' = volumetric heat generation rate, W/m³
r_{fp} = outer radius of pellet, m

If the pellet surface temperature is known, the central temperature may be determined from an appropriate expression for the thermal conductivity integral. For temperatures of 1650°C and below, the integral can be approximated by

$$\int_0^{T_c} \bar{k} \, dT = 40.4 \ln(464 + 7) + 0.027366 \exp(2.14 \times 10^{-3} T) - 248.02 \tag{7.52}$$

where T is in degrees Celsius. Cracking of the pellet due to the thermal gradient across the pellet decreases the thermal conductivity of the fuel somewhat. However, at the same time, the value of h_g increases and the net effect is to decrease center fuel temperatures slightly.

Fission gas release in the fuel rod. The release of fission product gases from the UO_2 fuel to the void spaces inside the fuel rod is a complicated function of gas bubble diffusion and gas trapping. Computer programs simulating the gas bubble behavior are used in the most sophisticated design procedures. However, in many cases simplified approaches are used. One approach assumes that a fraction, K_1, of the fission gas escapes without being trapped in the UO_2 and a second fraction, K_2, of the trapped gas is released gradually. If the fuel were operated from the time of its insertion to time t at a constant power, then r', the total fission gas released per unit volume during this time, is given by

$$r' = p'_{fg}\left\{t - \frac{1-K_1}{K_1 K_2}[1 - \exp(K_1 K_2 t)]\right\} \quad (7.53)$$

where p'_{fg} is the fission gas production rate per unit volume. By assuming that reactor power operation can be described by a series of constant power steps, total gas release can be determined. The number of moles per unit volume released, $\Delta r'_i$, during the ith time interval, Δt_i, is then given by

$$\Delta r'_i = r'_i - r'_{i-1} = p'_{fg,i}\left\{\Delta t_1 - \frac{1-K_1}{K_1 K_2}[1 - \exp(-K_1 K_2 \Delta t_1)]\right\}$$
$$+ (C_g)_{i-1}[1 - \exp(-K_1 K_2 \Delta t_1)] \quad (7.54)$$

where C_g is gas concentration in the fuel equal to $(p'_{fg}t - r')$. Since the total gas release from time zero is $\Sigma \Delta r'_i$, the fraction of gas produced that is released is given by

$$F'_{fg} = \frac{\sum_i \Delta r'_i}{\sum_i p_i \Delta t_i} \quad (7.55)$$

The constant K_1 and $K_1 K_2$ are functions of temperature and time at a given power level. For 95% dense fuel

$$K_1 = \exp\left(-\frac{12,450}{T} + 1.84\right)$$

$$K_1 K_2 = 0.25 \exp\left(-\frac{21,410}{T}\right)$$

where T is in degrees Rankine and t is in hours. The release fraction may then be used to determine the internal gas pressure in the fuel rod and the resulting elastic expansion of the fuel rod cladding. The internal pressure calculation is based on the

Sec. 7.4 Fuel Design and Analysis

ideal gas law,

$$p_i = \frac{n_{\text{gas}} R \overline{T}}{V_{\text{vs}}} \qquad (7.56)$$

where p_i = internal pressure
 n_{gas} = moles of gas that have been released from fuel plus moles of original fill gas
 R = ideal gas constant
 V_{vs} = total volume of the void space, cm^3
 $\overline{T}$ = average temperature of the void space in the rod, °K = $V_{\text{vs}}/[\Sigma_i(V_m)_i/T_i]$
 T_i = temperature of ith component of the void space, °K
 $(V_m)_i$ = volume of ith component of the void space, cm^3

In detailed calculations, one must also consider the change in internal void volume that occurs during operation due to thermal expansion and creep of the cladding.

Pellet-to-Clad Interactions

During reactor operation, the fuel pellets expand in both a radial and an axial direction. The fuel pellets tend to expand more than the cladding, because of the much higher temperatures that exist in the fuel. Thus the relative expansion between the fuel and cladding tends to reduce the initial diametral gap.

Since the external system pressure is higher than the internal gas pressure in the rod, the cladding gradually creeps down toward the fuel. At the same time, fuel swelling causes the fuel to expand farther outward in the radial direction. The combined effects of swelling and creep may eventually lead to hard contact between the fuel and cladding. Hard contact is defined as the condition where the calculated pellet diameter, without considering cracking, equals the cladding inner diameter. Under these conditions, a significant contact pressure exists at the fuel–cladding interface.

Once hard pellet-to-cladding contact is established, the cladding tends to follow the diametral change of the pellet. At LWR operating conditions, the fuel pellet is almost incompressible and therefore the pellet elastically expands the cladding to accommodate any increase in diameter due to a power increase. Creep of the cladding then causes this stress to relax to a level such that the outward creep of the cladding exactly equals the pellet expansion due to fuel swelling.

The strain on the cladding is the sum of the thermal expansion strain, elastic strain, and creep strain. Accordingly, the circumferential cladding strain, ε_0, is given by

$$\varepsilon_0 = \frac{1}{E}\left[\sigma_0' - \mu(\sigma_r' + \sigma_z') + \alpha_c'(\Delta T) + \varepsilon_c\right] \qquad (7.57)$$

where σ_r' = radial stress, N/m^2
 σ_0' = tangential stress, N/m^2
 σ_z' = axial stress, N/m^2

E = Young's modulus, N/m²
α'_c = cladding coefficient of thermal expansion, $\Delta L/L$(°C)
ΔT = increase in cladding temperature, °C
μ = Poisson's ratio
ε_c = creep strain, $\Delta L/L$

Calculation of the cladding stress requires that both the external system pressure, p_o, and the internal rod pressure, p_i, be known. We determine p_i from the contact pressure and fission gas pressure.

The contact pressure between the pellet and cladding can be estimated by first computing pellet and cladding dimensions while ignoring any elastic deformation of the cladding. We then determine the internal pressure required to produce an elastic deformation, ΔD_o, that would result in zero gap. This is given by

$$\Delta D_o = \frac{D_i}{E(y^2-1)} \{ p_i[(1-\mu)+(1+\mu)y^2] - 2y^2 p_0] \} \qquad (7.58)$$

where D_o, D_i = external and internal cladding diameters, respectively, m
p_i = internal rod pressure, N/m²
p_0 = external pressure, N/m²
E = Young's modulus for the cladding, N/m²
$y = D_o/D_i$
μ = Poisson's ratio for the cladding (~ 0.3)

Since the cladding and pellet are only in contact over a small fraction of the total area, it is assumed that the internal pressure force exerted on the cladding can be represented by $p_i = p_{\text{gas}} + p_{\text{contact}}$.

Although there are cladding stresses in the radial, circumferential, and axial directions, it is often assumed that friction between the pellet and cladding prevents significant axial strain or stress. Since the radial stress is always quite small, the only stress of significance is the circumferential stress, σ'_θ. This stress may be approximated by

$$\sigma'_\theta = \frac{\Delta p_c D_i}{2 l_c} \qquad (7.59)$$

where Δp_c = net pressure difference across cladding
D_i = internal diameter
l_c = wall thickness of cladding

Under the foregoing assumptions the circumferential creep rate may be determined by taking the circumferential stress, σ'_θ, as the effective stress to be used in the creep rate equation.

In order to determine the expected fuel rod behavior, a computer simulation is generally conducted. The fuel element expected to see the highest power is examined by considering the element to be made up of a series of axial segments. The fuel and cladding dimensions and temperature distribution are determined for each segment

Sec. 7.4 Fuel Design and Analysis

at the beginning of fuel life. A short time step is taken and with the temperatures just calculated, the fission gas release, fuel swelling, and cladding creep are estimated for the end of the time step. With these values, a new set of dimensions and temperatures is obtained. A second time step is then assumed, and the process is repeated for each axial segment. The stepwise process is continued, using the expected power history of the fuel element, until the end of the fuel element's core life is reached.

The computer simulation provides the total fission gas pressure, maximum contact pressure, and strain range (maximum outward strain—minimum inward strain) experienced by the cladding. Because of the Zircaloy cladding embrittlement that occurs in a reactor, this strain range must be limited to less than 0.01.

Even though the computer simulation indicates that the strain range is within allowable limits, fuel failures may be observed when the contact pressure is substantial. This is attributed to the fact that high local stresses at the junctions between the pellets may significantly increase the nominal calculated stress. A slight "hourglassing" of the pellets occurs during irradiation producing cladding ridges at pellet interfaces. This can sometimes be exacerbated by "cocking" of a pellet in the stack. The high local stresses coupled with the presence of halide fission products may sometimes lead to stress corrosion cracking of the clad. Fuel manufacturers have assembled considerable data on the frequency of fuel failure as a function of nominal stress and strain plus fuel exposure. By restricting the fuel operation to regions where failure probability is low, fuel failure rates can be kept to acceptable levels.

Example 7.3

A fuel sample of 95% dense UO_2 in a capsule is irradiated at a thermal flux of 6×10^{12} N/cm^2·s and generates 18.7 W/cm^3. The sample is a 1-cm pellet contained in a Zircaloy tube which has a wall thickness of 0.1 cm. The heat flux is 2×10^{13} N/cm^2·s. In the experiment, the outer wall of the cladding is maintained at 260°C. At the beginning of life, the fuel and cladding are just touching (zero contact pressure) at this operating condition. After an exposure of 30,000 MWd/metric ton at constant power, determine the outer diameter of the fuel rod and the circumferential stress.

Solution If the cladding and fuel are in contact at the beginning of life at operating conditions, they will remain in contact throughout life. We may therefore determine the final rod diameter by determining the increase in pellet diameter due to swelling and adding this to the original rod diameter. We obtain the change in fuel volume from Eq. (7.41).

$$\frac{\Delta V}{V_0} = d_e \alpha \left[\exp\left(-\frac{F_g \Delta V}{V_0} \right) - 1.0 \right] + 1.6 \times 10^{-6} (1-\alpha) B$$

Since fuel is 95% dense, $\alpha = 0.05$, and we are given $B = 30{,}000$ MWd/metric ton. Because $\Delta V/V_0$ is contained within the right-hand side of the swelling equation, a trial-and-error solution is required. We assume a series of values for $\Delta V/V_0$ and determine where our assumed value agrees with that calculated. This is easily done by plotting the assumed versus calculated value and obtain the solution when the curve crosses the 45° line. We find that $\Delta V/V_0 = 0.0078$. Since the coefficient of volume expansion is three times the coefficient of linear expansions, the ratio of the change in diameter, ΔD, to the outer diameter, D_o, is

$$\frac{\Delta D}{D_o} = \frac{0.0078}{3} = 0.0026$$

Hence the fuel clad diameter at the end of this burnup period is

$$D_{\text{final}} = D_o + \text{swelling} + \text{cladding thickness}$$

$$= 1.0 + \frac{0.0026}{1.0} + 0.1 \times 2 = 1.226 \text{ cm}$$

At the end of life the rate at which the fuel is swelling will be exactly equal to the rate at which the cladding creeps outward. We can rewrite our fuel swelling equation at the end of life as

$$\frac{\Delta V}{V_0} = \text{constant} + 1.6 \times 10^{-6}(0.95) B$$

or

$$\frac{d}{dB}\frac{\Delta V_0}{V_0} = 1.52 \times 10^{-6}$$

But since $\Delta D/D_o = \frac{1}{3}\Delta V/V_0$,

$$\frac{d(D/D_0)}{dB} = 0.507 \times 10^{-6}$$

To get the time rate of diameter change, we write

$$dB = \left(q''' \frac{\text{W}}{\text{cm}^3} \times \frac{\text{cm}^3}{\rho \text{ g}} \times \frac{\text{MW}}{10^6 \text{W}} \times \frac{10^6 \text{ g}}{\text{metric ton}} \right) dt$$

$$= 18.7 \frac{\text{W}}{\text{cm}^3} \times \frac{\text{cm}}{10.97 \text{ g}(0.95)}$$

$$= \frac{dB}{dt} = 1.77 \frac{\text{MWd/metric ton}}{\text{day}} = 2.06 \times 10^{-5} \frac{\text{MWd/metric ton}}{\text{s}}$$

$$\frac{d(D/D_o)}{dt} = 0.507 \times 10^{-6} D_o \times 2.06 \times 10^{-5} = 1.04 \times 10^{-11}$$

We now set the Zircaloy creep rate [from Eq. (7.45)] equal to the swelling rate:

$$1.04 \times 10^{-11} \text{ cm/cm} \cdot \text{s} = \dot{\varepsilon} = K\phi \left[\sigma'_{ts} + B \exp(C\sigma'_{ts}) \right] \exp(-1000/RT) t^{-1/2}$$

where in this case $\phi = 2 \times 10^{13}$ N/cm² · s × 10⁴ cm²/m². We solve for the stress, σ'_{ts}, by assuming various values of σ'_{ts} plotting the results and noting where the curve intersects the desired value.

7.5 THERMAL ANALYSIS

In establishing the adequacy of the thermal design of a nuclear reactor core,[†] the following criteria are generally used:

1. The maximum centerline fuel temperature must be safely below the centerline fuel melting.
2. The maximum cladding temperature must be safely below the temperature at which any fuel cladding reaction takes place and the temperature at which the cladding strength becomes inadequate.
3. The power and coolant flow to the core must be such that adequate cooling can be maintained during steady-state operation and all expected transients.
4. The steady-state power and coolant flow to the core must be such that during a hypothetical loss-of-coolant accident, the cladding temperature stays below some predetermined limit.

Fuel Centerline Temperature

In Section 7.4 we found that we could determine fuel temperatures using Eq. (7.51). We may use this equation to evaluate the centerline fuel temperature at the position of highest core power (maximum fuel temperature). This evaluation is usually carried out assuming that it is possible for the core to be operating somewhat above its nominal power (e.g., 105% nominal power) due to measurement errors, transients, and so on. We must make certain that the maximum temperature is well below center melting, as fuel performance with a molten center region is questionable. There is a significant volume expansion on fuel melting which could lead to high cladding stresses. In addition, it is considered possible that molten fuel could contact the cladding by flowing through a radial crack in the fuel.

Maximum Cladding Temperature

In light-water reactors (LWRs), the maximum cladding temperature seldom causes a problem. Although Zircaloy can interact with UO_2, this does not occur until temperatures above 675°C are reached. Since boiling occurs in the hot channel of both PWRs and BWRs, the temperature of the outer surface of the cladding, T_{co}, is close to the saturation temperature ($\sim 300°C$). The relationship between wall and saturation temperature in nucleate boiling is

$$T_{co} - T_{sat} = 25(q'')^{0.25} e^{-p/62} \qquad (7.60)$$

[†] We shall consider only cores fueled with rods consisting of UO_2 pellets in metallic cladding. HTGR cores are not discussed.

where T_{sat} = saturation temperature of coolant, °C
q'' = maximum heat flux, MW/m²
p = pressure, bar

Note that the foregoing is valid only for water at high pressures. The temperature at the inner surface of the cladding, T_{ci}, is then obtained from

$$T_{ci} = T_{co} + \frac{q'' l_c}{k_c} \qquad (7.61)$$

The value of T_{ci} is rarely above 400°C in a water reactor at steady state, and hence cladding fuel reaction is of concern only during accidents.

In a sodium-cooled fast breeder reactor, boiling does not occur. The cladding surface temperature is then related to the bulk temperature of the coolant, T_{bf}, by

$$T_{co} = T_{bf} + \frac{q''}{h_c} \qquad (7.62)$$

where h_c is the forced convection heat-transfer coefficient. The bulk temperature of the coolant continues to rise as it passes through the core. At any axial location z, measured from core centerline, we have

$$T_{bf} - T_{in} = \frac{4}{D_e G'_{mf} c_p} \int_{-z_c/2}^{z} q'' z \, dz \qquad (7.63)$$

where T_{in} = coolant inlet temperature
G'_{mf} = mass flow rate (mass/time area)
c_p = specific heat of coolant
D_e = equivalent channel diameter = $\frac{4 \, (\text{flow area})}{\text{wetted perimeter}}$
z_c = core length

Although the axial distribution of the heat flux is often close to being cosinusoidal with a maximum at the core centerline, the higher bulk temperatures in the upper portion of the core lead to a maximum cladding surface temperature between the core centerline and the core outlet. The value of T_{ci} is determined as a function of length and its maximum value compared to the allowable limit. In the LMFBR, the maximum cladding temperature is generally limited by cladding strength requirements (tensile strength decreases with temperature).

Adequate Core Cooling

For LWRs the requirement that the power and flow levels be such that adequate cooling is maintained requires a brief consideration of the boiling process. After boiling begins, only small differences between wall temperature and saturation temperature are seen [see Eq. (7.60)]. The bubbles found in the boiling process highly agitate the fluid adjacent to the cladding leading to very high rates of heat

Sec. 7.5 Thermal Analysis

transfer. However, once the heat flux goes above a critical value [called the critical heat flux (CHF)] there are so many bubbles at the wall that they coalesce into a single steam layer or film. Such a steam film acts to insulate the fuel rod and prevents transfer of heat from the fuel to the coolant. The fuel rod would then heat up rapidly and fuel and cladding would melt unless the reactor were shut down quickly.

It is obvious that the reactor must be designed to operate at heat fluxes well below the critical heat flux. The critical heat flux decreases when quality, x, is increased and when flow is decreased. In a PWR, high CHF values are assured by operating at high mass flows and the low qualities to which these lead. To determine the margin between operating conditions and the CHF, a number of empirical dimensional correlations are used. One empirical correlation used for PWR rod bundle design is

$$q''_{CHF} = \frac{(A' - bD_e)A'_1(A'_2G')^{A'_3 + A'_4(p-2000)} - A_9 G' h_{fg} x}{A'_5 (A_6 G)^{A_7 + A'_8(p-2000)}} \qquad (7.64)$$

where $q''_{CHF,U}$ = critical heat flux for uniformly heated rod, Btu/h·ft^2
G' = mass velocity, lb/h·ft^2
D_e = equivalent diameter, in.
x = quality
p = pressure, psi
h_{fg} = heat of evaporation, Btu/lb

$A' = 1.1551$	$b = 0.4070$	$A'_1 = 0.370 \times 10^5$
$A'_2 = 0.5914 \times 10^{-6}$	$A'_3 = 0.8304$	$A'_4 = 0.6848 \times 10^{-3}$
$A'_5 = 12.71$	$A'_6 = 0.30545 \times 10^{-5}$	$A'_6 = 0.30545 \times 10^{-5}$
$A'_7 = 0.7119$	$A'_9 = 0.2073 \times 10^3$	$A'_9 = 0.1521$

The correlation of Eq. (7.64) holds for uniformly heated rods only. For rods where there is an axial variation in the heat flux, the actual critical heat flux, $q''_{CHF,N}$ for the nonuniform heating is gotten from

$$q''_{CHF,N} = \frac{q''_{CHF,U}}{F(z)} \qquad (7.65)$$

where $F(z) = \dfrac{1.025C}{q''(z)[1 - \exp(Cz_U)]} \int_0^{CHF} q''(z) \exp[-C(z_U - z)] \, dz$

$$C = \frac{0.44(1-x)^{7.9}}{(G'/10^6)^{1.72}}$$

z_{CHF} = distance from inlet at which $q''_{CHF,U}$ being evaluated, in.
z_U = length which if heated at $q''_{CHF,U}$ would produce a quality, x, equal to the quality at the location being evaluated, in.
G' = mass velocity, lb/h·ft^2

The location of greatest concern is not known in advance. Values of $q''_{CHF,N}$ must therefore be computed for a series of axial locations downstream of the peak heat flux. For every such axial location, z, we determine the *departure from nucleate boiling ratio* (DNBR) from

$$\text{DNBR} = \frac{q''_{CHF,N}}{q''(z)} \qquad (7.66)$$

For safety sake, we require that the minimum value of the DNBR be greater then a preset limit (i.e., DNBR $\geq$ 1.25). We also require that the DNBR be significantly above 1.0 during expected operating transients.

A somewhat similar procedure is followed for BWRs, but the critical heat flux relationship used differs substantially from those applicable to PWR conditions.

Loss-of-Coolant Behavior

During a hypothetical *loss-of-coolant-accident* (LOCA), it is possible that the peak fuel rods will exceed the peak heat flux. Here the safety requirement is that the maximum cladding temperature remain below the temperature at which severe Zircaloy embrittlement can occur and at which Zircaloy and stainless steel can form a eutectic alloy ($\sim 1200°C$). An analytical simulation of expected accident sequences is required to demonstrate that this limit is met. Special emergency core cooling systems are provided to make certain that the core is adequately cooled during the postulated accident.

7.6 THE NUCLEAR FUEL CYCLE

The *nuclear fuel cycle* may be described as the flow path of the uranium fuel material used in nuclear power plants, from mining the ore to ultimate disposal of the spent fuel or waste material. All power reactors in the United States presently utilize the uranium fuel cycle. The fuel cycle can be either "open" or "closed," depending on future political, economic, and resource considerations. The traditional concept of a closed fuel cycle is shown in Fig. 7.21. The closed fuel cycle is distinguished by the reprocessing of the spent nuclear fuel, recovery and use of the plutonium, and disposal of the high-level waste fission products. Presently, due to the lack of reprocessing in the United States, the fuel cycle is considered to be "open" at the back end. This fuel cycle, shown in Fig. 7.22, is also referred to as the "throwaway" fuel cycle. The open-end fuel cycle is characterized by the one-time, no-reprocessing use of the nuclear fuel. The spent fuel elements would be removed from the reactor and stored at an above ground facility for a number of years. Ultimate disposal would consist of the disposal of the spent fuel assemblies in a below-ground geologically stable facility.

In this section we review briefly each of the major steps in the nuclear fuel cycle. We begin with an evaluation of the natural uranium resources.

Sec. 7.6 The Nuclear Fuel Cycle

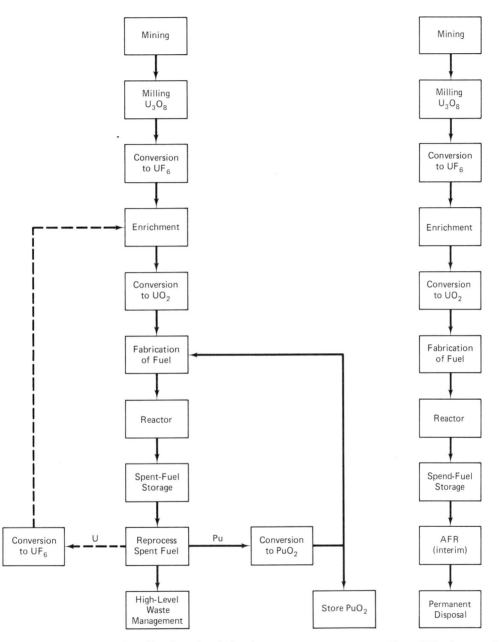

Figure 7.21 Closed uranium fuel cycle.

Figure 7.22 Open uranium fuel cycle.

Uranium Resources

Uranium occurs in significant concentrations in only a few countries of the world: the United States, USSR, Canada, Australia, and South Africa. In these countries, the concentration of uranium in the ore ranges from 0.1% to 3 or 4%. Deposits with ore concentrations of 0.001% or less are located in a number of countries. It is uneconomical, at present, to extract uranium from such ores.

The principal deposits in the United States lie in the West, primarily in the Colorado Plateau, which includes Arizona, Colorado, New Mexico, and Utah. Wyoming has deposits of medium-grade ore. Figure 7.23 shows these regions and their associated mineral belts. Much of the uranium ore comes from the Ambrosia Lake region of the uranium mineral belt shown on this figure.

The concentration of uranium in U.S. ores ranges from 0.1 to 0.4% of uranium. This is 1 to 5 kg of uranium oxide (U_3O_8) per 1000 kg of ore. This is considered a medium-grade ore concentration. As the concentration of uranium in the ore decreases, it becomes increasingly more expensive to produce the basic U_3O_8 that is sold as natural uranium.

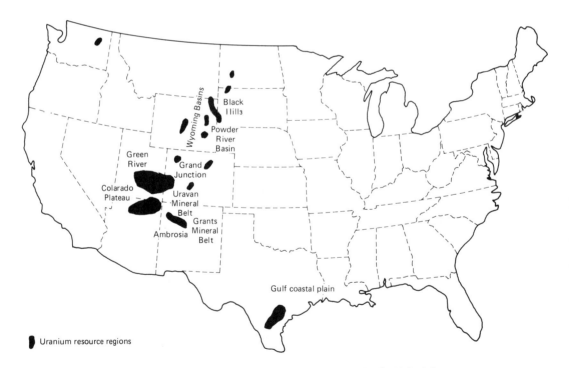

Figure 7.23 Major uranium ore locations in the United States.

TABLE 7.5 U.S. URANIUM RESOURCES (TONS OF U_3O_8)

Dollars per pound of U_3O_8	Reserves			
	Proven	Probable	Possible	Speculative
15	120,000	400,000	170,000	100,000
30	475,000	875,000	400,000	300,000
50	790,000	1,500,000	700,000	500,000
100	1,100,000	2,300,000	1,000,000	900,000

Source: Data from *U.S. Energy Supply Prospects to 2010*, National Academy Press, Washington, D.C., (1979).

Resource estimates are based on production costs, which are directly related to the uranium concentration. As the concentration decreases, the forward production costs increases. The forward production cost is the cost of mining and milling. It does not include exploration expenses. Resource estimates prepared by the U.S. Department of Energy under the National Uranium Resource Evaluation Program (NURE) are presented on this cost basis. Table 7.5 shows the U.S. uranium resource estimates on the basis of forward production cost. The values shown in the table are also tabulated in terms of proven reserves, with decreasing probability of potential resources.

Most experts believe that these resource estimates imply sufficient uranium to fuel all the power reactors in the United States that are presently on-line, under construction, or planned. Beyond these reactors, there is disagreement on the availability of natural uranium and the need for the introduction of fast breeder reactors to extend the fission reactor lifetime.

Uranium Mining

Approximately 85 to 90% of the uranium currently produced comes from the western states of New Mexico, Wyoming, Colorado, and Utah. About 50% of the uranium is surface mined in open pits and 50% is mined underground in deep mines. The mining techniques are very similar to those employed in coal mines.

The major difference between uranium mining and coal mining lies in the concentration of the useful product in the material mined. Coal constitutes a large fraction of the material removed from a coal mine. As we observed previously, the concentration of uranium is very low (1 to 5 kg per metric ton of material), and hence it constitutes a small fraction of the material mined. Thus an intermediate processing step, called *milling*, is used to increase the concentration of uranium in the product to be shipped from the mining region.

Uranium Milling

The uranium is concentrated from the 0.1 to 0.5% of the ore to nearly 95% by the chemical milling process shown schematically in Fig. 7.24. Approximately 90% of the uranium mills use this process. The remaining 10% use an alkaline leach method.

The ore is first crushed to a small particle size. The uranium ore is then leached from the bulk of the material by sulfuric acid. The pregnant leach liquor contains all the undissolved, nonmetallic constituents, which is nearly all the mass and volume of the original ore. The solids that remain in this liquor solution eventually form the mill tailings.

The separation of uranium from the nonmetallic components, as shown in Fig. 7.24, is basically a solvent extraction process. This process uses aqueous and organic liquids which are mutually immiscible. The sulfuric acid solution dissolves the metallic constituents of the ore, including the uranium. The acid solution containing the uranium is mixed with an organic solvent that has a selective affinity for uranium. The organic solution absorbs the uranium and is subsequently stripped of the uranium by yet another aqueous solution that has an even higher affinity for uranium (the eluent solution of Fig. 7.24). These operations can be repeated until up to 95% of the uranium is removed from the bulk ore.

The uranium is chemically precipitated from the eluent solution in the form of ammonium diurante, $(NH_4)_3U_2O_7$. This precipitate is yellow in color and is called *yellow cake*. Although the actual chemical content may vary, the uranium mill product prices are always quoted on the basis of the equivalent content of U_3O_8.

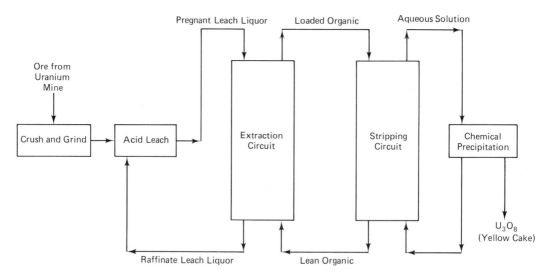

Figure 7.24 Simplified flow diagram for the uranium milling process.

Conversion to Uranium Hexafluoride

The mill product must be purified to reactor-grade material and converted to uranium hexafluoride, UF_6. At present, there are two methods that are used: a wet process, or solvent extraction, and a dry process, or hydrofluor process.

In the *hydrofluor process*, the yellow cake is ground to a small particle size and fed to a fluidized-bed reactor, where it is reduced by hydrogen gas and thermally cracked to UO_2. Successive hydrofluorination steps at elevated temperature produce an overall reaction of the form

$$UO_2 + 4HF \rightarrow 2H_2O + UF_4 \qquad (7.67)$$

In the final operation, the uranium tetrafluoride, UF_4, is reacted with fluorine gas at about 450°C to form uranium hexafluoride:

$$UF_4 + F_2 \rightarrow \underset{\substack{\text{uranium}\\\text{hexafluoride}}}{UF_6} \qquad (7.68)$$

Uranium hexafluoride is a solid at standard temperatures but sublimes to a gas at 56.5°C. The gaseous form is used as the feed material for the enrichment plants.

In the second process, the wet or *solvent extraction process*, the yellow cake is dissolved in a nitric acid solution. The resulting solution passes through a series of extraction, scrubbing, and stripping columns. The pumped uranyl nitrate solution is dried and calcined at 550°C to form uranium trioxide, UO_3. The UO_3 is converted to uranium hexafluoride by a process similar to the hydrofluor method described previously.

Uranium Enrichment

Natural uranium contains only three isotopes: $^{234}_{92}U$, $^{235}_{92}U$, and $^{238}_{92}U$. It contains only 0.711% of the fissile isotope ^{235}U, with the remainder of the natural uranium consisting mostly of the isotope ^{238}U. Only a small quantity of ^{234}U is found in natural uranium. Light-water power reactors require the fuel to be enriched in the isotope ^{235}U. As noted earlier, a typical enrichment would be in the range 1 to 4% ^{235}U. Special-purpose reactors, such as propulsion reactors, require enrichments as high as 93% ^{235}U.

The original and still dominant enrichment process is the gaseous diffusion process. All three currently operating U.S. enrichment plants use this process. The capacity of U.S. plants in 1982 was 4×10^6 kg per year of 3% enriched uranium. This is expected to be increased to 6×10^6 kg per year by 1985.

Centrifuge plants have been in use in Europe since 1975. The centrifuge enrichment process requires only 3 to 5% of the electrical power required for the gaseous diffusion process. The centrifuge enrichment process has been hampered by the need to develop a reliable ultra-high-speed (50,000 rpm) centrifuge machine.

Such machines now appear to be available. Other enrichment technologies are being developed.

Gas diffusion enrichment

Theory of Gas Diffusion. The gas diffusion process is based on the phenomenon of molecular effusion. According to the kinetic theory of gases, in a mixture of two gases, all molecules in thermal equilibrium have the same average kinetic energy. Since $KE = \frac{1}{2}mu^2$, where m is the mass of the molecule and u is the speed, the lighter molecules have a slightly higher velocity than that of the heavy molecules. The lighter molecules will thus strike the walls of the container more frequently than will the heavy gas molecules. If one outlet from the vessel is constructed with a barrier containing holes just large enough for passage of molecules (no laminar or viscous flow), the lighter gas molecules will strike the barrier more often than will the heavier gas molecules. Thus, more of the lighter gas molecules, in proportion to their concentration, will flow through the barrier holes. The flow of individual molecules through small holes is referred as *molecular effusion*.

The gas diffusion process for enriching uranium in the isotope ^{235}U employs a mixture of $^{235}UF_6$ and $^{238}UF_6$. The relative frequency with which the molecules enter and pass through a small hole is inversely proportional to the square root of their molecular weights. This ratio is known as the *ideal separation factor for gaseous diffusion*, α_0.

$$\alpha_0 = \sqrt{\frac{MW_1(^{238}UF_6)}{MW_2(^{235}UF_6)}}$$

$$= \sqrt{\frac{352}{349}} = 1.00429 \qquad (7.69)$$

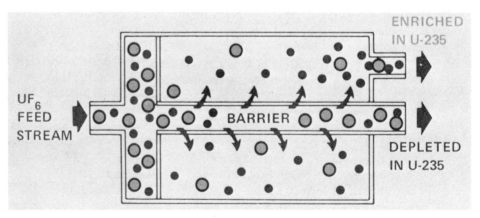

Figure 7.25 The separating unit of a gas diffusion cascade slightly enriches one of the streams in the isotope uranium 235. (Courtesy of the Union Carbide Corporation, Oak Ridge National Laboratory, Oak Ridge, Tenn.)

Sec. 7.6 The Nuclear Fuel Cycle

Figure 7.25 illustrates the molecular effusion of both types of UF_6 molecules through the porous tube barrier. The ideal stage separation factor, α_0, represents the best separation of the isotopes under ideal conditions. In practice, the conditions necessary for pure molecular effusion are not achieved; the actual stage separation factor is much less than 1.00429. The low value of this stage separation factor means that a large number of stages will be required to perform the isotope separation task.

Gas Diffusion Process Equipment. The smallest element of a diffusion plant that produces some separation is called a *separating unit*. A *stage* is a group of separating units connected in parallel. The overall enrichment plant, composed of a group of stages connected in series, is called a *diffusion cascade*.

The basic element of the gas diffusion plant is the stage. Each stage is composed of a diffuser vessel, gas compressor, motor control valve, and associated piping. Eight series-connected stages are shown in Fig. 7.26. An enrichment plant may contain up to 3200 stages in order to provide for the highest practical enrichment, 93 to 95%.

The diffuser vessel contains the barrier material and the gas cooler or heat exchanger used to control the process temperature. The UF_6 gas flows at high process pressure through the inside of the porous tubes as depicted in Fig. 7.25. About one-half of the gas effuses through the porous barrier tubes into the region of lower pressure on the outside of the tubes. The gas stream is slightly enriched in the isotope ^{235}U at this point in the process, but requires many additional stages of separation before the desired enrichment is achieved.

The axial flow compressors are used to compress the UF_6 gas to the required pressure. The electrical motors that power these compressors in U.S. plants are rated at 3300 hp. Because of the low stage separation factor, the gas diffusion process is characterized by a large internal flow rate of gas throughout the cascade. In most cases, the total internal flow rate may be around 7000,000 kg of UF_6 in order to obtain 1 kg of slightly enriched product.

Enrichment Cascade Equations. Consider a mass balance around the entire diffusion cascade, and a mass balance for the isotope ^{235}U. If, as indicated in Fig. 7.27, we let F designate the feed stream, P the product stream, and W the waste stream, we have

$$m_F = m_P + m_W \tag{7.70}$$

$$\chi_F m_F = \chi_P m_P + \chi_W m_W \tag{7.71}$$

where $m_F \equiv$ mass of feed material, kg
$\chi_F \equiv$ weight percent of ^{235}U in the feed material
$m_P \equiv$ mass of product material, kg
$\chi_P \equiv$ weight percent of ^{235}U in the product material
$m_W \equiv$ mass of the waste product, (tails flow), kg
$\chi_W \equiv$ weight percent of ^{235}U in the waste stream, called "tails assay"

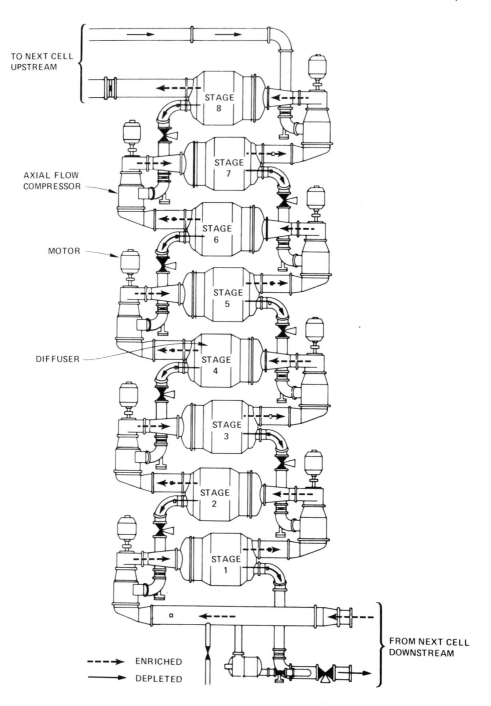

Figure 7.26 Stage arrangement in a gas diffusion cascade. (Courtesy of the Union Carbide Corporation, Oak Ridge National Laboratory, Oak Ridge, Tenn.)

Sec. 7.6 The Nuclear Fuel Cycle

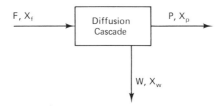

Figure 7.27 Diffusion cascade.

These equations can be solved to give the quantity of feed material required, m_F, to obtain a specified quantity of product m_P at enrichment χ_P.

$$m_F = \frac{m_P(\chi_P - \chi_W)}{\chi_F - \chi_W} \tag{7.72}$$

The waste product from this cascade would be

$$m_W = \frac{m_P(\chi_P - \chi_F)}{\chi_F - \chi_W} \tag{7.73}$$

Example 7.4

A fuel reload for a BWR will require 10,000 kg of 2.0% enriched ^{235}U. The U.S. Department of Energy (DOE) has specified the tails assay at 0.20%. Natural uranium will be supplied as the feed material. Calculate the mass of feed material that must be supplied to the DOE for this enrichment task.

Solution

$$m_P = 10{,}000 \text{ kg}$$
$$\chi_P = 0.02$$
$$\chi_W = 0.0020$$
$$\chi_F = 0.0071$$

Substituting these variables into Eq. (7.72), we have

$$m_F = 10{,}000\left(\frac{0.02 - 0.002}{0.0071 - 0.002}\right) = 10{,}000(3.46)$$
$$= 34{,}600 \text{ kg}$$

The waste produced can be found by direct substitution into Eq. (7.73) or simply by

$$m_W = m_F - m_P = 34{,}600 - 10{,}000 = 24{,}600 \text{ kg}$$

The minimum number of stages required for the diffusion occurs at a condition of "total reflux." This condition of operation describes a cascade where all product is retained and no product is being removed. Under such conditions, the minimum number of stages, n_{min}, is given by the *Underwood–Fenske equation*:

$$n_{min} = \frac{\ln[\chi_P(1-\chi_W)/(1-\chi_P)\chi_W]}{\ln \alpha_0} \tag{7.74}$$

Example 7.5

Using the data of Example 7.4 calculate the minimum number of stages required to enrich the natural uranium feed to 2.0% enrichment.

Solution Use Eq. (7.50) and the following data:

$$\chi_P = 0.020$$
$$\chi_F = 0.0071$$
$$\chi_W = 0.002$$
$$\alpha_0 = 1.0043$$

$$n_{min} = \frac{\ln[0.02(0.998)/0.98(0.002)]}{\ln 1.0043}$$

$$= 540 \text{ stages}$$

In an ideal cascade in which product is removed, certain modifications must be made to the Underwood–Fenske equation. The number of stages required is found to be twice the minimum, minus 1, or

$$n_{ideal} = 2n_{min} - 1 \tag{7.75}$$

Finally, the stage separation factor α is in practice around 1.003. Thus, for a somewhat practical ideal diffusion cascade, the number of stages on our example would be

$$n_{ideal} = 2\left\{\frac{\ln[0.02(0.998)/0.98(0.002)]}{\ln 1.003}\right\} - 1 = 1548 \text{ stages}$$

The number of stages required in the cascade increases sharply as the product enrichment increases. For 90% enriched product, with a 3% tails assay, the ideal number of stages required increases to 3738 stages. The interstage flow rate in the cascade for this high enrichment would be 41,636,000 mol of UF_6 pumped throughout the cascade for each mole of enriched UF_6 removed.

The measure of the work being done by a diffusion cascade in making m_P kilograms of product from m_F kilograms of feed is called the *separative work*, or *separative work units* (SWU).

$$\text{SWU} = m_P[\mathscr{V}(\chi_P) - \mathscr{V}(\chi_W)] - m_F[\mathscr{V}(\chi_F) - \mathscr{V}(\chi_W)] \tag{7.76}$$

and $\mathscr{V}(\chi_i)$ is the value function, defined as

$$\mathscr{V}(\chi_i) = (1 - 2\chi_i)\ln\frac{1-\chi_i}{\chi_i} \tag{7.77}$$

The separative work equation is generally evaluated in mass units such as kilograms. The enrichment charges are listed in "SWU units" or "kilogram SWU" units. U.S. charges in 1984 were just over $141/kg SWU, for long-term contracts.

Example 7.6

(a) Calculate the quantity of natural uranium feed material required to produce 10,000 kg of 3% enriched uranium.
(b) Calculate the kilogram SWU units for this enrichment task.
(c) Calculate the cost of enrichment.

Assume that the tails assay (value of χ_W in the plant waste stream) is 0.2%.

Solution

(a) Feed material required, m_F:

$$m_F = m_P \frac{\chi_P - \chi_W}{\chi_F - \chi_W} = 10,000 \left(\frac{0.03 - 0.002}{0.007 - 0.002} \right)$$

(b) Separative work required:

$$\text{SWU} = m_P[\mathscr{V}(\chi_P) - \mathscr{V}(\chi_W)] - m_F[\mathscr{V}(\chi_F) - \mathscr{V}(\chi_W)]$$

$$V(\chi_i) = (1 - 2\chi_i) \ln \frac{1 - \chi_i}{\chi_i}$$

$$V(0.03) = [1 - 2(0.03)] \ln \frac{1 - 0.03}{0.03} = 3.207$$

$$V(0.0071) = 4.869$$

$$V(0.002) = 6.188$$

On substitution, we obtain

$$\text{SWU} = 10,000(3.267 - 6.188) - 54,795(4.869 - 6.188)$$

$$= 43,065 \text{ kg}$$

(c) Cost of enrichment:

$$\text{cost} = 141.20 \frac{\$}{\text{SWU}} \times 43,065 \text{ SWU}$$

$$= \$6,080,000$$

Gas centrifuge enrichment. The principle of the gas centrifuge is based on the centrifugal force developed due to rotational motion. When UF_6 gas is fed into a centrifuge device, such as the one shown schematically in Fig. 7.28, the heavier $^{238}UF_6$ molecules are driven toward the outside of the device, while the lighter $^{235}UF_6$ molecules move toward the central axis of rotation. The gas is fed into the center of the device near the top. The high-speed rotational motion leads to the centrifugal force, which results in a pressure gradient, dp/dr, across the radius of the centrifuge:

$$\frac{dp}{dr} = \omega^2 r \rho \qquad (7.78)$$

where p = gas pressure
ω = angular velocity, rad/s
r = centrifuge radius
ρ = gas density

In a binary gas mixture, it can be shown that this leads to a stage separation factor of

$$\alpha_0 = \exp\left[\frac{(MW_2 - MW_1) \omega^2 (r_{oc}^2 - r_{ic}^2)}{2RT} \right] \qquad (7.79)$$

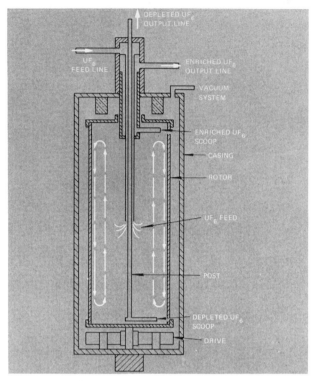

Figure 7.28 Gas centrifuge. (From *U.S. Gas Centrifuge Program for Uranium Enrichment*, Department of Energy brochure.)

where MW_2 = heavy molecular weight
MW_1 = lighter molecular weight
R = ideal gas constant = 8.315×10^7 ergs/g mol·°K
T = absolute temperature, °K
r_{oc} = outer radius of the centrifuge bowl, cm
r_{ic} = inner radius of the centrifuge bowl, cm

and ω has its previous meaning.

The separation factor in a gas centrifuge depends on the difference between the molecular weights, but in the gaseous diffusion process, it depends on the ratio of the two molecular weights. Thus gas centrifuge stage separation factors are considerably higher and fewer stages are required for the enrichment of ^{235}U.

The gas centrifuge machine introduces the UF_6 gas into the center of the device near the axis. The gas near the axis flows upward and becomes increasingly richer in the lighter isotope. A product line removes the slightly enriched gas near the top of the device, while a bottom scoop removes the waste stream (tails).

We have previously indicated that the gas centrifuge process consumes only about 4% of the power required by a gas diffusion plant. However, the capacity of each centrifuge is low and many centrifuge machines are required to form a cascade.

Fabrication of the Fuel Assemblies

The slightly enriched UF_6 is shipped as a solid from the enrichment facility to the fabrication plant. There it is sublimed and converted to UO_2 powder. A typical fabrication process for BWR fuel assemblies is shown in Fig. 7.29.

The UO_2 green ceramic is sintered to 95% of theoretical density. The sintered pellets are centerless ground to final dimensions, ± 0.01 mm. The pellets are inspected, weighed, and vacuum outgassed at high temperature to remove moisture and organic contaminants. The Zircaloy cladding tubes are cleaned, inspected, and cut to length. One end cap is welded onto the tube and the fuel pellets are loaded into the cladding tube. The rod is outgassed and backfilled with an inert gas such as helium.

There are a number of inspection and quality control checks throughout the fabrication process. Following the final-dimensional inspections on the fuel rods, the rods are assembled into fuel bundles.

Reprocessing and Waste Disposal

Transportation of spent fuel assemblies. The reactor fuel remains in the reactor core for three or four years, depending on the nuclear design of the core. (Newer fuel design may extend this time interval.) The fuel is removed from the core and stored in a water-filled basin called the *spent fuel pool*. As this on-site storage becomes full, the fuel must be shipped to an "away from the reactor" (AFR) storage facility or to a reprocessing plant.

The spent fuel is both thermally hot and radioactive. Shipments of spent fuel, including packaging, loading, inspection, documentation, and accident reporting, are regulated by the U.S. Department of Transportation (DOT). The design and approval of the shipping container, called a *shipping cask*, is regulated by the Nuclear Regulatory Commission (NRC). The applicable federal regulations are found in the Code of Federal Regulations under 49CFR100-199 and 10CFR71.

A typical spent fuel transport cash is shown in Fig. 7.30. The container is designed to withstand severe accident situations such as railroad accidents, petroleum fire, and immersion in water.

Reprocessing of nuclear fuel. The goals of fuel reprocessing are to recover the fissile uranium and plutonium from the spent fuel and to separate out the waste fission products for later conversion to a stable form and subsequent burial. Although there are now no commercial fuel reprocessing facilities in operation in the United States, there are several such facilities operating in other Western countries. Reprocessing facilities for defense purposes are in operation in the United States.

The reprocessing of nuclear fuel is a chemical process. The most widely used reprocessing method is the *Purex process*. This process is shown schematically in Fig. 7.31. The process begins with the mechanical disassembly of the fuel bundle to remove all the external hardware. The individual fuel rods are chopped into

334 Nuclear Power Stations Chap. 7

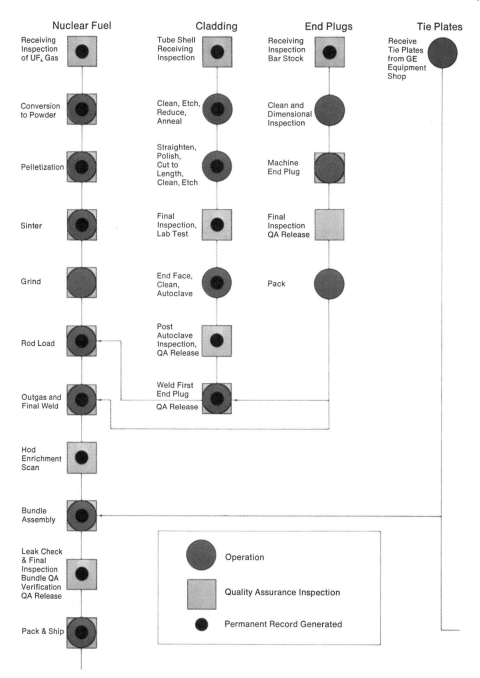

Figure 7.29 Flow process for BWR fuel rod fabrication. (Reprinted with the permission of the General Electric Company, Nuclear Energy Operations, San Jose, Calif.)

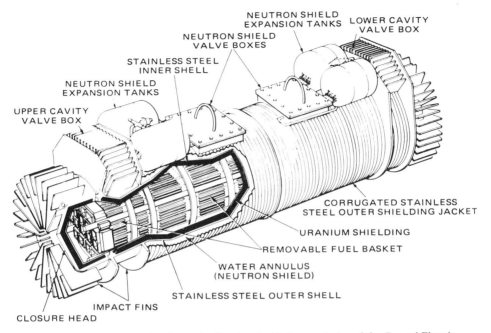

Figure 7.30 Spent fuel shipping cask. (Reprinted with the permission of the General Electric Company, Nuclear Energy Operations, San Jose, Calif.)

segments of 3 to 5 cm in length. These segments are placed in a basket device and immersed in a nitric acid solution. The fuel material is leached out of the clad by the reaction

$$3UO_2 + 8HNO_3 \rightarrow 3UO_2(NO_3)_2 + 2NO + 4H_2O \qquad (7.80)$$

Plutonium and fission products react similarly to form nitrates.

The Purex process requires that the chemical elements to be separated be in an aqueous solution in the nitrate form. The basic portion of the Purex process is a solvent extraction which removes the uranium and plutonium from the aqueous solution. The organic solvent used is tributyl phosphate (TBP) in kerosene.

Solvent extraction techniques can be used when one of the constitutients of an aqueous solution (in this case either uranium or plutonium) is considerably more soluble in a solvent than are the other elements. Since the uranium is more soluble in the organic solvent, it tends to pass into that phase, whereas the other elements (fission products) remain more concentrated in the aqueous solution.

This extraction process can be improved by the addition of a salting agent. This can be either an acid or a salt which has the same anion as the inorganic compound to be extracted. The salting agent, in this case nitric acid, enhances the extraction process.

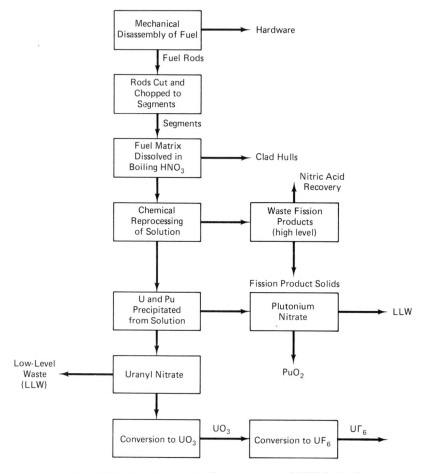

Figure 7.31 Flow diagram for the reprocessing of LWF fuel rods.

To further purify and decontaminate the uranium and plutonium streams, two or three extraction cycles are utilized. These are followed by an ion-exchange operation.

The useful product of the reprocessing plant is uranium in the form of UO_3 and UF_6 and plutonium in the form of Pu metal or PuO_2. There is also a high-level radioactive liquid waste, containing the waste fission products, that must be stored temporarily at the facility and various low-level radioactive products.

Fuel cycle waste management. Each step of the nuclear fuel cycle produces some type of radioactive waste. Compositions, quantities, and activity levels vary widely. For example, the ore material that remains from the first operation described in this section, mining and milling, is called the *mill tailings*. Some radon gas evolves from this bulk material. However, most of the concern over radioactive waste disposal centers on the management of the high-activity-level wastes produced by the reprocessing operation or storage of unprocessed radioactive fuel assemblies.

Once-Through Fuel Cycle Waste Management. The once-through fuel cycle will probably require the government to assume ownership of spent fuel at licensed or approved storage sites. This is considered an interim storage concept, and utilities would have the option of reclaiming the spent fuel if reprocessing becomes a reality. Transportation to the site and minimal storage costs for the spent fuel would be paid by the utilities.

Several concepts have been proposed for the interim storage of spent fuel. These include surface silos, water basins, and air-cooled surface vaults. All of the interim storage methods being investigated provide the option of later recovery of the fuel assemblies for reprocessing. The ultimate disposal of the spent fuel assemblies would be much the same as that of the high-level waste fission products from the reprocessing operation.

Closed-Fuel-Cycle Waste Management. As shown previously, the fission product wastes from the reprocessing operation generate high levels of heat and radioactivity. These high-level wastes (HLW) would be stored in liquid form at the reprocessing plant for about a two-year period. In one of the most promising methods being considered for disposal, the liquid wastes are processed to remove the liquid and the remaining material is converted to a solid oxide form by a process called *calcining*. This relatively inert calcined waste would then be incorporated into a host matrix of borosilicate glass, bitumen, concrete, or similar fixation-type material. This mixture would be cast into a cylindrical shape of approximately 30 cm (1 ft) diameter by 1 m in length (3 ft). The cylinder is finally encapsulated in a 1-in.-thick stainless steel jacket.

Ultimate disposal of spent fuel assemblies or encapsulated waste. Many varied concepts have been considered for the ultimate disposal of the waste fission products. All methods rely on the multibarrier concept: inert waste form, leaktight cannister, and stable disposal medium. At present, the preferred final disposal appears to be the burial of the waste in a geologically stable formation such as granite or salt. In the salt storage concept, the waste would be isolated deep underground in a bedded salt formation. The depth, thickness, and hydrology would be such that human or water intrusion would be very low probability events.

SYMBOLS

A	area
A_m	atomic mass number
ACT	activity, disintegrations/s
Av	Avogadro's number (6.03×10^{23})

B	fuel burnup, MWd/MTU
B_g	geometric buckling, cm^{-2}
c	velocity of light m/s
c_p	specific heat, J/kg · °K
C_g	fission gas concentration in the fuel
d_e	density correction factor [Eq. (7.41)]
D_e	equivalent diameter, m or in.
D_p	hot pellet diameter, m
D_i	internal clad diameter, m
D_o	external clad diameter, m
E	Young's modulus of elasticity, N/m^2
$\mathscr{E}$	energy, J
f_{th}	thermal utilization factor
F'	fission rate, fissions/s
F'_{fg}	fraction of fission gas produced/released
F_g	geometry factor
g^*	empirical function [Eq. (7.48)]
G'	mass velocity, lb/h ft^2
G_{mf}	mass velocity, mass/time area
h_{fg}	heat of evaporation; Btu/lb
h_c	convective heat-transfer coefficient, W/m^2 · °C
h_g	fuel pellet gap conductance, W/m^2 · °C
i_j	number of atoms of component i in molecule j
I_n	neutron beam intensity, n/cm^2 · s
k'_B	Boltzmann constant
k_c	thermal conductivity of cladding, W/m · °C
k_g	thermal conductivity of the gas in the gap, W/m · °C
$\bar{k}$	thermal conductivity of UO$_2$, W/m · °C
k_{eff}	effective neutron multiplication factor
k_∞	neutron multiplication factor of infinite system
KE	kinetic energy, J
KE(T)	neutron kinetic energy at temperature T, eV
KE(th)	thermal neutron kinetic energy, eV
l_0	rms value of fuel and cladding surface projections, m
l_c	clad thickness, m
l_{fc}	fuel clad-to-pellet gap thickness, m
i_t	material thickness, cm mass, kg
L_{th}	thermal diffusion length, cm

Chap. 7 Symbols

$\Delta L/L_0$	thermal expansion, m/m
m	mass, kg
$^A_Z m$	mass of atom, kg
m_e	mass of electron, kg
m_n	mass of neutron, kg
m_p	mass of proton, kg
M'_{He}	mole fraction of helium
MW	molecular weight
n_a	number of atoms
n_{gas}	total gram moles of gas in fuel rod
n_{ideal}	number of stages in an ideal cascade
n_{min}	minimum stages in cascade
n_n	number of neutrons in given atomic nucleus
n	neutron density, n/cm^3
N	atom density, atoms/cm^3
p	pressure, N/m^2, bar, or psi
p'_{fg}	fission gas production rate in fuel rod, atoms/fission
p_i	internal fuel rod pressure, N/m^2
p_o	external fuel rod pressure, N/m^2
Δp_c	net pressure differential across cladding, n/m^2
$\mathscr{p}_{\text{re}}$	resonance escape probability
$\mathscr{p}_{\text{th}}$	neutron leakage during slowing down
$\mathscr{P}_f$	fast nonleakage probability
$\mathscr{P}_{\text{NL}}$	neutron nonleakage probability
P_T	reactor thermal power, MW
$\mathscr{P}_{\text{th}}$	thermal nonleakage probability
q''	surface heat flux, W/m^2 or Btu/h·ft^2
q''_{CHF}	critical heat flux, Btu/h·ft^2
$q''_{\text{CHF,N}}$	critical heat flux, nonuniform heating, Btu/h·ft^2
$q''_{\text{CHF,U}}$	critical heat flux, uniform heating, Btu/h·ft^2
q'''	volumetric heat generation, W/m^3
Q_{NR}	energy value of a nuclear reaction, MeV
$\bar{r}$	atoms of given isotope/all atoms of given element
r'	gas release per unit volume
r_{fp}	fuel pellet radius, m
r_{ic}	inner radius of centrifuge, cm
r_{oc}	outer radius of centrifuge, cm
R	ideal gas constant

R_a	reaction rate, per second
R_c	radius of the active core, m
SWU	separative work units
t	time, s
t_a	average lifetime of prompt and delayed neutrons, s
t^*	mean neutron generation time, s
$t_{1/2}$	half-life of radioisotope, s
T	temperature, °K (unless specified otherwise)
$\overline{T}$	average temperature of void space in fuel rod, °K
T_{bf}	bulk fluid temperature, °C
T_c	center temperature of the fuel pellet, °C
T_{ci}	clad inner surface temperature, °C
T_{co}	clad outer surface temperature, °C
T_{gap}	temperature drop across the pellet to clad gap, °C
T_{in}	coolant inlet temperature, °C
T_s	surface temperature of fuel pellet, °C
T_{sat}	coolant saturation temperature, °C
u_n	neutron velocity, cm/s
V_j'	volume fraction of component j
V_m	volume, cm³
V_{vs}	total volume of void space, cm³
$\Delta V / V_0$	volumetric expansion, m³/m³
$\mathscr{V}(\chi_i)$	value function for argument χ_i
x	steam quality
y	D_o/D_c
z	axial position
z_c	height of the active core, m
z_U	length heated at $q''_{CHF,U}$ which would produce a quality equivalent to quality at location being evaluated
Z_e	atomic number (electrons)

Greek Symbols

α	void fraction
α_c'	cladding coefficient of thermal expansion, m/m · °C
α_0	ideal separation factor for gas diffusion
β_f	fuel swelling rate, % per MWd/MTU
β^*	delayed neutron fraction
Δ	mass defect, kg

Symbols

ε'	fast fission factor
$\dot{\varepsilon}$	transverse creep rate, m/m·s
ε_0	circumferential clad strain, m/m
ε_c	creep strain, m/m
η	thermal fission factor, neutrons released per thermal neutron captured in fuel material
λ	radioactive decay constant, s^{-1}
μ	Poisson's ratio, dimensionless
ν	number of neutrons produced per fission
ρ	density, kg/m^3 (unless specified otherwise)
ρ_r	reactivity
σ_i	microscopic cross section for ith reaction type, cm^2 or barns
$\sigma(KE)$	cross section at kinetic energy KE, barns
$\sigma(th)$	thermal cross section, barns
σ'_0	tangential stress, N/m^2
σ'_r	radial stress, N/m^2
σ'_{ts}	transverse stress, N/m^2
σ'_z	axial stress, N/m^2
σ'_θ	circumferential stress, N/m^2
Σ	macroscopic cross section, cm^{-1}
Σ_t	total macroscopic cross section, cm^{-1}
τ	Fermi age, cm^2
ϕ	neutron flux, N/cm^2·s
χ	weight percent of ^{235}U
ω	angular velocity, rad/s

Subscripts

a	absorption cross section
c	capture
f	fissionable, fission, or fuel
F	feed
P	product
t	total
th	thermal
W	waste or tails

PROBLEMS

7.1. Determine the total energy released when $^{235}_{92}\text{U}$ fissions to form $^{99}_{42}\text{Mo}$, $^{134}_{54}\text{Xe}$, and two neutrons. The isotopic masses of the elements involved in this reaction are

$$^{235}_{92}\text{U} \quad 235.117 \text{ amu}$$
$$^{99}_{42}\text{Mo} \quad 98.938 \text{ amu}$$
$$^{134}_{54}\text{Xe} \quad 133.948 \text{ amu}$$

7.2. A nuclear power reactor produces 800 MWe power and the plant has an overall efficiency of 31%. What is the total weight of the fission products produced by one year's continuous operation?

7.3. (a) Because the neutrons in a reactor have a range of energies, an average cross section, σ_{av}, which allows for this effect should be used. The average cross section may be defined as

$$\sigma_{av} = \frac{\int_0^\infty \sigma(\text{KE})\phi(\text{KE})\,d(\text{KE})}{\int_0^\infty \phi(\text{KE})\,d(\text{KE})}$$

If $\phi(\text{KE})$, the neutron flux per unit energy at energy level KE(T), is given by $\phi(\text{KE}) = C\,\text{KE}(T)\exp[-\text{KE}(T)/k'_B T]$, where $C = C'/(\pi k'_B T)^{3/2}$, C' is a constant, and $\sigma(\text{KE})$, the cross section at energy KE(T), is given by Eq. (7.24), show that

$$\sigma_{av} = \sigma(\text{th})[\text{KE}(\text{th})]^{1/2}\frac{0.5 k'_B T(\pi k'_B T)^{0.5}}{(k'_B T)^2}$$

(b) What effect, if any, would the use of averaged cross sections have on the results of Example 7.2 if μ_{1c} was assumed to be unchanged? Explain.

(c) If KE(th) and $\sigma(\text{th})$ are evaluated at $T = 293°\text{K}$ and $\text{KE}(T) = k'_B T$ write an expression for σ_{av} which is in terms of T and $\sigma(\text{th})$ only.

7.4. Determine the concentration of boron (as boric acid) that would have to be dissolved in the moderator of Example 7.2 in order to reduce k_{eff} to 1.0.

7.5. A nuclear power reactor core consists of the following:

Material	Density (g/cm³)	Volume fraction
UO₂	10.5	0.39
Water	0.81	0.525
Zirconium	6.52	0.085

The absorption cross section of zirconium is 0.185 barn for neutrons at a velocity of 2200 m/s. Other cross sections are given in Example 7.2. The reactor may be considered to be a right circular cylinder 10 ft in diameter and 12 ft high. Determine the enrichment of uranium required to provide a $k_{\text{eff}} = 1.04$ at the beginning of life with all control rods withdrawn. You may assume that $\mu_{\text{re}} = 0.9$.

7.6. A given power reactor must be shut down before either (1) the excess reactivity exceeds β^* ($\beta^* \approx 0.00266$) or (2) a power level exceeds 110% of nominal (level above which the fuel may be damaged). If the most reactive group of control rods can be withdrawn from the reactor at a rate such that the reactivity increase is 0.0008 per second, what is the maximum allowable time for the reactor shutdown (scram) system to be effective?

7.7. A 1-cm-diameter fuel rod contains 0.9-cm-diameter pellets of UO_2. The cladding is of Zircaloy. If the hot pellet operates at a power density of 500 W/cm and the initial pellet-to-cladding gap (at power) is 0.003 cm, how long will it take for the cladding to contact the fuel pellet? Assume that the external pressure is 135 bar and the internal pressure remains constant is 30 bar.

7.8. (a) Calculate the central temperature of the hot pellet described in Problem 7.7 after 100 h of operation. Assume (1) that pellet cracking leads to the pellet pieces lying against the clad at zero contact pressure, (2) the cladding surface temperature is 310°C, and (3) the rod is initially filled with helium.

(b) If the fuel temperature may be considered to remain constant, estimate the fraction of fission gas produced that is released after 2000 h.

7.9. For the conditions of Problem 7.7, estimate the contact pressure which will lead to an outward creep rate equal to the fuel swelling rate shortly after pellet-to-clad contact is established.

7.10. Derive an equation for determining the axial position of the maximum cladding temperature when the heat flux is cosinusoidal (maximum at core center) and there is no boiling in the core. Assume that h_c, G, D_e, T_{in}, and L are known.

7.11. Determine the minimum value of the critical heat flux for the following conditions:

$$T_{in} = 500°F \qquad p = 2000 \text{ psi}$$
$$G' = 1.5 \times 10^6 \text{ lb/h} \cdot \text{ft}^2 \qquad q''_{max} = 300{,}000 \text{ Btu/h} \cdot \text{ft}^2$$
$$D_e = 0.5 \text{ in}$$

Assume that the axial heat flux variation is cosinusoidal and that $F(z)$ may be taken as 1.0.

7.12. A reactor designer suggests that instead of using slightly enriched fuel for the reactor of Problem 7.5, 90% enriched uranium be used and that this be diluted in a special ceramic oxide. Assume that this oxide has the same thermal absorption cross section as $^{238}_{92}U$ but that there is essentially no resonance absorption. Determine the revised $^{235}_{92}U$ concentration required and the change in separative work units needed to supply a core loading.

BIBLIOGRAPHY

BENEDICT, M., ET AL., *Nuclear Chemical Engineering*, 2nd ed. New York: McGraw-Hill, 1982.

Brookhaven National Laboratory, *Neutron Cross Sections*, Report BNL 325. Upton, N.Y.: Cross Section Evaluation Center, Brookhaven National Laboratory, 1966.

COHEN, B., "The Disposal of Radioactive Wastes from Fission Reactors," *Scientific American*, Vol. 236, No. 6 (June 1977).

DUDERSTADT, J., and L. HAMILTON, *Nuclear Reactor Analysis*. New York: Wiley, 1976.

GLASSTONE, S., and A. SESONSKI, *Nuclear Reactor Engineering*. New York: Van Nostrand Reinhold, 1982.

KNIEF, R., *Nuclear Energy Technology*. New York: McGraw-Hill, 1981.

LAMARSH, J., *Nuclear Reactor Theory*. Reading, Mass.: Addison-Wesley, 1966.

MACDONALD, P. E., and L. B. THOMPSON, *MATPRO—A Handbook of Material Properties for Use in Analysis of Light Water Reactor Fuel Rod Behavior*, U.S. Nuclear Regulatory Commission Report ANCR-1263, Aerojet Nuclear Corp., 1976.

MACDONALD, P. E., and J. WEISMAN, "Effect of Pellet Cracking on Light Water Reactor Fuel Temperature," *Nuclear Technology*, Vol. 31, 357 (1976).

TONG, L., and J. WEISMAN, *Thermal Analysis of Pressurized Water Reactors*, 2nd ed. LaGrange Park, Ill.: American Nuclear Society, 1979.

WALKER, W., ET AL. (eds.), *Chart of the Nuclides*, 12th ed. San Jose, Calif.: Nuclear Energy Group, General Electric Co., 1977.

WEISMAN, J., *Elements of Nuclear Reactor Design*, 2nd ed. Melbourne, Fla.: R. E. Krieger, 1983, Chap. 6.

WEISMAN, J., and L. ECKART, "Basic Elements of Light Water Reactor Fuel Design," *Nuclear Technology*, Vol. 53, 326 (1981).

YU-LI, J. C., ET AL., "Some Considerations of the Behavior of Fission Gas Bubbles in Mixed Oxide Fuels," *Nuclear Technology*, Vol. 9, 188 (1970).

8

Production of Mechanical Energy

8.1 THE STEAM CYCLE

Basic Operating Parameters

The basic thermodynamics of modern steam power plants was reviewed in Chapter 2. In this section analysis techniques are developed for steam cycles typical of modern power plant operation.

As indicated in Fig. 8.1, a steam plant can be divided roughly into three segments: (a) the furnace/boiler and auxiliaries; (b) the turbine cycle (which includes, condenser, pumps, feedwater heaters, and heat rejection system) and (c) the electric generator.

Although segment (b) of Fig. 8.1 shows a Rankine cycle, a number of modifications and additions have been made to the basic cycle to improve the overall cycle efficiency. All sensible heat is utilized to the maximum practical extent. Feedwater flowing toward the boiler is first preheated in the economizer, which reclaims part of the heat in the flue gas. This reduces the temperature of the flue gas and decreases the heat that must be supplied to the boiler. Most of the superheated steam from the furnace flows directly to the turbine, but some is bled off for use in the feedwater heaters and additional steam may be used in the steam jet ejector, which maintains condenser vacuum. About 70% of the steam flows directly through the turbine and then through the exhaust ducts to the condenser.

As described in previous chapters, the condenser transfers the latent heat of the steam to water supplied from a river, lake, ocean, or cooling tower. The low-pressure condensate produced drains down to the condenser *hotwell*. A hotwell

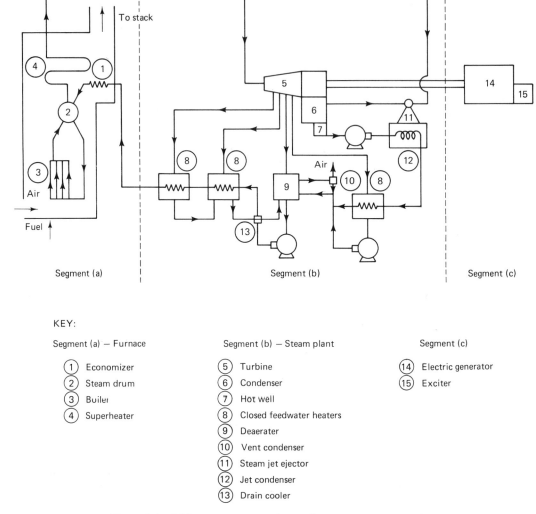

Figure 8.1 Major components of a modern steam power station.

pump takes the water from the hotwell and pumps it through the low-pressure portion of the feedwater heating system: low-pressure feedwater heaters and deaerator. In an actual system, there may be two to four tubular-type feedwater heaters prior to the deaerating heater.

In most plants, the boiler feed pump is connected to the discharge of the deaerator. The boiler feed pump raises the feedwater pressure to that of the boiler and pumps the feedwater through the high-pressure feedwater heaters. Although only two are shown in Fig. 8.1, an actual system would contain three to six such

Sec. 8.1 The Steam Cycle

heaters. The heaters are arranged in series and the liquid condensate collected in the steam trap of one heater is flashed to the pressure of the preceding heater. At the first (lowest steam pressure) heater the liquid from the steam trap is cooled by the water from the boiler feed pump (in drain cooler) so that it may enter the deaerator without flashing. The series of feedwater heaters raise the condensate temperature to the level desired at the economizer inlet. This temperature is in the range of 175 to 230°C (350 to 450°F).

Turbine inlet pressures vary from about 65 bar (950 psia) in nuclear plants up to 240 bar (3500 psia) in some fossil-fuel plants. Inlet temperatures vary from around 280°C in nuclear plants to over 600°C (1100°F) in many fossil-fuel plants. Turbine exhaust pressures range from 20 to 50 mmHg. The maximum steam temperature is normally limited by the superheater or high-pressure turbine materials. Although not shown in the schematic of Fig. 8.1, most plants will also include one or more stages of steam reheat. It was pointed out previously that reheat is usually desirable to prevent excessive end-point moisture. A reheat cycle is illustrated schematically in Fig. 2.9.

Actual Steam Cycles

In Chapter 2 we indicated how we could determine the output of a given turbine under ideal (isentropic) conditions. The theoretical efficiency of the steam cycle, η_R, is simply

$$\eta_R = \frac{W_t}{q_B} \tag{8.1}$$

where W_t = work done by the turbine assuming isentropic conditions
q_B = heat supplied by boiler or reactor

In actuality, a real turbine cycle is not entirely isentropic. Due to frictional effects, the entropy increases slightly in any expansion. This leads to a higher exit enthalpy and lower turbine work output. The ratio of the actual enthalpy change, Δh_a, to the ideal enthalpy change, Δh_i, under isentropic conditions is referred to as the turbine efficiency, η_T. That is, for any turbine section,

$$\eta_T = \frac{\Delta h_a}{\Delta h_i} \tag{8.2}$$

For any given turbine, the turbine efficiency remains nearly constant over a wide range of conditions. Hence the actual work, W_a, done by the turbine is given by

$$W_a = \eta_{th} \eta_T q_B \tag{8.3}$$

The turbine efficiency is a measure of how well a turbine performs for given steam cycle conditions.

An alternative measure of the merit of both the turbine and the steam conditions is given by the heat rate. This heat rate figure of merit can be further subdivided into two categories: a *turbine-cycle heat rate* (TCHR) and an *overall power plant heat rate* (OPPHR). The TCHR is defined as the heat required to be supplied to the steam per kilowatthour of electric power output. The OPPHR is defined as the heat required to be supplied to the power plant in the form of a basic fuel, per unit of kilowatthour of power output from the plant. Various other, similar heat rate definitions, such as the *boiler-turbine heat rate* (BTHR), follow from these concepts:

$$\text{TCHR} = \frac{\text{heat supplied to the turbine cycle/time}}{P_{GO}} \quad (8.4)$$

$$\text{BTHR} = \frac{\text{heat supplied to the turbine cycle/time}}{\eta_B \times P_{GO}} \quad (8.5)$$

$$\text{OPPHR} = \frac{\text{heat available from fuel/time}}{P_{GO}(1 - P_{ax}/P_{GO})} = \frac{\text{heat to the turbine/time}}{\eta_B(P_{GO})(1 - P_{ax}/P_{GO})} \quad (8.6)$$

where P_{GO} = gross generator output, kW
P_{ax} = auxiliary power consumed at the site, kW
η_B = boiler efficiency

The OPPHR is a measure of the fuel economy for power generation. Let us consider the determination of the turbine cycle heat rate for the simplified steam cycle shown in Fig. 8.2.

Assume that we are given, $\dot{m}_T$, the mass flow rate of steam to the turbine and the pressure and temperature of that steam. We are also told the pressures, p_1 and p_2, at which steam is extracted for heaters 1 and 2 and for the reheater. In addition, we are given the temperature increases in the regenerative feedwater heater and the steam temperature at the reheater exit. We are also given the final exhaust pressure and the turbine efficiency, η_F.

The actual work done by the turbine, per unit time W_a, is obtained from

$$W_a = \eta_T \{ (\dot{m}_T(h_1 - h_2')) + (\dot{m}_T - \dot{m}_1)(h_2' - h_3') \\ + (\dot{m}_T - \dot{m}_1 - \dot{m}_2)[(h_3' - h_{13}') + (h_{14}' - h_4')] \} \quad (8.7)$$

where h_j = actual enthalpy at location j
h_j' = enthalpy at location j determined assuming an ideal isentropic expansion in the turbine
$\dot{m}_T$ = mass flow rate entering turbine
$\dot{m}_1, \dot{m}_2$ = mass flow rate at positions 1 and 2

To compute W_a we must determine $\dot{m}_1$ and $\dot{m}_2$. To do this we must perform energy balances on the various segments of the system. If we make an energy balance around heater No. 1, we equate the energy transferred from the condensing steam to that absorbed by the feedwater. We then have

$$\dot{m}_T(h_{10} - h_9) = \dot{m}_1(h_2 - h_{12}) \quad (8.8)$$

We obtain h_{12} from the enthalpy of saturated liquid at the extraction pressure p_1. Note that here we must use actual enthalpies. We obtain h_2 from

$$\eta_T(h_1 - h_2') = h_1 - h_2 \quad (8.9)$$

Sec. 8.1   The Steam Cycle

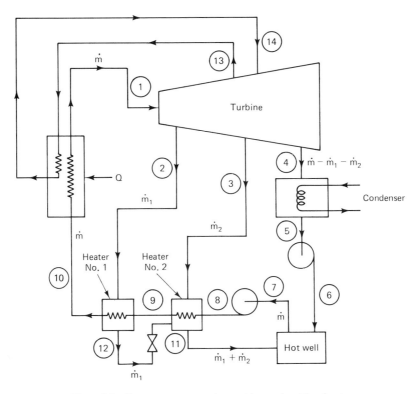

Figure 8.2 Two-stage regenerative rankine cycle with reheat.

Since all quantities in Eq. (8.8) are now known except $\dot{m}_1$, this quantity may be calculated.

We now obtain the value of $\dot{m}_2$ by making a heat balance around the system consisting of heater No. 2 and the hotwell. By equating the energy entering to that leaving, we get

$$\dot{m}_2 h_3 + \dot{m}_1 h_{12} + (\dot{m}_T - \dot{m}_1 - \dot{m}_2) h_5 = \dot{m}_T h_9 \tag{8.10}$$

if we assume that pump work is negligible. We may again obtain h_3 by noting that

$$\eta_T (h_2' - h_3') = (h_2 - h_3) \tag{8.11}$$

and can then compute $\dot{m}_2$ since all other parameters in Eq. (8.10) are known.

Once we specify the amount of reheat $(h_{14} - h_{13})$ desired, we calculate q_B from

$$q_B = \dot{m}_T (h_1 - h_{10}) + (\dot{m}_T - \dot{m}_1 - \dot{m}_2)(h_{14} - h_{13}) \tag{8.12}$$

We obtain the turbine cycle heat rate from

$$\text{TCHR} = \frac{\dot{m}_T (h_1 - h_{10}) + (\dot{m}_T - \dot{m}_1 - \dot{m}_2)(h_{14} - h_{13})}{W_a \eta_G} \tag{8.13}$$

where η_G is the generator efficiency (kilowatts of electric power produced/kilowatt of mechanical power supplied).

The heat load on the condenser per unit time, q_C, may be obtained from

$$q_C = (\dot{m}_T - \dot{m}_1 - \dot{m}_2)(h_{14} - h_5) \tag{8.14}$$

We must again use the actual stream enthalpies. It should be noted that we could have obtained q_C from an energy balance around the entire system.

$$q_C = q_B - W_a \tag{8.15}$$

Example 8.1

A turbine is fed with steam at 110 bar and 450°C. Determine the actual enthalpy and quality of the steam extracted at 14 bar if the turbine stage efficiency is 0.92.

Solution

Enthalpy of steam entering stage = 3222 kJ/kg

Entropy of entering steam = 6.355 kJ/(kg · °K)

Under isentropic conditions $s_{\text{extracted}} = 6.355$ kJ/(kg · °K)

At 14 bar:

$$s_f = 2.284 \qquad s_g = 6.469 \text{ kJ(kg} \cdot \text{°C)}$$

$$h_f = 830.0 \qquad h_g = 2790 \text{ kJ/kg}$$

We determine x under isentropic conditions from

$$6.355 = (1 - x_s)2.284 + 6.469 x_s$$

$$x = 0.975$$

Therefore,

$$h_s = 0.025(830.0) + 0.975(2790) = 2740 \text{ kJ/kg}$$

$$\Delta h_s = 3222 - 2740 = 482 \text{ kJ/kg}$$

$$\Delta h_a = 0.92(482) = 443$$

actual enthalpy = $h_a = 3222 - 443 = 2779$ kJ/kg

We obtain x_a from $2779 = (1 - x_a)(830) + x_a(2790)$:

$$x_a = 0.995$$

In the initial design phase, the problem just examined would have been stated differently. If we know the power that we wish to generate, we would ask what steam flow rate would be required to achieve this power. If the turbine is 100% efficient, the steam flow rate for the cycle described is

$$\dot{m}_T = \frac{W_a}{(h_1 - h_2') + (1 - f_1')(h_2' - h_3') + (1 - f_1' - f_2')[(h_3' - h_{13}') + (h_{14} - h_4)]} \tag{8.16}$$

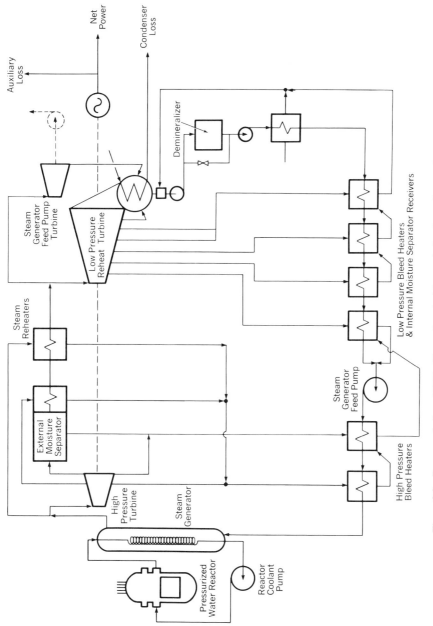

Figure 8.3 Power cycle for a PWR. (Reproduced with permission from *Steam, Its Generation and Use*, The Babcock & Wilcox Company, Lynchburgh, Va., 1978.)

where f_1, and f_2, are the fractions of turbine inlet flow going to heaters 1 and 2, respectively. $\dot{m}_T$ is known as the *theoretical steam rate*. The *actual steam rate* $\dot{m}_a$ is simply obtained from

$$\dot{m}_a = \frac{\dot{m}_t}{\eta_T} \qquad (8.17)$$

The steam cycle calculations described in this section are still somewhat idealized. For example, the final stages of all modern turbines contain some means for internal water separation. Provision is made to remove droplets slung by the blades to the turbine casing. Hence the mass flow rate and moisture in the latter stages of the turbine tend to be overestimated.

In Chapter 2 it was observed that in a nuclear power station there is no opportunity to provide much steam reheat. Nevertheless, the end-point moisture problem remains severe. In earlier plants this was dealt with by incorporating a mechanical moisture separator. The steam leaving the moisture separator was dry and saturated vapor and was fed to the next turbine stage. The saturated liquid from the drain of the moisture separator would be returned to the feedwater train at an appropriate place. More recently it has been customary to couple a moisture separator with a little reheat. Some steam from the steam generator is used to reheat the returning steam slightly. A steam cycle for a modern PWR is shown diagramatically in Fig. 8.3.

8.2 GAS TURBINE POWER PLANTS

The fundamental aspects of a simple Brayton cycle were presented in Chapter 2. In this section the Brayton cycle analysis techniques are extended to practical systems and applied to gas turbine performance for power applications.

Gas turbines for power applications range from 1 MW of rated output to a few hundred megawatts of rated output. In Chapter 2 it was pointed out that gas turbines have been widely used by public utilities for meeting peak power demand. Gas turbines are characterized by low capital costs, high operating costs, and quick startup capability.

Closed Brayton Cycle Analysis Assuming Ideal Gas Behavior

Both the open and closed Brayton cycles were described in Chapter 2. For simplicity, we begin our examination with the closed cycle and assume ideal gas behavior.

Figure 8.4 illustrates the behavior in a closed cycle where a gas such as helium, air, or carbon dioxide (see Table 8.1 for properties of some possible working fluids) is compressed from p_1 to p_2, heated from T_2 to T_3, and then expanded from p_3 (same as p_2) to p_4. The solid lines represent ideal behavior where compressors and turbines are 100% efficient and compressions and expansions are isentropic.

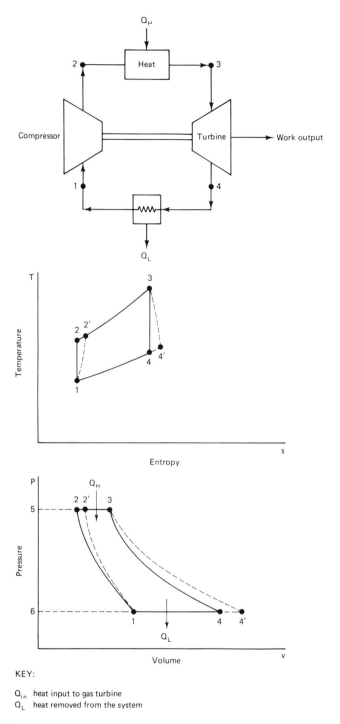

KEY:

Q_{in} heat input to gas turbine
Q_L heat removed from the system

Figure 8.4 $T–s$ and $p–v$ diagram for a closed-cycle gas turbine system.

TABLE 8.1 PROPERTIES OF GAS COOLANTS AT LOW PRESSURES AND TEMPERATURES[a]

Gas	Molecular weight	c_p (kJ/kg · °C)	$c_p/c_v = \gamma$ (dimensionless)	k (kJ/h · m · °C)	μ (kg/h · m)
H_2	2.016	14.322	1.405	0.77879	0.0357
He	4.003	5.233	1.659	0.56073	0.0744
Steam	18.016	1.842	1.335	0.08100	0.0446
CO_2	44.01	0.846	1.29	0.08100	0.0670
Air	28.97	1.005	1.4	0.09345	0.0684
N_2	28.02	1.040	1.4	0.11214	0.0714

[a] c_p, specific heat at constant pressure; c_v, specific heat at constant volume; k, thermal conductivity; μ viscosity.

Real turbines are, of course, not 100% efficient. The nonideal Brayton cycle is shown by the dashed lines and primes (') in Fig. 8.4. Both the compression process with fluid friction (1–2') and the expansion process with fluid friction (3–4') show an increase in entropy. Pressure drops during the heat addition process (2–3) and the heat rejection process (4–1) have still been neglected. These pressure drops need be included only for low-pressure ratio cases.

The efficiency for the compression and expansion processes are now defined by the following expressions:

For the compressor:

$$\eta_{cp} = \frac{\text{ideal work}}{\text{actual work}} = \frac{h_2 - h_1}{h_2' - h_1} \simeq \frac{T_2 - T_1}{T_2' - T_1} \qquad (8.18)$$

For the turbine:

$$\eta_T = \frac{\text{actual work}}{\text{ideal work}} = \frac{h_3 - h_4'}{h_3 - h_4} \simeq \frac{T_3 - T_4'}{T_3 - T_4} \qquad (8.19)$$

The temperature relationships assume constant specific heats.

The net work output of the cycle is still the difference between the turbine work output and the work input required by the compressor. By following the approach used in Section 2.3, we have

$$W_T = (\dot{m} c_p T_3)\left(1 - \frac{1}{r_p^{(\gamma-1)/\gamma}}\right)\eta_T$$

$$W_c = \frac{1}{\eta_{cp}}\left[\dot{m} c_p T_2 \left(1 - \frac{1}{r_p^{(\gamma-1)/\gamma}}\right)\right]$$

$$= \frac{1}{\eta_{cp}}\left[\dot{m} c_p T_1 r_p^{(\gamma-1)/\gamma}\left(1 - \frac{1}{r_p^{(\gamma-1)/\gamma}}\right)\right] \qquad (8.20)$$

Sec. 8.2 Gas Turbine Power Plants

Since the net work, W_{net}, equals $(W_T = W_c)$, we have, after simplification,

$$W_{net} = \dot{m} c_p T_1 \left[\left(\eta_T \frac{T_3}{T_1} - \frac{r_p^{(\gamma-1)/\gamma}}{\eta_{cp}} \right) \left(1 - \frac{1}{r_p^{(\gamma-1)/\gamma}} \right) \right] \quad (8.21)$$

The specific power of the cycle can be used as one measure of cycle performance. We define

$$\text{specific power} \equiv \frac{W_{net}}{\dot{m}} \quad (8.22)$$

The specific power is dependent on the specific heat of the coolant gas used in the cycle and there is a pressure ratio at which an optimum specific power is reached.

The heat added to the cycle, q_A, is given by

$$q_A = \dot{m} c_p (T_3 - T_2') \quad (8.23a)$$

$$q_A = \dot{m} c_p \left[(T_3 - T_1) - T_1 \frac{r_p^{(\gamma-1)/\gamma} - 1}{\eta_{cp}} \right] \quad (8.23b)$$

The efficiency can be determined from the standard expression for the overall thermal efficiency, η_o:

$$\eta_o = \frac{W_{net}}{q_A} \quad (8.24)$$

which is nothing more than division of Eq. (8.21) by Eq. (8.23b).

The thermal efficiency is sensitive to turbine and compressor efficiencies. A simple example will illustrate this sensitivity.

Example 8.2

A Brayton cycle operating with air as the working fluid has a turbine inlet temperature, T_3, of 850°C (1502°F) and a compressor inlet temperature, T_1, of 10°C (41°F). The pressure ratio across the compressor r_{cp} is 4.25. Compare the overall system efficiency for turbine and compressor efficiencies of (a) 100%, (b) 90%, and (c) 80%.

Solution

(a) For 100% efficiency:

$$\eta_{cp} = \eta_T = 1.0$$

By using Eqs. (8.21) and (8.23b), we have

$$\eta_{th} = \frac{T_1 \left[\eta_T (T_3/T_1) - r_p^{(\gamma-1)/\gamma}/\eta_{cp} \right] \left(1 - 1/r_p^{(\gamma-1)/\gamma} \right)}{(T_3 - T_1) - T_1 \left(r_p^{(\gamma-1)/\gamma}/\eta_{cp} \right)} \qquad \begin{array}{l} T_3 = 850 + 273 = 1123°K \\ T_1 = 10 + 273 = 283°K \end{array}$$

$$= \frac{283 \left\{ (1)(1123/283) - \left[(4.25)^{0.4/1.4}/1 \right] \right\} \left[1 - (1/4.25)^{0.4/1.4} \right]}{(1123 - 283) - 283 \left[(4.25)^{0.4/1.4}/1 \right]}$$

$$= 0.338$$

(b) For 0.9 efficiency:
$$\eta_{cp} = \eta_T = 0.9$$
$$\eta_{th} = 0.263$$

(c) For 0.8 efficiency:
$$\eta_{cp} = \eta_T = 0.8$$
$$\eta_{th} = 0.185$$

This example demonstrates the rapid decrease in thermal efficiency with a relatively small decrease in the compressor and turbine efficiency.

The efficiency of the cycle will, of course, always increase when T_3 is increased. However, turbine materials limit the maximum value of T_3. For any given value of T_3 there is a pressure ratio that will provide the maximum thermal efficiency. A simple expression for the optimum value of r_p may be obtained if we make the approximation that the heat supplied is $(h_3 - h_2)$ instead of $(h_3 - h_{2'})$. If we do so,

$$Q_h \simeq c_p(T_3 - T_2) = c_p T_1 \left(\frac{T_3}{T_1} - r_p^{(\gamma-1)/\gamma} \right) \tag{8.25}$$

and

$$\eta_{th} \simeq \frac{(T_3/T_1)\eta_T - \left(T_3\eta_T/T_1 r_p^{(\gamma-1)/\gamma}\right) - \left(r_p^{(\gamma-1)/\gamma}/\eta_{cp}\right) + (1/\eta_{cp})}{(T_3/T_1) - r_p^{(\gamma-1)/\gamma}} \tag{8.26}$$

After differentiating with respect to r_p, equating the differential to zero, and simplifying the result, we have

$$r_{p(opt)} \simeq \frac{T_3/T_1}{1 + \sqrt{[(T_3/T_1) - 1][(1/\eta_T\eta_{cp}) - 1]}} \tag{8.27}$$

The efficiency of a Brayton cycle can be increased by adding a regenerator to the system. In these systems, the turbine exhaust gases are used to preheat the compressed gas before it enters the combustor. If the regenerator had an efficiency of 100%, as shown in Fig. 8.5, the turbine exhaust would be cooled to $T_{2'}$ and the coolant temperature entering the combustor would be raised from $T_{2'}$ to $T_{2''}$. Although the net work of the cycle remains the same, the net heat added to the cycle can be reduced from $\dot{m}c_p(T_3 - T_{2'})$ to $\dot{m}c_p(T_3 - T_{2''})$. Thus the cycle efficiency will increase correspondingly.

An actual regenerator operates somewhat below 100% efficiency with the turbine exhaust being cooled only to $T_{4'''}$ and the actual temperature increase of the compressed gas going from $T_{2'}$ to $T_{2'''}$ rather than $T_{2''}$. The efficiency of a regenerator is the ratio of actual heat transferred to the theoretical amount of heat that could be transferred. With the terminology of

$$\eta_r = \frac{h_{2'''} - h_{2'}}{h_{4'} - h_{2'}} \tag{8.28}$$

Sec. 8.2   Gas Turbine Power Plants

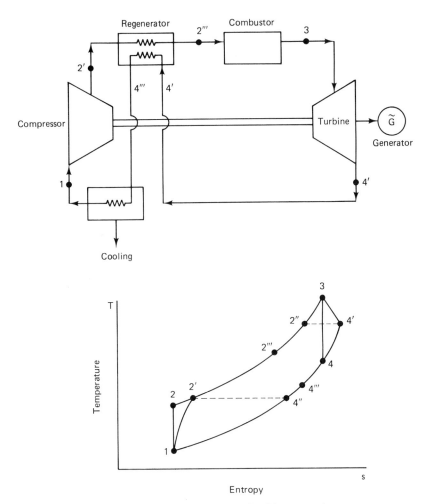

Figure 8.5 Closed gas turbine cycle with regeneration.

For a constant specific heat, Eq. (8.28) can be expressed as

$$\eta_r = \frac{T_{2'''} - T_{2'}}{T_{4'} - T_{2'}} \tag{8.29}$$

When regeneration is used, there are other additions that can be made to increase cycle efficiency further. From basic thermodynamics, we know that isothermal compression requires less work than does adiabatic compression. Adiabatic compression can be made to approach the isothermal process if we compress the gas in several stages and cool the gas between stages. If there is no regeneration, the gain obtained by intercooling is partly offset by the additional heat which must be supplied at low temperature. There is little, if any, improvement in thermal efficiency from intercooling unless intercooling is used together with regeneration.

The expansion process may be divided into stages. If reheating of the gas is provided between each stage, the system efficiency can be increased further when intercooling is used. The net work of a system containing a series of compressors and turbines may be found by algebraically summing the work done in each stage. For the simple case in which all turbines in the system have the same efficiency (η_T), and the compressors have efficiency η_{cp} and the same pressure ratio (r_p), the net work done is

$$W_{net} = \dot{m}\left[\eta_T T_3(N_T + 1)\left(1 - \frac{1}{r_p^{(\gamma-1)/\gamma}}\right) - \frac{N_c + 1}{\eta_{cp}}T_1\left(r_{cp}^{(\gamma-1)/\gamma} - 1\right)\right] \quad (8.30)$$

The total heat input to the system would be

$$q_A = \dot{m}\left\{c_p T_3\left[\left(1 - \frac{T_1}{T_3}r_{cp}^{(\gamma-1)/\gamma}\right) + N_T\left(1 - \frac{1}{r_{cp}^{(\gamma-1)/\gamma}}\right)\right.\right.$$

$$\left.\left. -\eta_r\left(\frac{1}{r_p^{(\gamma-1)/\gamma}} - \frac{T_1}{T_3}r_{cp}^{(\gamma-1)/\gamma}\right)\right]\right\} \quad (8.31)$$

where r_p = pressure ratio across the turbines
r_{cp} = pressure ratio across the compressors
η_r = regenerator efficiency
N_T = number of turbine stages
N_c = number of compressor stages

The efficiency of the cycle increases as the number of reheat and intercooling stages are added. This increase in efficiency diminishes with each successive stage addition, while the capital cost for each additional stage remains relatively constant. Thus the net gain in efficiency for each stage addition must be evaluated against the increase in capital cost.

Open Brayton Cycle Analysis Assuming Ideal Gas Behavior

In an open Brayton cycle that utilizes a fossil fuel as the primary source of heat, the mass flow through the system is not constant and the previously developed equations must be modified to account for the variation in mass flow rates. Figure 2.5(a) illustrates an open-cycle gas turbine power system. The compressor and turbine work are now obtained from

$$W_c = \dot{m}_1 c_p T_2\left(1 - \frac{1}{r_p^{(\gamma-1)/\gamma}}\right) \quad (8.32)$$

$$W_T = (\dot{m}_1 + \dot{m}_2)c_{pm}T_3\left(1 - \frac{1}{r_p^{(\gamma_m-1)/\gamma_m}}\right) \quad (8.33)$$

We see that with the relatively low top temperature which is common, combustion gases make up only a small fraction of the total flow through the turbine. Some of the air required to dilute the combustion gases is provided as excess air in combustion and the additional air is added to the gases leaving the combustor. The temperature of the air leaving the combustor may be calculated by assuming an adiabatic process and using the calculational techniques developed in Chapter 4.

In view of the relatively small change in mass flow and properties through the system, the approximate analyses developed previously can also be considered to be generally applicable to the open cycle. Less equipment is needed since no cooler is required, but with the same turbine and compressor conditions and efficiencies, essentially the same cycle efficiency is obtained.

It should be noted that regeneration is still valuable in the open cycle despite the need to dilute the gases leaving the combustor. By preheating the air entering the combustor, less heat is required to reach the allowable turbine inlet temperature. Since gas turbine plants are usually used for peaking plants where low capital cost is of prime importance, use of compressor intercooling and turbine reheating for increasing efficiency is not widespread.

Accurate Analysis of the Open Brayton Cycle

While the assumption that the working fluid in the Brayton cycle is an ideal gas allows us to obtain simple analytical expressions for cycle behavior and draw conclusions as to the effect of cycle conditions, it does not lead to an accurate evaluation of compressor and turbine work. For such an evaluation we should, as we did for steam turbines calculations, use gas enthalpies and entropies evaluated from appropriate thermodynamic tables or charts. Such tables and charts are available for air and mixtures of air with various levels of combustion gases.

In most open cycles, the maximum pressure will be only several atmospheres. In this range, the variation of specific heat with pressure is small. Although gas thermodynamic properties are functions of both temperature and pressure, it is possible to develop reasonably accurate tables, for the low-pressure conditions common to Brayton cycles, which are functions of temperature only. We know from basic thermodynamics that at constant pressure

$$dh = c_p \, dT \tag{8.34}$$

Although pressure has little effect on the c_p of air and combustion gases over the range of interest, c_p does vary with T. By taking c_p independent of pressure, we may write

$$h_2 - h_1 = \int_{T_1}^{T_2} c_p \, dT \tag{8.35}$$

Sec. 8.2 Gas Turbine Power Plants

The values of c_{pm} and γ_m should be properly evaluated for the appropriate mixture (indicated by the subscript m) of primary fluid and combustion gases. However, in many approximate analyses this complication is ignored and air properties as well as a constant mass flow are assumed.

The approximation that the air alone is being circulated is not unreasonable for the usual conditions. For most gas turbines, the inlet temperature must be kept below about 1000°C unless special turbine blade cooling is provided. In order to keep the turbine inlet temperature below this value, a large amount of air in excess of that required for combustion must be provided. The combustion gases are then only a small fraction of the total gas flow. This is illustrated in the following example.

Example 8.3

Assume the following open-cycle system parameters:

Working fluid: air
$c_p = 1.005$ kJ/kg · °C
$\gamma = 1.4$
T_1 = air inlet temperature = 300°K
T_3 = maximum gas temperature = 1150°K
Inlet pressure = 1 atm
r_p = pressure ratio = 4
η_{cp} = compressor efficiency = 80%
η_T = turbine efficiency = 85%
HV = heating value of fuel = 41,860 kJ/kg
Heater losses = 10% of fuel heating value

We are to determine the mass of fuel required per unit mass of air.

Solution In the compressor, the work is given by

$$\frac{W_c}{\dot{m}} = \frac{(T_1 r_p^{(\gamma-1)/\gamma} - T_1)c_p}{\eta_{cp}} = \frac{500(1.486 - 1)(1.005)}{0.8} = 183.2 \text{ kJ/kg}$$

The heat supplied by the heater is given by

$$\frac{q_A}{\dot{m}_{air}} = (h_3 - h_2)_{air}$$

If we can assume that fuel mass flow is negligible, we have

$$h_2 = c_p T_1 + \frac{W_c}{\dot{m}_{air}} = (1.005)(300) + 183.2 = 484.7 \text{ kJ/kg}$$

$$\frac{q_A}{\dot{m}_{air}} = (c_p T_3 - h_2)_{air} = 1.005(1150) - 484.7 = 671.1 \text{ kJ/kg air}$$

= energy per unit mass of air required to heat air to exit temperature
net heat added per unit mass of fuel = 0.9 × HV = 37,680 kJ/kg fuel

$$\text{fuel/air ratio} = \frac{671.1 \text{ kJ/kg air}}{37,680 \text{ kJ/kg fuel}} = 0.0178 \frac{\text{kg fuel}}{\text{kg air}}$$

Sec. 8.2 Gas Turbine Power Plants

When we define $h = 0$ at a base temperature T_0, we have

$$h_2 = \int_{T_0}^{T_2} c_p \, dT \tag{8.36}$$

and can then tabulate h as a function of temperature.

To obtain entropy as a function of temperature, we recall that for a reversible process

$$T \, ds = dh - v \, dp \tag{8.37}$$

where v is the specific volume. After substitution of Eq. (8.34), we have

$$ds = \frac{c_p \, dT}{T} - \frac{v \, dp}{T} \tag{8.38}$$

We now make use of the fact that although c_p varies with temperature, air and combustion gases show only small deviations from the ideal gas law ($pv = RT$) over the range of interest. Hence

$$ds = \frac{c_p \, dT}{T} - \frac{R \, dp}{p} \tag{8.39}$$

and

$$s_2 - s_1 = \int_{T_1}^{T_2} \frac{c_p \, dt}{T} - R \ln \frac{p_2}{p_1} \tag{8.40}$$

Note if British units are used, R must be divided by J, the mechanical equivalent of heat. By defining $\bar{\phi}$ for any temperature T' as

$$\bar{\phi} = \int_{T_0}^{T'} \frac{c_p \, dT}{T} \tag{8.41}$$

where T_0 is our arbitrary base temperature, we can write

$$s_2 - s_1 = \bar{\phi}_2 - \bar{\phi}_1 - R \ln \frac{p_2}{p_1} \tag{8.42}$$

We may now tabulate $\bar{\phi}$ as a function of temperature along with h and subsequently compute entropy changes from the tabulated values of $\bar{\phi}$ and the change in pressure.

The utility of the tabulation of $\bar{\phi}$ and h as T may be enhanced by considering fluid behavior under reversible, adiabatic conditions. In going from condition 1 to 2 we will be moving along an isentropic line; hence $s_1 = s_2$ and for this condition

$$\bar{\phi}_2 - \bar{\phi}_1 = R \ln \frac{p_2}{p_1} = R(\ln p_2 - \ln p_1) \tag{8.43}$$

Since $\bar{\phi}_2$ depends only on T_2, and $\bar{\phi}_1$ depends only on T_1, Eq. (8.43) is satisfied by any two pressures having the ratio p_2/p_1 provided that the temperatures T_2 and T_1 are fixed. We may then define an arbitrary set of relative pressures, $(p_{re})_1$ and $(p_{re})_2$, such that

$$\frac{(p_{re})_2}{(p_{re})_1} = \frac{p_2}{p_1} \tag{8.44}$$

For Eq. (8.43) to hold between any two points on an isentropic line, each value of p_{re} must be a function of temperature alone. We may therefore tabulate the function $(p_{re})_1$ computed from an arbitrary base as a function of gas temperature.

Effectively, such a tabulation of p_{re} values relates the change in temperature to the change in pressure along an isentropic path. This is equivalent to the relationship between temperature and pressure ratios along an isentropic path defined by Eq. (2.30) for an ideal gas.

Values of h, $\bar{\phi}$, and p_{re} have been tabulated for air and for air–combustion gas mixtures resulting from the combustion of saturated straight-chain hydrocarbons with various quantities of excess air (added in original combustion or as a subsequent diluent). Appendix Table A.5 gives these data for air. Because of the large amount of excess air added to keep exit-gas temperatures in an acceptable range, the thermodynamic properties of the actual combustion gas mixture differ little from those of air.

The use of Table A.5 may be clarified by the following example.

Example 8.4

Consider the problem of finding the work done when 1 kg of air and combustion product mixture (combustion products + 400% theoretical air) is expanded from 4 bar and 800°C to 1 bar. Assume a 90% efficient turbine.

Solution At the initial state, $T_3 = 1073°K$, and we find from Table A.5 that $h_3 = 1130$ kJ/kg and $(p_{re})_3 = 151.36$. Since the expansion ratio is 4, $(p_{re})_4 = 151.36/4 = 37.84$. We now enter Table A.5 with this value of p_{re} and find by interpolation that $h = 765.4$ kJ/kg. The work done per kilogram is then

$$\frac{W_T}{\dot{m}} = \eta_T(h_3 - h_4) = (1130.15 - 765.4) \text{ kJ/kg} \times 0.9 = 328.4 \text{ kJ/kg}$$

If we had used the actual properties of the combustion gas–air mixture, we would have found that the total work done was 327.2 kJ/kg. The error in the results because of the assumption that the gas is all air is therefore negligible. A greater error is obtained if we assume that the working fluid is an ideal gas with a value of γ equal to that of air. In that case we would have

$$\frac{W_T}{\dot{m}} = \eta_T c_p T_3 \left(1 - \frac{1}{r_p^{(\gamma-1)/\gamma}}\right)$$

$$= 0.9 \times 1.005 \text{ kJ/kg} \cdot °K \, (1073°K)\left(1 - \frac{1}{4.285}\right) = 317.4 \text{ kJ/kg}$$

When greater accuracy is needed, we can construct thermodynamic tables or charts for any given exit-gas composition from the properties of O_2, N_2, CO_2, and H_2O vapors. Since gas mixtures are ideal solutions which exhibit no heat of mixing, we can readily calculate molal properties from

$$h_m = \sum_i y_i h_{m_i}$$

$$s_m = \sum_i \left(\phi_i - R \ln \frac{y_i p}{p(0)}\right) = \sum_i \left(\phi_i - R \ln y_i - R \ln \frac{p}{p(0)}\right) \quad (8.45)$$

where h_m, s_m = molal enthalpies and entropies, respectively, of given mixture at T
h_{m_i}, s_{m_i} = molal enthalpies and entropies, respectively, of ith component at given T
y_i = mole fraction of component i
$p(0)$ = initial pressure

8.3 COMBINED CYCLES

High-pressure steam turbines have been built to operate at pressures as high as 340 bar (5000 psi) and temperatures of about 670°C. However, most of the industry has apparently concluded that better overall economy (considering initial, maintenance, and outage costs as well as fuel costs) is obtained with pressures below 235 bar and temperatures below 570°C. Hence, most steam turbines do not take full advantage of the high flame temperatures that are produced in the combustion process.

In contrast to the relatively low temperature operation of steam turbines, low-pressure gas turbines may be operated with temperatures as high as 1000°C. They thus offer the possibility of achieving higher thermodynamic efficiencies. It is theoretically possible to devise a Brayton cycle having an efficiency close to that of a Rankine cycle operating over the same temperature range. However, this requires a large number of intercoolers between compression stages and a large number of reheat stages in the turbine in addition to regenerative heating. Such a turbine–compressor set would be very costly. A more attractive possibility is to use a relatively simple Brayton cycle as a "topping" cycle. The exhaust gases from the gas turbine would then flow to a steam generator and a conventional Rankine cycle would follow.

Figure 8.6 provides a schematic diagram of the combined cycle. The waste gases from the gas turbine may be joined by additional hot gases produced by combustion. The combined gas flow generates steam in the conventional manner. Although not shown in the diagram, the steam cycle would have the usual steam reheaters, feedwater heaters, and economizer.

Combined cycles of the type illustrated are not feasible when pulverized coal is used as the fuel. The fly ash from the coal causes an unacceptable rate of turbine blade erosion. Such cycles would be feasible with oil or natural gas as fuel. However, this is not attractive because of the high cost and limited supply of such fuels. Combined cycles seem most attractive when medium-Btu gas produced by coal gasification is the fuel.

In one process a slurry of coal and water is fed to a gasifier. There it is reacted exothermically with oxygen at high pressure and temperature (40 atm and 1400°K) to produce a mixture of mostly CO and H_2 which has a heat of combustion of about 11×10^6 J/m³ (300 Btu/standard ft³). After ash removal by both a settling chamber and wet scrubber, sulfur compounds (H_2S and COS) are removed by absorption. The chemical absorption process used produces a concentrated stream of H_2S and absorbant for recycle. The H_2S is then converted to elemental sulfur

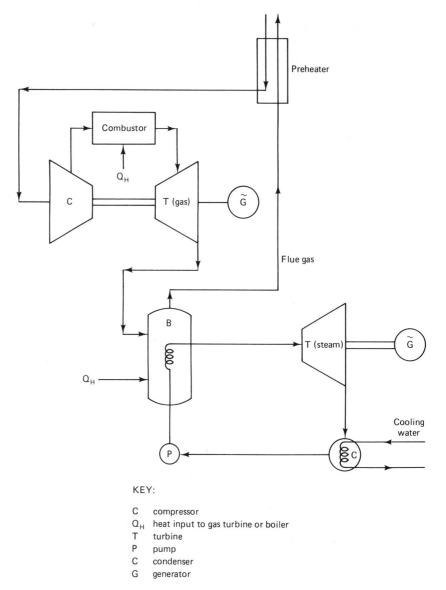

Figure 8.6 Combined gas turbine/steam cycle.

KEY:
- C   compressor
- Q_H heat input to gas turbine or boiler
- T   turbine
- P   pump
- C   condenser
- G   generator

which can be sold. Note that removal of sulfur from the fuel eliminates the need for large and costly flue gas scrubbers.

The purified synthesis gas is burned in a conventional gas turbine. Since the heat of combustion per unit mass of fuel is much lower than for natural gas or petroleum fractions, considerably less excess air is required in order to keep combustion exit temperatures from exceeding those allowed in the turbine.

Sec. 8.4   Steam and Gas Turbine Configurations

The oxygen required for production of the fuel gas is obtained by distillation of liquefied air. The compressed air used by the oxygen plant is supplied by a compressor driven by a small steam turbine which receives a bleed stream from the steam generator.

Efficiency of Combined Cycles

The interest in a combined cycle stems from the potential for higher overall cycle thermal efficiency. Assume that the two cycles, cycle 1 and cycle 2, have thermal efficiencies given by η_{c1} and η_{c2}. The combined cycles are connected together such that the heat exhausted from cycle 1 becomes the heat source for cycle 2. The heat flow is thus arranged in series between the two cycles. If the heat input to cycle 1 is q_1 and if in addition to using the heat rejected from cycle 1, there is heat input of q_2 to cycle 2, the combined cycle efficiency, η_{co}, would be

$$\eta_{co} = \eta_{c2} + \frac{q_1 \eta_{c1}(1 - \eta_{c2})}{q_1 + q_2} \tag{8.46}$$

where η_{c1} = efficiency of cycle 1
η_{c2} = efficiency of cycle 2

If the second cycle can be designed to operate only on the heat rejected from the first cycle, the additional heat supplied to the second cycle is, of course, zero ($q_2 = 0$). Note that the actual net efficiency of the system shown in Fig. 8.6 would be less than given by Eq. (8.44) since we have not accounted for the energy required for the oxygen plant compressor.

Combined Cycle with a Magnetohydrodynamic Generator

An alternative means of increasing thermal efficiency is to pass the hot combustion gases through a magnetohydrodynamic (MHD) generator before they are used to produce steam. In an MHD generator, electricity is produced directly without the need for a thermal-to-mechanical energy conversion device. We discuss MHD, and combined cycles in which it is used, in Chapter 9.

8.4 STEAM AND GAS TURBINE CONFIGURATIONS

Basic Concepts

The steam or gas turbine is one of engineering's outstanding achievements. Because of their efficiency and simplicity, turbines are utilized for electric power production, propulsion, and a variety of small shaft horsepower applications.

Steam power in the eighteenth and early nineteenth centuries was produced entirely by steam engines of steadily improving design. Although many inventors recognized the desirability of using steam directly to power a device similar to a water turbine (available since 1835) it was not until 1884 that Charles A. Parsons demonstrated the first steam turbine. Parsons later added a condenser to improve the efficiency of the device. In 1891, a steam turbine was first used to generate electricity, and in 1897 the first steam turbine powered ship was built.

A steam or gas turbine converts the internal energy of the working fluid into mechanical work. Such a turbine consists of sets of stationary blades (or nozzles) and sets of rotating blades. The gas or steam expands as it flows through a set of stationary blades or nozzles, converting a portion of the enthalpy of the gas or steam into kinetic energy. This increased kinetic energy is a result of the increased velocity of the steam or gas. The high-velocity fluid is directed over the rotating turbine blades, causing a change in momentum of the fluid. This rate of change of the momentum of the fluid cause a force to be exerted on the rotating blades, thus producing mechanical work.

Steam and gas turbine design is based on two concepts: reaction blading and impulse blading. Reaction blading utilizes the fact that a reaction force develops whenever a velocity is generated. In a pure reaction turbine the steam or gas would enter the moving blading without a significant velocity and drop in pressure so that a substantial velocity would be generated.

An impulse force arises whenever a jet is bent from its original direction or diminished in velocity. In a pure impulse turbine, the working fluid drops in pressure when it passes through the nozzle or stationary blades and acquires a velocity. The high-velocity fluid then enters the rotating blading, where some of the kinetic energy is absorbed, and it exits with a diminished velocity. In a pure impulse turbine there would be no pressure drop across the rotating blading.

Multistage Turbines

In its most basic form, a turbine would consist of a single set of nozzles or stationary blades followed by a set of rotating blades. Such an arrangement is satisfactory only for very low power applications, since only a very limited amount of power can be produced. In addition, the system would be inefficient since the exit fluid would retain most of the thermal energy of the entering stream. Therefore, all turbines for power production consist of a number of sets, or "stages," of nozzles and blades. Rows of specially designed rotating blades project outward from a cylindrical rotor. Between each row of these rotating blades is a row of fixed (stationary) blades that project inward from the cylindrical shell. The flow paths between the stationary blades are arranged such that the fluid increases in velocity as it flows across.

Perhaps the simplest form of turbine staging is the *pressure staging* used in the Rateau impulse turbine. This design consists of many rows of simple impulse turbines in a series arrangements. The pressure drop occurs primarily through these stationary blades or nozzles. Since there are usually many sets of these stationary nozzles, the pressure loss across any set is small. The velocity acquired by the steam as it enters the moving blades is low, resulting in low blade velocity and correspondingly low exit pressure losses from each stage. The blades can also be designed such that the steam leaving one set of blades is directed onto the next set of stationary nozzles, thereby further minimizing kinetic energy losses.

Reaction staging turbine design consists of alternating sets of moving and stationary blades in a converging passageway. Both the stationary and moving blades are designed in such a manner that passageways through both sets of blades converge. Each set of blades, moving or stationary, acts as a group of convergent nozzles. Pressure drop occurs across both stationary and moving blades. The pressure drop and the velocity increase across the stationary blade. The steam leaving the stationary blade is directed onto the moving blades, which absorb part of the kinetic energy. The steam entering any set of blades has a considerable amount of kinetic energy. Thus, although referred to as a reaction turbine, the blading is not really pure reaction blading. In many reaction turbines, the impulse effect is actually greater than the reaction effect.

A third type of turbine staging, called *velocity staging* or *Curtis staging*, is often used for turbines having only a few stages. A Curtis stage consists of a set of nozzles and two sets of rotating blades separated by a set of stationary reversing blades. It is used for single-stage turbines when the exit velocity is high. The set of stationary blades redirects the flow, without significant pressure change, to the second row of moving blades, where the excess kinetic energy is absorbed. The efficiency of a Curtis impulse stage is less than a Rateau impulse stage, and it is not generally used to central station power applications.

A velocity–pressure diagram is often used to illustrate the flow–pressure characteristics through a turbine stage. Figure 8.7 illustrates the behavior in typical impulse and reaction stages. In Fig. 8.7(a) we see the behavior in an impulse stage. The fixed blades accelerate the steam or gas until the fluid is about double the blading velocity. At the same time the pressure drops significantly. In the rotating blades, the velocity drops to its original value with only small changes in pressure.

In the reaction stage [Fig. 8.7(b)], the steam velocity is again increased and pressure decreased in the stationary blade. However, the velocity increase is only about half of that seen in the impulse stage. In the rotating blades, both the absolute velocity and pressure are decreased.

In turbine design, the energy transferred to the rotor in each stage is proportioned to the change in absolute steam velocity in the direction of rotation. The energy transferred per stage is therefore about twice as high in impulse blading as in reaction blading. Practically, this means that for a given power output, an impulse turbine will require fewer stages than will a reaction turbine. This advantage in performance is offset by practical design and manufacturing considerations. The shaft seal design is much more difficult for impulse turbines. Also, in reaction turbines, stationary and rotating blades have the same shape, thus simplifying design

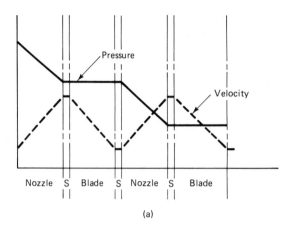

(a)

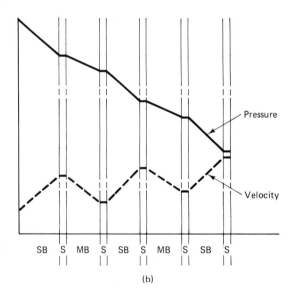

(b)

KEY:
SB   stationary blade
S   space between blades
MB   moving blade

Figure 8.7 Pressure and velocity changes through the stages of (a) an impulse and (b) a reaction turbine.

Sec. 8.4 Steam and Gas Turbine Configurations

and reducing manufacturing cost. The blade concept used makes very little difference in overall turbine efficiency. Turbine efficiencies are generally above 90% for each concept.

We have noted previously that a practical turbine design cannot be pure reaction staging. In a large axial-flow machine one cannot have a pure impulse turbine. The annulus following the nozzle exit is filled with steam flowing with a high tangential velocity in a vortex that exists between inner and outer boundaries. For equilibrium, there must be a static pressure gradient from a low value at the inner boundary to a high value at the interboundary. A single pressure condition can only be approximated for very small length blades. The deviation is measured by the radius ratio, r_o/r_i, the ratio of the outer radius to the inner radius. Only at a radius ratio near 1.0 would a nearly uniform pressure condition exist across the turbine blades. All axial-flow turbine stages (where radius ratio $\gg 1.0$) tend to have significant impulse components at the inner radius and very significant reaction components at the outer radius.

Working-Fluid Flow and Expansion

The working fluid enters the turbine at a substantially higher pressures than that at which it leaves. This is particularly true of a steam turbine, where the first stage may be at a pressure greater than 160 atm while the last stage is at fractions of an atmosphere. To accommodate the large increase in gas volume that accompanies the pressure reduction, the flow area must be increased progressively for each stage. The substantial increase in flow area accounts for the use of the truncated cone as the traditional flow diagram symbol for a steam turbine.

As the pressure decreases, it is often not possible to keep within allowable blade size limits using only a single flow path. When this occurs, the flow is taken to a second, "double-flow" unit. Here the vapor enters at the center of the unit and flows toward each end.

Steam Turbine Configurations

Steam turbines are descriptively classified by the following:

1. Steam supply and exhaust conditions
2. Casing or shaft arrangement
3. Number of exhaust stages in parallel
4. Type of stage design; impulse or reaction
5. Direction of steam flow
6. Single stage or multistage

Steam supply and exhaust conditions indicate whether the turbine is condensing or noncondensing (noncondensing turbines may be used in some small industrial applications) and whether the back pressure is above or below the atmospheric pressure. Other subclassifications under this category include regenerative, extraction, and reheat conditions.

The casing or shaft arrangement describes the manner in which these components are designed: for example, single casing, tandem compound, or cross compound. A *tandem compound* turbine has two or more casings with the shafts coupled together in line. A *cross compound* shaft arrangement refers to a turbine with two or more shafts that are not in line. In such arrangements the shafts often have different speeds.

The number of exhaust stages describes the turbine steam exhaust arrangement. There may be two or three steam flow exhaust stages, which are referred to as double-flow or triple-flow exhaust.

The direction of steam flow describes the manner in which the steam flows through the turbine. Three flow paths are used: axial flow, radial flow, and tangential flow. In the United States, only axial-flow turbines are used for large units. Some small turbines operate on the tangential-flow principle. Radial-flow steam turbines are sometimes used in Europe. All large steam turbines are multistage. Single-stage turbines are only used for very low power applications.

In a large power plant, a turboset usually consists of three steam turbines: high pressure, intermediate pressure, and low pressure. All three turbines are generally connected to the same shaft, which is coupled to the electric generator shaft. A turboset for a large power plant, with the outer turbine casing removed, is shown in Fig. 8.8. In this, as in most very large plants, it is necessary to split the steam flow between two low-pressure turbines in order to keep the last-stage blade diameter within acceptable limits.

A tandem-compound turbine might typically operate with steam inlet conditions of 16,650 kPa (2400 psi), 540°C (~ 1000°F), and reheat to 548°C. The

Figure 8.8 Large modern steam turbine. (Courtesy of General Electric Company, Schenectady, N.Y.)

Sec. 8.5 Turbine Analysis

left-hand turbine serves as the high-pressure portion of the unit. The central unit receives reheated steam at a central post and the steam flows toward both ends of the turbine. The exhaust from the intermediate unit feeds the low-pressure unit at the center and again flows to both ends. The double-flow intermediate and low-pressure units lead to near-zero axial thrusts and thus minimize the difficulty in designing axial thrust bearings for these large units.

Modern steam turbines are usually designed to meet the power plant output specifications with a single turboset. Thus a single turboset may be rated at up to 1300 MW.

Gas Turbine Configurations

Gas turbines operate with much smaller pressure ratios than those employed with large steam turbines. In addition, most gas turbines are of relatively low power, generally 100 MW or less. For these reasons the volumetric flow at the low-pressure stage is low enough so that a turboset generally has but a single turbine and compressor. Both turbine and compressor, which are mounted on a single shaft, are almost universally axial-flow machines with unidirectional flow.

8.5 TURBINE ANALYSIS

The large turbines used for central station power applications are axial-flow machines where impulse provides the major thrust in the high-pressure stages. For the sake of simplicity, this section will be limited to a discussion of impulse blading.

From the previous discussion of impulse turbines, we know that the kinetic energy of the working fluid is increased at the beginning of each stage as it flows through a nozzle. Since the high-velocity fluid leaving the nozzle is directed toward the moving turbine blades where some portion of the kinetic energy of the fluid is converted to mechanical energy of the rotating turbine, our task is to determine the portion of available kinetic energy that is converted to useful work. The fraction of kinetic energy so converted may be determined by a knowledge of the stage efficiency.

The *nozzle-bucket efficiency* of a stage is defined as the ratio of the energy per kilogram of steam converted into mechanical work to the kinetic energy produced per kilogram of steam by isentropic expansion across the stage nozzle.

It is convenient to consider the overall efficiency of a stage, η_s, as having two components. The nozzle efficiency, η_n, represents the ratio of the kinetic energy (per unit mass) of the fluid actually leaving the nozzle to the kinetic energy the fluid would have been obtain in a frictionless isentopic expansion. The blade efficiency, η_{be}, is the ratio of the mechanical work produced per unit mass of fluid to the kinetic energy per unit mass actually produced in the nozzle. Therefore, in the absence of leakage, windage, and moisture losses, the overall stage efficiency, η_s, is the same as the nozzle-bucket efficiency and is given by

$$\eta_s = \eta_{be}\eta_n \qquad (8.47)$$

Theoretical Blade Efficiency

The ideal or theoretical blade efficiency for a single-row nozzle blade system may be calculated with the use of the vector diagram shown in Fig. 8.9. As indicated in the diagram, the steam velocity leaving the nozzle is u_1 and the rotating turbine blade has a velocity u_b. The relative velocity between the steam and blade is $u_1 - u_b$. From Newton's second law, the force in the blade is

$$F = m\frac{d(\Delta u)}{dt} \qquad (8.48)$$

where Δu is the total change in steam velocity relative to the blade, $2(u_1 - u_b)$. Thus

$$F = 2\dot{m}_{sc}(u_1 - u_b) \qquad (8.49)$$

where $\dot{m}_{sc}$ is the flow rate of working fluid (mass/time). The power produced by the steam in this stage is

$$P = 2\dot{m}_s(u_1 - u_b)u_b \qquad (8.50)$$

The theoretical stage efficiency is then found by the ratio of actual power to the power that would be obtained if all the kinetic energy were converted to mechanical energy. That is, we assume that the fluid entering the nozzle had zero velocity (first stage of turbine); then the kinetic energy produced by the nozzle is $\frac{1}{2}mu_1^2$:

$$\eta_{te} = \frac{2\dot{m}_{sc}(u_1 - u_b)u_b}{\frac{1}{2}\dot{m}_{sc}(u_1^2)}$$

$$\eta_{te} = \frac{4u_b(u_1 - u_b)}{u_1^2} \qquad (8.51)$$

Although this simple analysis applies only to a single-row nozzle blade system, some important observations can be made from this equation. The equation can be

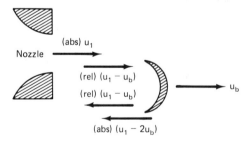

Figure 8.9 Vector diagram for an ideal turbine–nozzle bucket arrangement.

Sec. 8.5 Turbine Analysis

rewritten in terms of the ratio of the blade velocity u_b, to the fluid velocity, u_1:

$$\eta_{te} = 4\left[\frac{u_b}{u_1} - \left(\frac{u_b}{u_1}\right)^2\right] \qquad (8.52)$$

With this equation it is easy to show that the theoretical stage efficiency reaches a maximum of 100% when the blade speed is one-half the fluid velocity (ratio $u_b/u_1 = 0.50$). The absolute velocity of the exit fluid is than zero and all the energy in the steam is transferred to the blade.

This same analysis approach can be extended to the interior rows of a multistage turbine where the inlet fluid velocity is not zero. If the inlet velocity is u_{in}, the increase in fluid kinetic energy imparted by the nozzle is $\frac{1}{2}(u_1^2 - u_{in}^2)$ and the simple stage efficiency becomes

$$\eta_{te} = \frac{4u_b(u_1 - u_b)}{u_1^2 - u_{in}^2} \qquad (8.53)$$

Actual Blade Efficiency

In a real turbine, the stage inlet and outlet fluid velocity vectors are not horizontal or in the plane of the rotating turbine wheel. In an actual nozzle–blade system for an impulse turbine, the velocity vector diagram would be represented by the vectors shown in Fig. 8.10. The symbols indicated in Fig. 8.10 are defined as follows:

u_0 = theoretical fluid velocity leaving the nozzle

u_1 = actual fluid velocity leaving the nozzle, $u_1 = K_1 u_0$

K_1 = nozzle velocity coefficient

u_2 = velocity of the entering fluid relative to the turbine blade

α_b = angle at which the fluid leaves the nozzle relative to the plane of rotation of the blade

β_b = relative angle of entrance of the fluid to the blade

γ_b = relative angle of exit of the fluid from the blade

u_3 = relative exit velocity of the steam, $u_3 = K_2 u_2$

K_2 = blade velocity coefficient

u_4 = absolute velocity of the steam leaving the blade

u_{ta}, u_{tb} = tangential components of the steam velocity relative to blade

u_b = absolute blade velocity

The power output is evaluated from the same principles that were used to determine the theoretical efficiency of an ideal stage. The force on the blade is

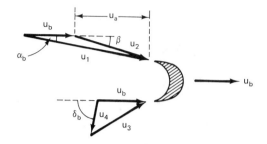

Figure 8.10 Velocity vector diagram for a real turbine.

determined by the change in the tangential component of the fluid velocity relative to the blade. The tangential component of the fluid velocity relative to the blade is u_{ta} entering and u_{tb} leaving the stage. The total change in tangential velocity is therefore $u_{ta} + u_{tb}$. The tangential force on the turbine blade is then

$$F = \dot{m}_{sc}(u_{ta} + u_{tb}) \tag{8.54}$$

The actual power produced by this set of blades is

$$P = \dot{m}_{sc}(u_{ta} + u_{tb})u_b \tag{8.55}$$

The blade efficiency is the ratio of actual power that could be extracted if all the kinetic energy imparted to the nozzle were converted to useful work. Hence we have

$$\eta_{be} = \frac{2u_b(u_{ta} + u_{tb})}{u_1^2 - u_{in}^2} \tag{8.56}$$

where u_{in} is the inlet velocity to the nozzle.

With blade angles α_b, β_b, and γ_b specified, we obtain u_{ta} from

$$u_{ta} = u_1 \cos \alpha_b - u_b \tag{8.57a}$$

and u_{tb} from

$$u_{tb} = u_3 \cos \gamma_b \tag{8.57b}$$

The value of u_3 would equal u_2 if the system were frictionless, but as this is not the case,

$$u_3 = K_2 u_2 \tag{8.58}$$

where K_2 is the blade velocity coefficient. Hence

$$u_{tb} = K_2 u_2 \cos \gamma_b = K_2 \frac{u_{ta}}{\cos \beta_b} \cos \gamma_b \tag{8.59}$$

Nozzle Efficiency

The flow process through an ideal nozzle can be approximated by a constant entropy process, in which the enthalpy change of the steam is converted to kinetic energy. If we assume that the inlet velocity to the nozzle is $u_{\dot{m}}$ and equate the kinetic energy

Sec. 8.5　Turbine Analysis

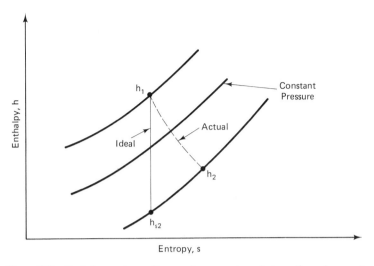

Figure 8.11 Enthalpy–entropy diagram for expansion of steam through a nozzle.

change to the change in enthalpy, we have

$$u_0^2 - u_{\text{in}}^2 = -2(h_2 - h_1) \qquad (8.60)$$

where u_0 = theoretical nozzle exit velocity
u_{in} = inlet velocity to nozzle

Note that if British units are used in Eq. (8.60), the left-hand side must be divided by J, the mechanical equivalent of heat.

The actual expansion through a nozzle is not isentropic, due primarily to frictional losses. On an h–s diagram, the expansion follows the path h_1 to h_2 as shown in Fig. 8.11. The nozzle efficiency, η_n, is the ratio

$$\eta_n = \frac{h_1 - h_2}{h_1 - h_{s2}} = \frac{\Delta h_a'}{\Delta h_s} \qquad (8.61)$$

Once values of η_n have been obtained experimentally, we may calculate u_1, the actual nozzle exit velocity, by again equating the kinetic energy change to the enthalpy change:

$$u_1^2 - u_{\text{in}}^2 = 2\Delta h_a = 2\eta_n \Delta h_s \qquad (8.62\text{a})$$

or

$$u_1 = \sqrt{2\eta_n \Delta h_s + u_{\text{in}}^2} \qquad (8.62\text{b})$$

Actual Stage Efficiencies and Configuration

When test measurements are made, actual stage efficiencies are found to be less than that predicted by the product of η_{be} and η_n. This arises since there are several effects that are not accounted for by η_{be} and η_n alone. These include:

1. *Windage*: rotational losses (found proportional to fluid density, diameter, and cube of the speed)
2. *Leakage losses*: bypassing of steam through the seals between the nozzle disk and rotating shaft
3. *Moisture losses*: mechanical interaction of slow-moving moisture droplets, in last stages of turbine, with buckets

It is common practice to lump all of these miscellaneous loss effects into a single efficiency factor which we shall designate as η_l. The total stage efficiency is then given by

$$\eta_s = \eta_{be}\eta_n\eta_l \tag{8.63}$$

It is generally found that for given turbine design parameters, η_s remains constant throughout nearly all of the turbine. Although a good approximation of the turbine efficiency, the assumption of constant stage efficiency does not apply to the last stage of a turbine. Here we encounter *exhaust losses*, caused by (1) the unrecovered energy in the exit steam (no stage follows the last stage to recover this energy) and (2) in a steam turbine, the loss resulting from the pressure drop through the exhaust hood to the condenser. Although the exhaust hood pressure losses are of the order of $\frac{1}{2}$ cm Hg, this pressure drop is responsible for a significant reduction in the energy available since the pressure drop is a sizable fraction of the system pressure at that point.

Empirical data are required to evaluate exit losses. The losses, reported in terms of Δe, the loss in energy per unit mass of steam flow at the exit, are generally given as a function of the mass flow at the exit stage. If we take the stage efficiency as essentially constant, then the total turbine efficiency, η_T, may be approximated as

$$\eta_T = \eta_s - \frac{\Delta e}{(\Delta \mathbf{h}_s)_{tot}} \tag{8.64}$$

where $(\Delta \mathbf{h}_s)_{tot}$ is the total enthalpy change across the turbine under ideal isentropic conditions (kJ/kg).

In actual practice, based on analysis and experience, certain nozzle/bucket angles are frequently employed. For a nozzle angle, α_b, of 12 to 14°, the optimum blade entrance angle is 27°. In this design, the blade exit angle will often be 24°, resulting in a slightly smaller exit area and a slight increase in the steam velocity leaving the stage. For this type of blade design, 27°/24° or some variation, such as 28°/23°, is often used in commercial practice.

Experience shows that the length or radial height of the turbine blade is also an important performance parameter. While for blade heights below about 25 cm there is an appreciable falloff in efficiency, the effect of blade height is small for blades

Sec. 8.5 Turbine Analysis

larger than 25 cm. This indicates that some losses are nearly constant and that their effect on efficiency is reduced in larger turbines.

For the 60-Hz current produced in the United States, most turbines are operated at 3600 rpm, although some older machines were designed for 1800 rpm. For the 50-Hz current utilized in the USSR, 3000-rpm machines have generally been used, but some of the largest Soviet machines operate at 1500 rpm.

Stage Reheat

We have pointed out that the actual expansion process in the nozzles is not isentropic. Some of the energy that would have been converted into kinetic energy of the steam is converted by friction into heat, which in turn raises the enthalpy of the steam (see Fig. 8.11). In addition, there are friction losses in the blading. The difference between the isentropic and actual enthalpy is known as *reheat*.

The major portion of the reheat occurs in the nozzle. This reheat may be readily determined from the nozzle efficiency:

$$\Delta h_i - \Delta h_a = \Delta h_s (1 - \eta_n) \tag{8.65}$$

Because of friction effects, there will always be some pressure drop across the blades and some additional reheat there. The entire expansion process in a stage is illustrated in the h–s diagram of Fig. 8.12. In the ideal process we would expand from 1 to $2s$. If the blades were frictionless, the working fluid would expand along the line 1–2. Because of the blade friction, the working fluid actually expands along line 1–2'.

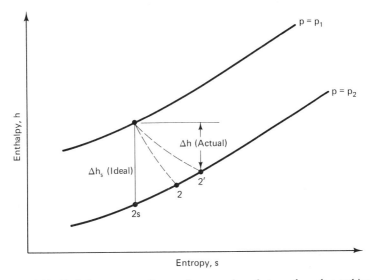

Figure 8.12 Enthalpy–entropy diagram for expansion of steam through a turbine.

We define the *stage reheat factor* (SRF) as

$$\text{SRF} = \frac{\Delta h_s}{\Delta h_a} \tag{8.66}$$

Hence

$$\text{SRF} = \frac{1}{\eta_s}$$

Considering the entire turbine staging, the ideal entropy change across the turbine would be given as

$$(\Delta \mathbf{h}_s)_{\text{tot}} = \sum_{N_T} \Delta h_s \tag{8.67}$$

where N_T is the number of turbine stages.

If the constant-pressure lines were parallel and exit losses negligible, we could write

$$\eta_s (\Delta h_s)_{\text{tot}} = \eta_s \sum_{N_T} \Delta h_s \tag{8.68}$$

However, the divergence of the constant-pressure lines causes the sum of the entropy changes across the individual stages to be greater than predicted. That is,

$$\eta_s \sum_{N_T} (\Delta h_s)_i > \eta_s (\Delta \mathbf{h}_s)_{\text{tot}} \tag{8.69}$$

Moving to the right in the h–s diagram gives larger changes in enthalpy for a given pressure change (see Fig. 8.13). On this basis, an *overall reheat factor* (RF) is often

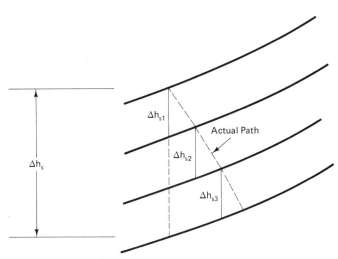

Figure 8.13 Enthalpy–entropy diagram for expansion of steam through three turbine stages.

Sec. 8.5 Turbine Analysis

defined such that

$$\sum_{N_T} (\Delta h_s)_i = RF \times (\Delta h_s)_{tot} \tag{8.70}$$

The reheat factor, RF, can be used with the ideal Δh_s across a stage to obtain the number of stages. Since the reheat factor is close to 1.0, as a first approximation, the number of stages, N_T, required in the turbine is

$$N_T = \frac{(\Delta h_s)_{tot}}{\Delta h_s} \tag{8.71}$$

Although the effect of reheat may be ignored in obtaining the approximate number of turbine stages, it should be considered in computing condenser and feedwater heater behavior. Actual steam enthalpies should be used in all such heat balances.

Preliminary Turbine Design

The preceding equations can be used to perform a preliminary analysis of the turbine. Assume that turbine design conditions such as the turbine wheel speed, blade velocity, nozzle angle, ratio of turbine blade speed to steam velocity, and the nozzle efficiency, η_n, were given. For a given steam flow (kg/h) and steam conditions (temperature and pressure), the procedure would be as follows:

1. Use the blade speed and the ratio u_b/u_1 to calculate the steam exit velocity from the nozzle. This is also the blade inlet velocity.
2. Use the results from step 1 and the blade angles to get relative velocities and stage efficiency.
3. Use the results from step 2 to calculate the isentropic enthalpy decrease across a stage.
4. Calculate the number of stages required from Eq. (8.71).
5. Calculate the overall turbine efficiency, neglecting the effects of minor losses, as

$$\eta_T = \eta_{be}\eta_n\eta_l \tag{8.72}$$

Example 8.5

Given the following turbine design data:

Turbine speed = 1800 rpm
Steam flow = 0.5×10^6 kg/h
Inlet steam at 6.9 MPa (~ 1000 psi), saturated
Outlet steam at 2.76 MPa (~ 400 psi)
Nozzle – nozzle velocity coefficient = 0.97
Blading $\alpha_b = 13°$, $\gamma_b = 24°$

$$\frac{u_b}{u_1} = \frac{\cos^2\alpha_b}{2}$$

Blade velocity coefficient $K_2 = 0.92$
KE of steam leaving last stage is lost
Blade velocity $u_b = 70$ m/s
Blading is symmetrical and mean diameter of all stages is constant
Windage and leakage losses = 3.5%

Determine (a) the blading efficiency, (b) the stage efficiency, (c) the number of stages required, and (d) the actual power output.

Solution

(a) Determination of the blading efficiency:

$$\eta_{be} = \frac{2u_b(u_{ta} + u_{tb})}{u_1^2 - u_{in}^2}$$

where u_{in} = inlet velocity to the nozzle $\simeq$ outlet velocity from the blade $\simeq u_4$
$u_{ta} = u_1 \cos \alpha_b - u_b$
and

$$\frac{u_b}{u_1} = \frac{\cos^2 \alpha_b}{2} = 0.4747$$

$$u_1 = \frac{u_b}{0.4747} = 147.5 \text{ m/s}$$

Since $u_{ta} = u_1 \cos \alpha_b - u_b$, we have

$$u_{ta} = 147.46(\cos 13°) - 70$$
$$= 73.68 \text{ m/s}$$

From the vector diagram, Fig. 8.10, we find that

$$u_2 \sin \beta_b = u_1 \sin \alpha_b = 33.17 \text{ m/s}$$

and since

$$\tan \beta_b = \frac{u_2 \sin \beta_b}{u_2 \cos \beta_b} = \frac{33.17}{73.68} = 0.45$$

we have

$$\beta_b = 24.23°$$

We can now solve for u_2:

$$u_2 = \frac{u_1 \sin \alpha_b}{\sin \beta_b} = \frac{33.17}{0.4104} = 80.8 \text{ m/s}$$

We obtain u_3 from

$$u_3 = K_2 u_2 = 0.92 \times 80.82 \text{ m/s} = 74.44 \text{ m/s}$$

From our vector diagram (Fig. 8.10) we have

$$u_4 \sin \delta_b = u_3 \sin \gamma_b = 30.242 \text{ m/s}$$
$$u_4 \cos \delta_b = u_b u_3 \cos \gamma_b$$

Sec. 8.5   Turbine Analysis

or
$$u_4 \cos \delta_b = u_3 \cos \gamma_b - u_b$$
$$u_4 \cos \delta_b = -2.074 \text{ m/s}$$

Therefore,
$$\tan \delta_b = \frac{u_4 \sin \gamma_b}{u_4 \cos \delta_b} = \frac{30.242}{-2.074} = -14.58$$

and thus
$$\delta_b = 93.950$$

Hence
$$u_4 = \frac{u_3 \sin \gamma_b}{\sin \delta_b} = \frac{30.242 \text{ m/s}}{0.9976} = 30.31 \text{ m/s}$$
$$u_{tb} = u_3 \cos \gamma_b = 67.926$$

We can now evaluate η_{be}:
$$\eta_{be} = \frac{2u_b(u_{ta} + u_{tb})}{u_1^2 - u_{im}^2} = \frac{u_b(u_{ta} + u_{tb})}{u_1^2 - u_4^2}$$

We have previously evaluated u_{ta}, u_{tb}, u_4, and u_1 and were given u_b. By using these values, we have
$$\eta_{be} = \frac{2(70)(73.68 + 67.93)}{(147.5)^2 - (30.31)^2} = 0.952$$

(b) Determination of the stage efficiency:
$$\eta_s = \eta_{be} \eta_n \eta_l$$

where η_{be} = blade efficiency = 0.952
η_n = nozzle efficiency
$\eta_l = 1 - $(windage and leakage losses) $= 1 - 0.035 = 0.965$

Now
$$\eta_n = \frac{h_1 - h_2}{h_1 - h_{s2}} = \frac{u_1^2 - u_{in}^2}{u_0^2 - u_{in}^2}$$

where u_0 = theoretical nozzle exit velocity
u_1 = actual nozzle exit velocity
$u_1 = K_1 u_0 = 0.97 u_0$

Therefore,
$$u_0 = \frac{u_1}{0.97} = \frac{147.46}{0.97} = 152.0$$

and thus
$$\eta_n = \frac{u_1^2 - u_{in}^2}{u_0^2 - u_{in}^2} = \frac{(147.5)^2 - (30.31)^2}{(152.01)^2 - (30.31)^2} = 0.938$$

and
$$\eta_s = \eta_b \eta_{ne} \eta_l = (0.952)(0.938)(0.965) = 0.862$$

(c) Determination of the number of stages required:

$$N_T = \text{number of stages} = \frac{\Delta h_s}{\Delta h_s}$$

where $\Delta h_s = h_{in} - h_{out}|_s$
$s_{in} = s_{out} = 5.82$
$s_{out} = s_g + (1-x)s_{fg} = 6.21 - (1-x)3.616$

Therefore,
$$x = 0.89$$
$$h_{in} = 2773$$
$$h_{out} = h_g - (1-x)h_{fg} \quad \text{where } h_g = 2804 \text{ and } h_{fg} = 1817$$
$$= 2804 + 0.11(1817) = 2604$$

and thus
$$(\Delta h_s)_{tot} = h_{in} - h_{out}$$
$$= 2773 - 2604 = 169 \text{ kJ/kg}$$
$$\Delta h_s = \frac{u_1^2 - u_{in}^2}{2} = \frac{(152.0)^2 - (30.31)^2}{2(1000)} = 11.1 \text{ kJ/kg}$$
$$= \frac{(\Delta h_s)_{tot}}{\Delta h_s} = \frac{169.37}{11.19} = 15.26$$

Since fractional stages are impossible, we conclude that 16 stages are required.

(d) Determination of the actual power output:
$$(\Delta h_a)_{tot} = \eta_T (\Delta h_s)_{tot}$$

where
$$\eta_T = \eta_{be} \eta_n \eta_l - \frac{\Delta e}{(\Delta h_s)_{tot}}$$

where Δe is the kinetic energy loss in last stage; that is,
$$\Delta e = \frac{u_4^2}{2} = \frac{(30.134)^2}{2(1000)} = 0.454$$

Therefore,
$$\eta_T = \eta_{be} \eta_n \eta_l - \frac{\Delta e}{(\Delta h_s)_{tot}} = \eta_s - \frac{\Delta e}{(\Delta h_s)_{tot}} = 0.862 - \frac{0.454}{169.37} = 0.859$$
$$(\Delta h_a)_{tot} = \eta_T (\Delta h_s)_{tot} = 0.859 \times 169.37 = 145.5 \text{ kJ/kg}$$
$$\text{power} = \dot{m}(\Delta h_a)_{tot} = 0.5 \times 10^6 \text{ kg/h} \times 145.5 \text{ kJ/kg}$$
$$= 72.77 \times 10.9 \text{ J/h}$$
$$= 20.2 \times 10^6 \text{ J/s} = 20.2 \text{ MW}$$

Steam-Turbine Blade Problems

There are two materials problems of concern to the steam turbine designer: moisture erosion and stress corrosion cracking of turbine blades. Although these problems have been identified for a number of years, work remains to be done to understand and resolve them completely.

Moisture erosion. Although moisture erosion of turbine blades has caused difficulties for some time, the wet steam conditions that prevail in many nuclear power stations have exacerbated the difficulties. The problem has been aggravated further by increases in turbine tip speeds.

Research into the mechanism of turbine erosion has shown that water droplets are responsible for most of the erosion process in turbine blades. The erosion of the turbine blades, especially the last-stage blades, depends on the moisture content of the steam entering the preceding stationary blades. Although the steam velocity increases with tip speed, the relative velocity of the water particles leaving the stationary blades back edge is very low. The rapid acceleration of the water drops by the rotating blades is believed to cause blade erosion. The erosion increases directly as a function of tip velocity.

An empirical equation has been developed to assess the relative risk of, or exposure to, erosion for a given set of turbine conditions. The *erosion parameter*, EP is

$$\text{EP} = (1-x)^2 \left(\frac{D_b \omega_r}{3600}\right)^3 \frac{1}{p_{\text{st}}} \qquad (8.73)$$

where x = steam quality ahead of last stage
D_b = blade diameter of last stage, in.
ω_r = revolutions per minute
p_{st} = steam pressure at last stage, in. Hg (absolute)

A higher value for the erosion coefficient implies a more severe erosion problem. Good practice sets the erosion coefficient at less than 3600.

Erosion of turbine blades can be reduced by several techniques. A larger clearance can be included in the design between the stationary and rotating blades. This would allow the steam to accelerate the water particles prior to entering the rotating blades. Alternatively, steam reheating can be used to virtually eliminate erosion problems by reducing the moisture content of the steam entering the last stage. Finally, better material selection for the turbine blades or hardening of the leading edge can improve erosion resistance.

Stress corrosion cracking. There is no doubt that stress corrosion cracking (SCC) has caused failure of turbine blades. While stress corrosion cracking occurs in pure, wet steam, it is accelerated in a caustic environment. Numerous instances of SCC have been reported in the literature for both fossil and nuclear turbine units. Chromium-molybdenun alloy and chromium-nickel alloy (stainless) steels are vulnerable to SSC.

SCC occurs primarily in a wet steam environment. The crack usually begins in a highly stressed crevice or in a region of high tensile stress. The severity of the cracking is highest in high-temperature regions of the turbine disks. There appears to be a relationship between the growth rate of the cracks and steam temperature.

A number of cracked blades have been inspected very carefully. The metallurgical structure of the blade material appeared to have very little effect on the SCC phenomenon. Cracking is predominantly intergranular, originating at defect locations, such as nonmetallic inclusions, corrosion pits, and machining defects of heavily worked metals. Crack growth rates increase with increasing steam temperature.

Although much work remains to be performed to better understand and predict SCC, some conclusions can be drawn regarding this phenomenon. These conclusions are:

1. CrMo and NiCrMoV steels are susceptible to SCC.
2. $3\frac{1}{2}\%$ Ni steel alloys are not susceptible to SCC.
3. Cracking occurs only in wet steam in regions of high tensile stress.
4. Cracking is intergranular.
5. Cracking may occur in high-purity wet steam, but is enhanced by hydroxides or sulfides.
6. The time required for a crack to develop is dependent in the system water purity level and the water chemistry control.
7. No correlation exists between alloy grain size and SCC.
8. No correlation exists between temper embrittlement of the steels used and SCC.

Vibration failures. Design of an acceptable blade for the last stages of large turbines is a complex fluid dynamics problem. Blade angles and twist must be carefully selected to achieve good efficiencies and to avoid periodic shedding of strong vortices. Such vortex shedding can lead to turbine blade vibration and subsequent fatigue failure of the blades. The failure probability is substantially increased if the vibration frequency is close to the natural frequency of the blade. Blade failures due to vibration-caused fatigue have been observed. Experimental tests of proposed blade designs are generally required.

8.6 OFF-DESIGN TURBINE ANALYSIS

Rankine Cycle Operation at Part Load

At part load, the mass flow to the turbine will be reduced. The lower mass flow means that the pressure drop across the turbine must also be reduced. Since the condenser pressure is maintained nearly constant, the pressure at the first stage must

Sec. 8.6 Off-Design Turbine Analysis

be reduced. The inlet steam pressure may be reduced by throttling valves or by admission of steam to only some of the first-stage nozzles. With the latter, and more common, procedure, critical flow of superheated steam across the inlet nozzles in use is to be expected. A pressure discontinuity will exist and there is then an isenthalpic (throttling) expansion to the lower pressure of the turbine body.

The reduced pressure drop across an impulse turbine may be estimated by noting that essentially all of the turbine pressure drop is across the nozzles. With the assumption of critical flow across all of the nozzles, it may be shown that the mass flow through the turbine, $\dot{m}_T$, is related to inlet pressure by

$$\dot{m}_T \simeq C_{tc} \sqrt{\frac{p_{in}}{v_{in}}} \qquad (8.74)$$

where p_{in} = inlet pressure
v_{in} = specific volume of steam at inlet
C_{fc} = flow coefficient for turbine

By use of the design conditions, the value of C_{fc} may be determined. Since the steam expands isenthalpically from the known inlet line conditions, v_{in} is known for any given value of p_{in}. By assuming a series of inlet pressures with appropriately corresponding p_{in}, the value of p_{in} that corresponds to the reduced mass flow may be estimated.

The inlet pressure and enthalpy establishes the inlet entropy. The value of Δh_s from the inlet pressure to the position of the first extraction stage may be computed once the pressure of the first extraction stage is known. Determination of this pressure again requires use of Eq. (8.74). However, we must now use a revised C_{fc} which is determined by using the design flow, pressure, and specific volume at the first extraction stage in Eq. (8.74). With the revised C_{fc}, a trial-and-error calculation finds the p and related v (v determined by either assuming isentropic expansion, or if a more accurate calculation is desired, assuming Δh_s appropriately reduced for turbine efficiency) at the reduced flow rate.

Each section of the turbine is treated in a similar manner. At the last stage of the turbine an allowance must be made for the increased exit losses seen. At low loads, exit losses increase sharply. Such data would be obtained from a graph, showing Δe versus mass flow or percent of design flow, provided by the manufacturer. In addition, the condenser pressure should be corrected for the change from design conditions. Since the overall UA of the condenser is essentially fixed, the lower mass flow will allow a somewhat lower back pressure to be maintained.

The mass flows and enthalpies to each of the feedwater heaters must also be reevaluated. The enthalpies of the extraction steam will correspond to the new conditions at the extraction points. The mass flows are determined by assuming UA is unchanged from design conditions, all the entering steam is condensed and the exit fluid leaves through the steam trap at the enthalpy of saturated liquid, h_f. Hence

$$\dot{m}_{out} = \frac{UA(\Delta T)_{LM}}{h_{in} - h_f} \qquad (8.75)$$

Although the value of the log mean temperature difference is unknown, it may be determined readily by consideration of the overall heat balance on the feedwater heater.

Example 8.6

Feedwater entering at 200°C is heated by saturated steam at 300°C in a heat exchanger for which the overall UA is 5×10^4 kJ/(h · °C). Determine the outlet temperature of the feedwater if the feedwater flow is 10,000 kg/h.

Solution From an overall heat balance we know that

$$\dot{m}_F c_p (T_{in} - T_{out}) = \dot{m}_{ext}(h_{in} - h_f)$$

where $\dot{m}_F$ = feedwater flow rate
T_{in}, T_{out} = inlet and outlet feedwater temperatures, respectively

Hence

$$\dot{m}_F c_p (T_{in} - T_{out}) = UA(\Delta T)_{LM} = UA \left[\frac{(T_s - T_{in}) - (T_s - T_{out})}{\ln \frac{T_s - T_{in}}{T_s - T_{out}}} \right]$$

$$\dot{m}_F c_p (T_{in} - T_{out}) = UA \left(\frac{T_{in} - T_{out}}{\ln \frac{T_s - T_{in}}{T_s - T_{out}}} \right)$$

$$\dot{m}_F c_p = \frac{UA}{\ln \frac{T_s - T_{in}}{T_s - T_{out}}}$$

$$\ln(T_s - T_{out}) = \ln(T_s - T_{in}) - \frac{UA}{\dot{m}_F c_p}$$

All the quantities needed are available except c_p. We find $c_p = 5.65$ kJ/kg by evaluating h_f at two temperatures in the region of 300°C and dividing by the temperature difference. On numerical substitution

$$\ln(300 - T_{out}) = \ln(300 - 200°C) - \frac{5 \times 10^4 \text{ kJ/h} \cdot °C}{10^4 \text{ kg/h} \times 5.65 \text{ kJ/kg} \cdot °C}$$

$$= 4.606 - 0.885$$

$$T_{out} = 259.2°C$$

Brayton Cycle Operation at Part Load

In electric utility application, the turbine must operate at a constant speed despite load changes. Since the compressor and turbine are on the same shaft, the compressor also operates at constant speed. Further, it operates at essentially the same total gas flow as the turbine. In an open cycle, the only means of control available is to vary the amount of fuel fed to the combustor. This sets the turbine inlet temperature and the torque produced.

Sec. 8.6   Off-Design Turbine Analysis   387

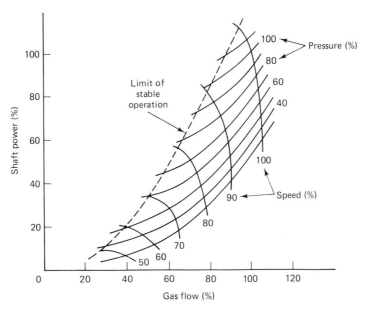

Figure 8.14 Typical gas compressor characteristics.

The net work produced depends on the point at which the given turbine and compressor can operate together at a set fuel supply. The common operating point depends on the characteristics of both turbine and compressor. Compressor performance is commonly presented on a plot of shaft power versus percent of design flow with lines of constant speed, constant pressure ratio, and efficiency (see Fig. 8.14 for typical curve). Examination of the 100% speed curve, which is the only curve of concern, shows that the total flow increases only slightly, while efficiency decreases slightly as the pressure ratio is decreased. If specific compressor characteristics are unavailable, it is generally adequate for preliminary analysis to assume the total gas flow is unchanged.

If a specific pressure ratio is assumed, the turbine inlet pressure is specified. Over the range of pressure ratios expected in part load operation, the flow through the turbine, $\dot{m}_T$, can generally be approximated by

$$\dot{m}_T \simeq K \frac{p_{\text{inlet}}}{\sqrt{T_{\text{inlet}}}} \tag{8.76}$$

If we know the design conditions, then

$$\dot{m}_T \simeq \left(\frac{\dot{m}\sqrt{T_{\text{inlet}}}}{p_{\text{inlet}}} \right)_{\text{design}} \left(\frac{p_{\text{inlet}}}{\sqrt{T_{\text{inlet}}}} \right)_{\text{part load}} \tag{8.77}$$

We find $\dot{m}_T$ for the assumed pressure ratio from the compressor characteristics curves, or, in the absence of specific data, assume the equality of $\dot{m}_T$ and $\dot{m}_{\text{design}}$.

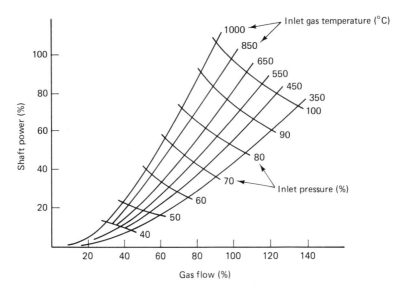

Figure 8.15 Typical turbine characteristics.

We may then solve for the required turbine inlet temperature and from this the fuel needed may be obtained.

At constant speed, the turbine efficiency changes little with load. The turbine power output may then be computed from the inlet temperature and pressure ratio in the usual manner. The net work is obtained by subtracting the compressor work obtained from the compressor diagram at the interaction of the design speed and assumed pressure ratio. By repetition of the process for several pressure ratios below the design value, the full range of part load behavior can be determined.

When a specific turbine has been selected, more accurate predictions of part load behavior can be obtained using the manufacturer's data on turbine performance. These data are generally presented as plots of output power versus gas flow with inlet temperature and inlet pressure as parameters (see Fig. 8.15 for typical behavior). From the compressor characteristic curve, we determine the compressor flow for any assumed pressure ratio. With this flow and turbine inlet pressure, we determine the required turbine inlet temperature and available shaft power from the chart of turbine characteristics.

8.7 TURBINE CONTROL

The actual demand for electric power is constantly varying. The load demand curve (Fig. 1.9) illustrated the gross daily variations. There is, however, a constantly changing demand for electric power caused by routine industrial, commercial, and residential variations. These minute-by-minute changes in demand must be met by

Sec. 8.7 Turbine Control

the overall power station control system. This section focuses on turbine control, which is a fundamental part of the overall control requirement.

Steam Turbine Control

The overall power plant station controls must sense and respond to changes in system load. The goal is to maintain bus-bar voltage and frequency constant. Consider a small system supplied by a single turbine generator and assume that a large industrial motor is switched on in the system. This reduces the overall electrical resistance of the system, allowing a larger current to flow through the generator armature circuit. This increase in current increases the turning resistance of the field rotor through an increase in armature magnetic field strength. The increased resistance slows the turbine–generator shaft rotational speed.

The reduction in the turbine shaft speed is sensed by the turbine governor. The governor signals the turbine steam valves to open, increasing the steam flow to the turbine. The increased steam flow to the turbine causes a pressure drop in the steam supply system. If the power plant is a fossil system, the pressure reduction signals the automatic combustion control system to increase the forced-draft and induced-draft fan speeds. The pressure reduction also signals the combustion control system to increase the firing rate. In a nuclear power system, the reduction in pressure would require an increase in the power level of the reactor. This would be achieved through a slight withdrawal of the shim control rods.

Power plant steady-state conditions are again achieved as the increased heat rate produces additional steam that again balances the system pressure. The governor reaches a new balance point which maintains the higher load requirement, allowing an increased steam flow to bring the turbine–generator shaft speed back to design conditions.

In an actual system, there will be a number of turbine–generators in parallel. When load and generation are in balance, all of the generators are locked into the speed which produces the desired system frequency (synchronous speed). Since the generators are operated in parallel, all generators must always produce a common voltage and frequency under all circumstances.

The load that a turbine can deliver is determined by its load–speed characteristic. At a given position of the turbine speed controller, this load-speed characteristic is fixed. When additional load is suddenly added to the system, the only way that this can be immediately supplied is for the system frequency to drop slightly. Thus, if we have two turbines in parallel, with the load–speed characteristics shown in Fig. 8.16, the speed (which is proportional to frequency) is c, which is below the speed c' needed to produce the desired system frequency. To restore the system frequency, the load–speed characteristics of one of the turbines must be changed. This is accomplished by moving the turbine's speed controller in the direction that would increase speed. In a steam turbine, the speed–load curve is changed and the turbine admission valves are opened further.

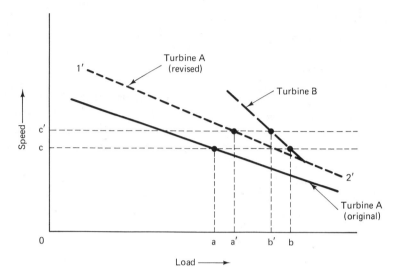

Figure 8.16 Turbine load–speed characteristic curve.

Assume that in the example of Fig. 8.16, turbine A is to take the increased load. As shown in the figure, the load on turbine A would be increased until the system was operating at the desired frequency. At that time the new load–frequency curve of unit A would be 1′–2′ and it would have increased its share of the load to 0–a′ while turbine B would have reduced its load to 0–b′.

If the load on the system were to decrease, the frequency would rise as both turbines increase their speed slightly. The speed–load characteristics of turbine A would then be changed by closing the turbine admission valves. A new equilibrium position would then be reached when the system frequency is restored to its desired value.

Behavior of a system containing a number of turbine-generators in parallel is essentially the same. Increasing the system load causes the speed of all turbine generators to decrease slightly. The system dispatcher, or more likely the dispatch computer, determines which turbine–generator will be used to take the load. The turbine admission valves to the selected turbine are opened, its speed–load characteristic is changed, and the system reaches a new equilibrium at a slightly higher frequency. The loads on the other turbines in the system will be decreased since they will be operating at a slightly higher speed. However, since there are normally a considerable number of turbine generators in parallel, the load change or any turbine, except the one designated to take the load, is quite small. The reverse behavior would be seen on a load decrease. Note that the operation of this control scheme requires that all turbine generators have a speed–load characteristic curve with a reasonable negative slope. A nearly flat speed–load curve would result in unstable operation since such a curve would lead to large changes in load for small changes in speed.

Sec. 8.7 Turbine Control

Mechanical–hydraulic speed control systems. Older turbines are generally controlled by a mechanical–hydraulic governor. A typical governing system of this type is shown in Fig. 8.17. In this system the flow of oil back to the reservoir is opposed by the pressure built up by the governor impeller attached to an extension of the turbine shaft. The pressure created is proportional to the square of the speed of the turbine shaft. Variations in impeller oil pressure are reflected in the governing force amplifier (C). This consists of a bellows which is attached to a cup valve and a spring. Changes in primary governing pressure cause movement of the spring-controlled cup valve, which alters the rate of leak-off in the secondary system (D).

Variations in the secondary system pressure are magnified by relay piston (E). This piston causes the pilot valve (F) to move there, changing the pressure above and below the valve positioning piston (G), which moves the steam control valve (H).

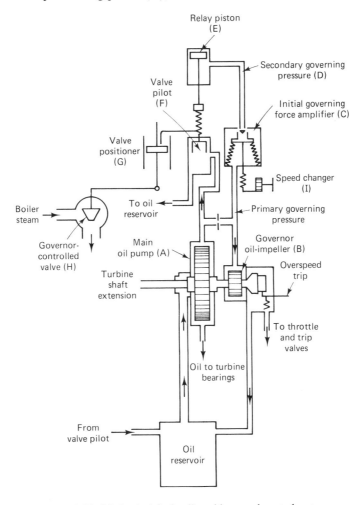

Figure 8.17 Mechanical–hydraulic turbine speed control system.

At constant load, a speed change may be introduced by the speed changer (I). This unit may be considered as a spring of variable tension which opposes the pressure generated by the governor. At constant load, an increase in spring tension would cause an increase in speed. When the turbine is being operated in parallel with others, it is locked into the synchronous speed of the system. An increase in spring tension causes the governor valve to open wider (or more valves to open) and power output is increased with little speed change.

Electrohydraulic speed control systems. The modern electrohydraulic control system provides the same basic control actions as the mechanical–hydraulic system just described. However, the basic signals received, and their initial amplification, are electrical rather than mechanical. The speed signal is generally a voltage proportional to turbine speed generated by a small alternator attached to the turbine shaft. The speed signal is electrically compared to the set point and an error signal produced. After appropriate amplification, the signal operates a servomotor which produces the hydraulic pressure needed to reposition the control valves. The system differentiates the speed signal and uses this to limit the acceleration of the turbine during startup.

As indicated previously, the turbine speed is locked into the synchronous speed of the system when it is operated in parallel with other turbine generators. To obtain a change in load, an electrical signal, which is equivalent to a speed error signal, is introduced (see Fig. 8.18). The load error electrical signal added is proportional to the difference between the desired load and actual load (load torque). The amplified load error signal operates the servomotor, which in turn generates the hydraulic pressure to reposition the steam control valves.

In addition to the normal operating controls, there is also an auxiliary pressure control that will close the control valves if the boiler pressure decreases below some predetermined value (such as 90% of design pressure). This system also regulates the control valves to limit pressure increases to less than some predetermined value, such as 110% of design rated pressure.

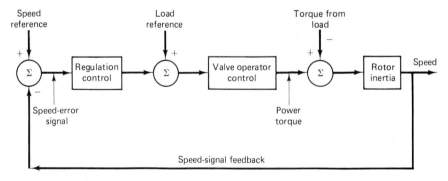

Figure 8.18 Electrohydraulic turbine speed control.

Sec. 8.8 Hydraulic Turbines and Pumped Storage

Turbine overspeed protection is provided by an overspeed trip reference signal. If the turbine speed exceeds the overspeed trip reference speed, the stop valves are immediately closed.

There are several other signals that can shut down the turbine to prevent damage. These include excessive vibration, low pressure, bearing failure, and high temperature.

Gas Turbine Control

Gas turbines are controlled in essentially the same manner as steam turbines. Mechanical–haydraulic or electrohydraulic speed control systems may be used. However, these units alter the speed–load curve by varying the quantity of fuel that is supplied to the combustion chamber instead of changing the position of steam admission valves.

8.8 HYDRAULIC TURBINES AND PUMPED STORAGE

A hydraulic turbine converts the potential energy of water into the rotational kinetic energy of a turbine. The water has potential energy by virtue of its height above the turbine. The quantity of power available from a hydraulic turbine can be estimated as the product of the hydraulic head and volumetric flow times the turbine efficiency. Before efficiency corrections, the power extracted from a stream equals the negative of the pumping power. Hence

$$P(\text{kW}) = \frac{Q_W H'_{\text{net}} \rho g \eta_h}{g_c} \quad (8.78)$$

where g = gravitational acceleration 9.8 m/s²†
g_c = gravitational conversion factor (unity in SI units†)
Q_W = water flow through the turbine, m³/s†
H'_{net} = net hydraulic head, m†
ρ = water density, kg/m³ †
η_h = overall hydraulic turbine efficiency, fraction (~ 0.85 to 0.9)

The net hydraulic head is the static head (in meters) minus losses in the system due to pressure drops in the piping and valves (usually around 5% of original head).

In a manner consistent with our previous turbine analyses, hydraulic turbines may be divided into two main categories, impulse turbines and reaction turbines. Impulse turbines are generally used for high-hydraulic-head applications, while reaction turbines are generally best suited for low-hydraulic-head applications (10 to

---

†If British units are used, $\rho = \text{lb}_m/\text{ft}^3$, $Q_w = \text{ft}^3/\text{s}$, $H'_{\text{net}} = \text{ft}$, $g_c = (\text{lb}_m/\text{lb}_f)32.2 \text{ ft/s}^2$, $g = 32.2$ ft/s², and $P = \text{ft} \cdot \text{lb}_f/\text{s}$.

500 ft). Since most of the useful high-hydraulic-head sites have been developed, the current emphasis in the power industry is on the development of low-hydraulic-head, small hydroelectric applications. Since reaction turbines are used for these applications, reaction turbine designs are of current interest.

Impulse Turbines

Most high-hydraulic-head sites use impulse turbines. A stream of high-velocity water is accelerated through a nozzle onto a set of scoop-shaped buckets which form the rotor. The nozzle forms a sharp angle with the rotor such that the impulse of the fluid jet on the buckets accelerates the wheel causing it to rotate. The static pressure at the rotor outlet is identical to the rotor inlet; thus the energy transferred from the water to the rotor is entirely kinetic.

Impulse turbines are best suited to applications where the hydraulic head ranges from a few hundred feet to a few thousand feet. The power output of an impulse turbine unit may range from 400 kWe to 110,000 kWe. Impulse turbines are further characterized by high reliability and low maintenance costs. Efficiency frequently exceeds 90% for these devices.

Figure 8.19(a) illustrates the common arrangements for hydraulic impulse turbines. A jet from a single nozzle is directed against a wheel with a horizontal shaft. Although a single wheel may drive a generator, it is quite common to have two wheels drive a single horizontal shaft generator (one wheel on each side of generator).

Present impulse turbines are almost always *Pelton wheels*. Each bucket consists of two scoops with a "splitter" in the middle. The splitter splits the jet from the nozzle into two streams, which exit on opposite sides of the wheel [see Fig. 8.19(b)].

Although less common, vertical-shaft Pelton wheels are also used. These units tend to be large for hydraulic turbines (e.g., 50 MW and above) and use multiple jets (up to six) on a single wheel. It is possible to use impulse turbines for low-head applicators, although it is not very common. A cross-flow turbine (water flows across entire turbine wheel before exiting) has been developed for this purpose.

The basic principle behind the operation of the impulse hydraulic turbine is the same as that behind the impulse steam turbine. When the entering jet is deflected by the turbine blade surface, a change in momentum occurs and a force is exerted on the blade. Since the blade is allowed to move, the force acts through a distance producing work.

By making use of the vector diagram of Fig. 8.19(c) and the principles of Section 8.5, we find that the horizontal force, F_x, on the bucket is

$$F_x = (u_j - u_t)(1 - \cos\beta_u)\frac{Q_w \rho}{g_c} \qquad (8.79)$$

where u_j = jet velocity
u_t = peripheral velocity of wheel
Q_w = volumetric flow of water

Sec. 8.8 Hydraulic Turbines and Pumped Storage

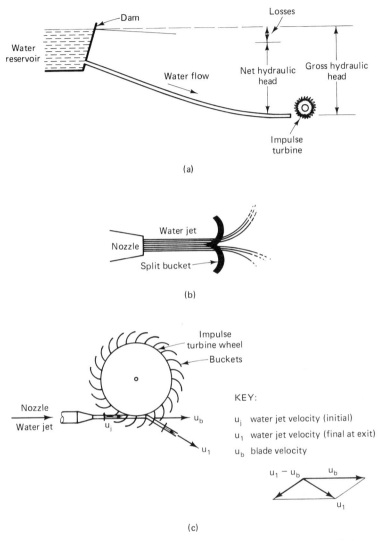

Figure 8.19 (a) Overall arrangement of a hydraulic plant; (b) jet splitting by bucket; (c) detailed arrangement and vector diagram for an impulse turbine.

For a frictionless mass, the power, P, obtained is

$$P = (u_j - u_t)(1 - \cos\beta_u)\frac{Q_w \rho u_t}{g_c} \tag{8.80}$$

As found for steam turbine behavior, P is a maximum when $u_t g_c = u_j/2$; hence

$$P_{\max} = \frac{Q_w \rho}{g_c}\frac{u_j^2}{4}(1 - \cos\beta_u) \tag{8.81}$$

The efficiency of a frictionless wheel, η_w, will be the ratio of P_{max} to the kinetic energy dissipated by the jet per unit time.

$$\eta_w = \frac{Q_w \rho}{4g_c} \frac{u_j^2 (1 - \cos \beta_u)}{Q_w \rho u_j^2 / 2g_c} \tag{8.82}$$

If $\beta_u = 180$, the theoretical efficiency could be 100% ($\cos \beta_u = -1$). However, practical blade angles are about 165°, giving a maximum efficiency of 98%. The actual efficiency of the wheel is further reduced by the fact that the peripheral wheel speeds are only about 48% of the jet velocity and that the wheel is not frictionless. Actual peak efficiencies are in the vicinity of 90%.

In an impulse turbine, nearly all of the pressure loss takes place across the nozzle. The pressure head, from the reservoir, less frictional losses in the piping and nozzle, is converted to kinetic energy of the exiting jet. If the head on the entrance to the nozzle is H'_{in}, then by Bernouli's theorem,

$$H'_{in} = f_h + \frac{u_j^2}{2g} \tag{8.83}$$

where f_h is the frictional head loss in nozzle, or

$$H'_{in} = (1 + C_{nl}) \frac{u_j^2}{2g} \tag{8.84}$$

where C_{nl} is the nozzle loss coefficient (typical value $\simeq 0.04$).

Reaction Turbines

In contrast to an impulse turbine, the major portion of the pressure loss in a reaction turbine takes place in the rotating wheel. To keep these pressure losses at a minimum, it is necessary for all passages to be full of liquid and the entire circumference of the turbine wheel be utilized. Fluid is therefore admitted around the entire circumference.

All present-day hydraulic reaction turbines may be regarded as variants of the Francis turbine first constructed in 1849. The basic operational principles may be understood by reference to Fig. 8.20. Water from the supply pipe enters a spiral scroll case which surrounds the turbine. The water flows inward from the scroll case through a set of guide vanes. The vanes exert a torque on the fluid and give the fluid a tangential velocity component. The fluid then flows through the moving runner (turbine wheel), which decreases the tangential velocity. This produces a torque on the runner blades. Since the runner blades are moving, work is done. The water exits through the central pipe.

It may be shown that the torque exerted on the runner is equivalent to that exerted by two single forces: one force concentrated at the entrance and the other concentrated at the exit. The entrance force is $\rho Q (u_j)_{in}$ and the exit force is $\rho Q_w (u_j)_{out}$ with $(u_j)_{in}$ and $(u_j)_{out}$ being the entrance and exit fluid velocities,

Sec. 8.8 Hydraulic Turbines and Pumped Storage

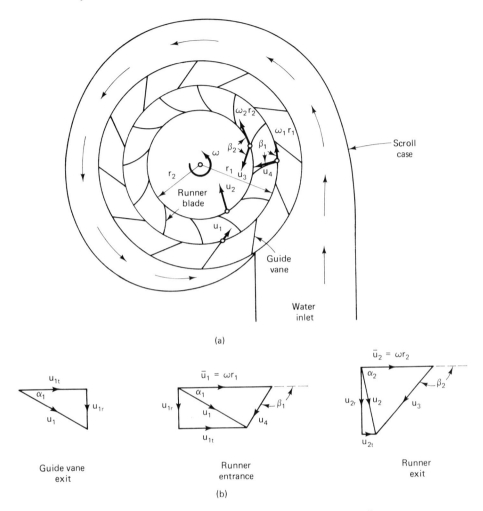

Figure 8.20 (a) Hydraulic reaction turbine; (b) velocity vector diagram.

respectively. By referring to Fig. 8.20, we evaluate the momentum and find the total torque, T_0, to be

$$T_0 = \frac{\rho Q_w}{g_c}\left[r_1(u_j)_{\text{in}}\cos\alpha_1 - r_2(u_j)_{\text{out}}\cos\alpha_2\right] \quad (8.85)$$

where the angles α_1 and α_2 and the runner blade radii, r_1 and r_2, are defined in Fig. 8.20. By multiplying the torque by the angular velocity, $\omega(\text{rad/s})$ of the runner, we have the rate at which work is being done. Hence for a frictionless machine, the power, P, produced is

$$P = \frac{\omega\rho}{g_c}\left\{Q_w\left[r_1(u_j)_{\text{in}}\cos\alpha_1 - r_2(u_j)_{\text{out}}\cos\alpha_2\right]\right\} \quad (8.86a)$$

or

$$P = \frac{\rho Q_w}{g_c}\left[(u_t)_{in}(u_j)_{in}\cos\alpha_1 - (u_t)_{out}(u_j)_{out}\cos\alpha_2\right] \quad (8.86b)$$

where $(u_t)_{in}$ and $(u_t)_{out}$ are the peripheral speeds at outer and inner radii of runner, respectively.

Examination of the velocity diagrams shown in Fig. 8.20 will show that there is a significant increase in the radial velocity, u_r, as the fluid passes through the runner. However, the absolute and radial velocities at the guide exit should be the same as the runner exit. The runner blades must be appropriately shaped to accomplish this. That is, the runner vane angle, β_1, which is fixed by construction, should be the same as the β_1 determined by the velocity diagram (Fig. 8.20). If $(u_t)_{in}$ and $(u_j)_{in}$ are fixed, then

$$\cot\beta_1 = \frac{(u_j)_{in}\cos\alpha_1 - (u_t)_{in}}{(u_j)_{in}\sin\alpha_1} \quad (8.87)$$

At other values of β_1 there will be a shock loss which decrease the turbine efficiency.

The head loss H'_L, through a turbine is found to be proportional to $u^2/2g$; hence we may write

$$(u_j)_{in} = C_1^*\sqrt{2gH'_L} \qquad (u_j)_{out} = C_2^*\sqrt{2gH'_L} \quad (8.88)$$

where C_1^* and C_2^* are the empirical loss coefficients.

Further, it is customary to express the peripheral speed in terms of $\sqrt{2gH'_L}$; that is,

$$(u_t)_{in} = \phi^*\sqrt{2gH'_L} \quad (8.89)$$

where ϕ^* is a constant for a given machine. We can now rewrite Eq. (8.87) in dimensionless form as

$$\cot\beta_1 = \frac{C_1^*\cos\alpha_1 - \phi^*}{C_1^*\sin\alpha_1} \quad (8.90)$$

We may also express the efficiency of our turbine in dimensionless form. From Eq. (8.78), we have

$$\eta_h = \frac{g_c P}{Q_w H'_L \rho g} \quad (8.91)$$

By making use of Eqs. (8.86b), (8.88), and (8.89),

$$\eta_h = \frac{(u_t)_{in}(u_j)_{in}\cos\alpha_1 - (u_t)_{out}(u_j)_{out}\cos\alpha_1}{gH'_L} = 2\phi^*\left(C_1^*\cos\alpha_1 - \frac{r_2}{r_1}C_2^*\cos\alpha_2\right) \quad (8.92)$$

For the efficiency to be a maximum, α_2 will be close to 90°. If for simplicity we take

Sec. 8.8 Hydraulic Turbines and Pumped Storage

$\alpha_2 = 90°$, then

$$\eta_h = 2\phi^* C_1^* \cos \alpha_1 \qquad (8.93)$$

In a reaction turbine, less then half of the net head is converted into kinetic energy leaving the guide vanes and entering the runner, and C_1^* is of the order of 0.6. For Francis turbines, typical values of ϕ^* are of the order of 0.7 to 0.85 for maximum efficiency.

The relationship between volumetric flow and runner dimension is obtained by recognizing that

$$Q_w = A_c u_r \qquad (8.94)$$

where A_c = circumferential flow area at the OD of the runner
u_r = radial fluid velocity at the OD of the runner

The circumferential flow area is obtained from

$$A_c = f'' \pi D_r L_b = f'' \pi m_b D_r^2 \qquad (8.95)$$

where f'' = fraction of flow area that is free space
D_r = outer runner diameter
L_b = blade height
$m_b = L_b / D_r$

Since $u_r = (u_j)_{in} \sin \alpha_1$, then

$$Q_w = f'' m_b \pi D_r \sin \alpha_1$$

or

$$Q_w = f'' \pi m_b D_r^2 C_1^* \sqrt{2gH_L'} (\sin \alpha_1) \qquad (8.96)$$

Substitution of the foregoing into Eq. (8.78) yields

$$P = \sqrt{2} f'' \pi m_b D_r^2 (gH_L')^{3/2} \rho \eta_h \sin \alpha_1 \qquad (8.97)$$

Configurations. Figure 8.21 shows a typical configuration for the standard Francis turbine. Such turbines are typically used for heads ranging from 40 to 400

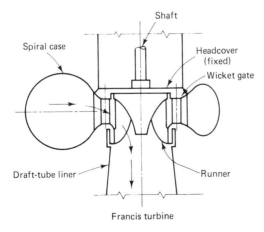

Figure 8.21 Francis hydraulic turbine. (Reprinted with permission from *Power*; copyright 1981 McGraw-Hill, Inc., New York.)

ft. Flow rate is adjusted by movable wickets or gates at the entrance to the guide vanes. The flow exits through a central "draft tube," which is a conduit with a very gradually enlarging cross section. The draft tube converts some of the discharge velocity head to pressure head. This allows the turbine to be set at an elevation above that of the discharge basin without loss in net head. By reducing exit velocity, the draft tube also reduces the head loss at the submerged discharge.

Propeller turbines are used in the lowest head range, from 10 to 60 ft. If the blades are adjustable, the turbine is called a *Kaplan turbine* (Fig. 8.22). The number of blades varies with the hydraulic head; the higher the head, the more blades in the runner.

Blade pitch in Kaplan turbines is automatically adjusted by means of an oil-driven piston generally located in the main shaft. The controls are designed such that the blade angle varies automatically with the wicket-gate opening to produce a maximum possible turbine efficiency per any given flow and head condition. Note that the optimum value of β_1 in Eq. (8.87) depends on the flow to the turbine.

The fixed-propeller machines have good efficiency at the optimum design point, but efficiency decreases sharply with reduction in flow. Because of the

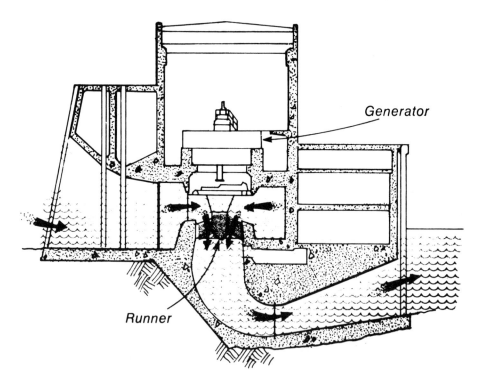

Figure 8.22 Kaplan reaction turbine. (Reprinted with permission from *Power*; copyright 1981 McGraw-Hill, Inc., New York.)

Sec. 8.8 Hydraulic Turbines and Pumped Storage

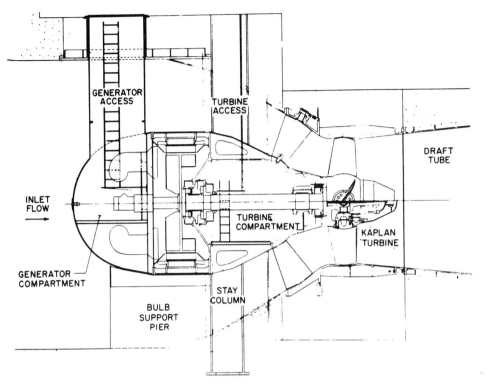

Figure 8.23 Bulb-type hydraulic turbine generator. (Reproduced with permission from *Proceedings of the American Power Conference*, Vol. 41, 1979.)

adjustable blade feature, the Kaplan turbine has a relatively flat efficiency curve over a wide range of flows.

An alternative to the Kaplan turbine for low-head applications is the axial-flow bulb-type turbine. A typical installation is shown in Fig. 8.23. The generator is enclosed in a streamlined watertight housing placed right in the water flow passage. Seal problems do not exist, but generator cooling and maintenance are difficult.

Pumped-Water Energy Storage

This is an energy storage technique rather than a primary energy resource, such as hydroelectric power. Less costly, off-peak electrical power is used to pump water from one reservoir to another. Usually, there is an upper reservoir, located about 90 to 360 m (300 to 1200 ft) above the pump–turbine. When there is a peak demand for electric power, the water from the upper reservoir is allowed to run down through the turbines. The electrical power output can be estimated from Eq. (8.78).

The electrical power so generated is usually less expensive than that produced from other types of peaker units, such as gas turbines.

Low capital cost is achieved by using the same unit as pump-motor and turbine-generator. When current is supplied to the generator it acts as a synchronous motor driving the turbine as a pump by rotating it in the reverse direction.

Aboveground pumped hydro. Presently, all pumped hydro plants are the conventional aboveground type of facility. The upper reservoir is usually constructed on top of a high plateau and the lower reservoir in a valley. Some facilities have a head of 2000 ft (670 m). New facilities and improvements in the pump–turbine design are expected to improve the capability up to 3500 ft of head (1070 m), with plant output of 2000 MW.

The typical system is shown schematically in Fig. 8.24. The facility consists of eight pump–turbines, which generate 60 MW each. The net hydraulic head is 162 ft with a maximum discharge of 6000 ft^3/per unit. Francis-type pump–turbines are used in the facility shown.

Underground pumped hydro. Conventional pumped hydro storage facilities cause extensive environmental changes to both the topology of the region and the fish and wildlife. Appropriate governmental approval of such projects is therefore often difficult to obtain. The goal of underground pumped hydro (UPH) is to provide an energy storage system at competitive cost with other types of peaker units, while minimizing the environmental impact of the project.

The UPH concept utilizes an underground reservoir located in suitable bedrock 3000 to 5000 ft (1900 to 1500 m) below ground surface level. The power output from any hydro unit is the direct product of head times flow rate [Eq. (8.78)]. Because the UPH cost is heavily dominated by the volume of the cavern, the water

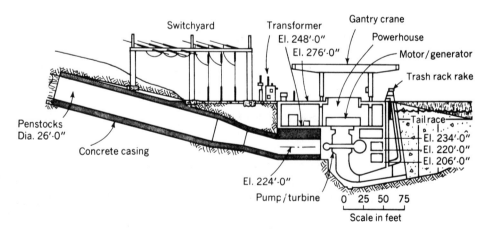

Figure 8.24 Pumped hydro facility. (From the August 1, 1977 issue of *Electrical World*; © copyright 1977, McGraw-Hill, Inc.; all rights reserved.)

flow must be minimized. This can be done by maximizing the head, or the depth of the lower reservoir. Heads of 5000 ft depth appear desirable.

For a 5000-ft head facility, three types of reversible pump-turbine designs are considered feasible: single-stage (Francis type), multiple-stage pump-turbine, and a Pelton impulse turbine with a separate centrifugal pump.

An UPH facility would operate as follows. When peak power is needed (generating mode), water is released from the surface reservoir. The water falls through the penstocks to the turbine–generator units. The turbine discharge is through the draft tube tunnels to the lower underground reservoir.

During off-peak periods, the process is reversed. Low-cost, off-peak electrical power is used to pump the water from the lower reservoir back to the surface reservoir or to discharge it to a stream or river. In either case, the lower reservoir is pumped out during off-peak times.

SYMBOLS

A	area, m^2
A_c	circumferential flow area at runner O.D.
c_p	specific heat at constant pressure, J/kg·°K
$(c_p)_{ave}$	specific heat of the gas mixture, J/kg·°K
c_v	specific heat at constant volume, J/kg·°K
C^*, C_{fc}	flow coefficient for turbine
C_{nl}	nozzle loss coefficient
D_b	blade diameter, m
D_r	outer diameter of runner, m
Δe	energy loss per unit mass of steam, J
EP	erosion parameter [Eq. (8.73)]
f''	fraction of flow area to free area
f_h	frictional head loss in nozzle
F	force, N
g	gravitational acceleration, m/s^2
g_c	gravity conversion constant, 9.8 m/s^2 (unity in SI units; 32.2 ft/s^2 in British units)
h	specific enthalpy, J/kg
h'	specific enthalpy (isentropic turbine expansion), J/kg
Δh	enthalpy change, J/kg
$\mathbf{\Delta h}$	enthalpy change across entire turbine, J/kg
h_f	enthalpy of the saturated liquid, J/kg
h_g	enthalpy of steam or vapor, J/kg

h'_m	molal enthalpy, J/mol
H'	hydraulic head, m
H'_L	hydraulic head loss, m
H'_{net}	net hydraulic head, m
HV	heating value of fuel, kJ/kg
J	mechanical equivalent of heat
k	thermal conductivity, W/m² · °C
K_1	nozzle velocity coefficient
K_2	blade velocity coefficient
L_b	blade height, m
m	mass, kg
m_b	ratio of blade height to blade diameter
$\dot{m}$	mass flow rate, kg/s
$\dot{m}_a$	actual steam rate, kg/s
$\dot{m}_{air}$	mass flow rate of air, kg/s
$\dot{m}_F$	feedwater flow rate, kg/s
$\dot{m}_{sc}$	mass flow rate of the working fluid, kg/s
$\dot{m}_T$	mass flow rate of steam entering the turbine, kg/s
$\dot{m}_{T'}$	mass flow rate through the turbine (the actual steam rate), kg/s
N_c	number of compressor stages
N_T	number of turbine stages
p	pressure, Pa
$\hat{p}_{st}$	steam pressure, Pa
p_{in}	inlet pressure to the turbine, Pa or atm
$p(0)$	reference or initial pressure, Pa or atm
p_{re}	relative pressure
P	power, W
P_{ax}	auxiliary power, kW
P_{GO}	gross generator output, kW
q_1	heat added to cycle 1, J/s
q_2	heat added to cycle 2, J/s
q_A	net heat added to the cycle, J/s
q_B	heat supplied to boiler or reactor, J/s
q_C	heat load on the condenser, J/s
Q_h	heat flow or heat supplied, J/kg
Q_w	water flow through turbine, m³/s
r	radius, m
r_{cp}	pressure ratio across compressor

Chap. 8   Symbols

r_p	pressure ratio across the turbine
R	ideal gas constant
s	specific entropy, J/kg · °K
s_f, s_g	specific entropy of saturated liquid and vapor, respectively
s_{fg}	entropy change on evaporation $= s_g - s_f$
s_m	molal entropy, J/mol · °K
T	temperature, °K
To	torque, N · m
$(\Delta T)_{LM}$	log mean temperature difference, °C
u	velocity, m/s
u_0	theoretical fluid velocity leaving nozzle, m/s
u_1	actual fluid velocity leaving nozzle, m/s
u_2	velociity of fluid relative to turbine blade, m/s
u_3	relative exit velocity of steam, m/s
u_4	absolute velocity of steam leaving blade, m/s
u_b	blade velocity, m/s
u_{in}	inlet velocity to nozzle, m/s
u_j	jet velocity, m/s
u_r	radial fluid velocity at O.D. of runner, m/s
u_t	peripheral velocity of the wheel, m/s
u_{ta}, u_{tb}	tangential components of steam velocity relative to blade, m/s
Δu	velocity change, m/s
U	overall coefficient of heat transfer, W/m² · °C
v_{in}	specific volume of steam at inlet, m³/kg
v_m	specific volume of steam at inlet (ratio)
W_a	actual turbine work output, J/s
W_c	work done by the compressor, J/s
W_{net}	net work of the turbine, J/s
W_t	turbine work (isentropic), J/s
W_T	work done by the turbine, J/s
x	steam quality, mass fraction of vapor in a vapor–liquid mixture
y_i	mole fraction of component i

Greek Symbols

α_b	angle of fluid leaving nozzle relative to the plane of the rotating turbine blade, degrees
β_b	relative angle of entrance of the fluid from the blade, degrees

β_1	runner vane angle
γ	c_p/c_v
γ_b	relative angle of exit of fluid from the blade, degrees
δ_b	blade design angle, degrees
η_B	boiler efficiency
η_{be}	blade efficiency
$\eta_{c0}, \eta_{c1}, \eta_{c2}$	efficiencies of combined cycle, cycle 1, and cycle 2, respectively
η_{cp}	compressor efficiency
η_G	generator efficiency
η_h	overall hydraulic turbine efficiency
η_i	stage efficiency factor
η_n	nozzle efficiency
η_o	overall component or system efficiency
η_r	regenerator efficiency
η_R	theoretical efficiency of the steam cycle
η_s	overall stage efficiency
η_{th}	thermodynamic efficiency
η'_T	turbine efficiency (isentropic)
η_T	turbine efficiency
η_{te}	theoretical stage efficiency
η_w	efficiency of a frictionless wheel
μ	viscosity, Pa · s
ρ	density, kg/m^3
$\overline{\phi}$	material parameter
ϕ^*	flow coefficient
ω	angular velocity, rad/s
ω_r	angular velocity, rpm

Subscripts

0	initial condition
1, 2, ...	state or condition at location 1, 2, ...
a	actual
m	mixture of gases
i	ideal
in	inlet condition
out	outlet condition

s	under isentropic conditions
sat	saturation or steam
tot	total change across turbine or turbine section
x	x coordinate

PROBLEMS

8.1. The inlet steam to the high-pressure section of a steam turbine is at 155 bar and 510°C and the outlet is at 35 bar. At 120 bar and 60 bar, 2% of the steam flow is bled to regenerative feedwater heaters.

  (a) Determine turbine work done per unit mass of steam flowing if the turbine stage efficiency is 93%.

  (b) What is the feedwater temperature rise across the two regenerative feedwater heaters?

8.2. A steam turbine is to produce (at full load) 500 MW of mechanical power using inlet steam at 180 bar and 250°C superheat. The condenser operates at 0.15 bar. Steam is withdrawn for regenerative heaters at 130 bar, 70 bar, 15 bar, and 5 bar. The feedwater exits from each regenerative heater at 12°C below the saturation temperature corresponding to the steam feeding the heat exchanger. After leaving the regenerative heaters, the steam goes directly to the boiler. A reheater is incorporated at a pressure of 20 bar and the steam is returned to the turbine with 100°C superheat. Determine the total steam flow required if the turbine stage efficiency is 0.925.

8.3. The boiler feeding the turbine described in Problem 8.2 has a boiler efficiency of 87%. The auxiliary power load is 15 MW. For the conditions of Problem 8.2, calculate the TCHR, BTHR, and OPPHR.

8.4. At part-load conditions, the turbine described in Problem 8.2 is supplied with 70% of the normal steam flow. Assume that steam conditions (temperature and pressure) leaving the boiler have not changed and condenser pressure changes are very small. Estimate the mechanical power produced by the turbine at the reduced flow. Assume that the change in exit loss is negligible.

8.5. An open Brayton cycle turbine may be considered to operate with an ideal diatomic gas. The turbine inlet temperature is 725°C and the gas enters the compressor at 20°C. Turbine and compressor efficiencies are 89%. The turbine operates at the optimum pressure ratio.

  (a) Determine the cycle efficiency for a simple open cycle with no regeneration.

  (b) Determine the cycle efficiency when a regenerator is added and the entering gas is heated to 25°C below the gas exhaust temperature before going to the combustor.

  (c) Determine the turbine efficiency if air is taken to be the turbine gas, actual air properties are used, and conditions of part (b) apply.

  (d) Assume that fuel having a heating value of 18,000 kJ/kg and containing 2.1 atoms of H per atom of C is used. With the conditions of part (b), determine the rates at which fuel and air must be supplied to produce 25 MW of power.

8.6. A direct Brayton cycle turbine may be considered to use air as the working fluid. The inlet gas to the turbine is at 8 bar and 600°C. The outlet gas exhausts to the atmosphere. The turbine may be considered to be of the Rateau impulse type. The nozzles placed between each stage increase the gas velocity from 135 to 335 m/s. The nozzle efficiency is 0.95 and the stage efficiency is 0.93. Determine:

(a) The mechanical work done per unit mass of air.

(b) The number of stages required.

(c) The flow areas required for the inlet and outlet stages.

8.7. Draw a scale velocity diagram and determine the stage efficiency for a steam turbine stage in which the following conditions hold:

$\alpha_b = 15°$

$\gamma_b = 22°$

Blade velocity coefficient = 0.91

Average blade velocity = 350 m/sec

Turbine speed = 3600 rpm

Nozzle velocity coefficient = 0.96

You may assume that the ratio of turbine blade velocity to vapor velocity is set at the ideal value $[(\cos^2\alpha)/2]$.

8.8. Assume that turbine stages having the characteristics described in Problem 8.7 are used in the high-pressure stages of the steam turbine in Problem 8.2. You may assume that the blading is symmetrical and that the mean diameter of all stages is constant. Determine:

(a) The total number of stages required.

(b) The blading height at stages following the extraction points and at the turbine exit.

8.9. A simple single-jet Pelton wheel is operated from a reservoir providing a head of 200 m. The nozzle loss coefficient is 0.41 and the peripheral wheel speed is 47% of the jet velocity. The blade angle is 165°. Determine the total water flow required to generate 80 MW of mechanical power if the wheel may be considered to be frictionless.

8.10. A designer is considering a Francis reaction turbine as an alternative to the Pelton wheel for the condition of Problem 8.8. For the class of machine the designer is considering, $C_1 = 0.6$, ϕ^* should be approximately 0.8 and ω_r is set at 300 rpm. Determine a set of design parameters that will provide the requisite power. Will the reaction turbine require a greater or lesser total water flow than the impulse turbine of Problem 8.8?

BIBLIOGRAPHY

CULP, A., *Principles of Energy Conversion*. New York: McGraw-Hill, 1979.

EGGENBERGER, M., *Basic Elements of Control Systems*, Report GET-3096E. Schenectady, N.Y.: General Electric Company, 1977.

HODGE, J., ET AL., "U.K. Experience of Stress-Corrosion Cracking in Steam Turbine Disks," *Proc. Institute of Mechanical Engineers*, Vol. 193, 93–109 (1979).

IRVINE, T. F., and J. P. HARTNETT (eds.), *Steam and Air Tables in SI Units*. New York: Hemisphere, 1976.

ROBERTS, J., *Structural Materials in Nuclear Power Systems*. New York: Plenum, 1981.

SALISBURY, J., *Steam Turbines and Their Cycles*, 2nd ed. Melbourne, Fla.: R. E. Krieger, 1974.

WEEKS, J., *Stress Corrosion Cracking in Turbine Rotors in Nuclear Power Reactors*, Nuclear Regulatory Commission Report NUREG 22689-R, Brookhaven National Laboratory, 1978.

9

Production of Electrical Energy

9.1 PRINCIPLES OF ALTERNATOR DESIGN AND OPERATION

At present, essentially all central power stations produce alternating current and do so by means of alternators (also called synchronous generators). The alternator converts mechanical energy into electrical energy by the Faraday effect. The basic principle of operation is that the movement of a conductor perpendicularly through a magnetic field produces a voltage gradient. The magnitude of the voltage gradient, $d\mathbf{V}/dx$, is determined by the vector cross product of flux density and conductor velocity. That is,

$$\frac{d\mathbf{V}}{dx} = \mathbf{u}_c \times \mathbf{B}_m \tag{9.1}$$

where $d\mathbf{V}/dx$ = induced voltage gradient, V/m
$\mathbf{u}_c$ = conductor velocity, m/s
$\mathbf{B}_m$ = magnetic flux density, T (tesla)

Virtually all large central power stations utilize a synchronous 60-Hz three-phase alternator. Large new units have efficiencies in excess of 90%.

Alternator (or Synchronous Generator) Operation

The simplest form of an alternator is a rotating coil between two poles of a magnet [see Fig. 9.1(a)]. As the coil rotates thus in counterclockwise direction, with uniform velocity, a varying voltage is induced across the ends of the coil. When the coil is in position 1, no voltage is generated as the conductor is moving parallel to the lines of

Sec. 9.1 Principles of Alternator Design and Operation 411

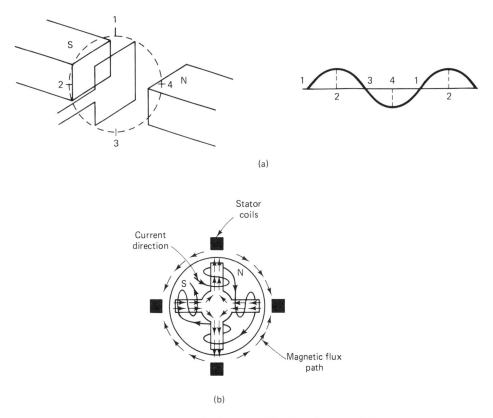

Figure 9.1 (a) Simple alternator; (b) salient alternator design.

magnetic flux ($\mathbf{u}_c \times \mathbf{B}_m = 0$). As the coil moves toward position 2, the voltage increases, reading a maximum at position 2, where flux and conductor velocity are mutually perpendicular. As the coils move toward position 3, the voltage again decreases, and reads a zero value at position 3. When the coil moves beyond position 3, the direction of the voltage in the conductors will be reversed as each conductor is under a pole of opposite region. The voltage increases until it reaches a negative maximum at position 4 and then goes back to zero at position 1. The variation in voltage produced is shown in the graph in the right-hand portion of Fig. 9.1(a). We see that we obtain a sinusoidally varying voltage.

In a practical alternator, we must produce large currents at high voltage. If rotating coils were used, this would require a substantial number of large coils rotating at high speeds. The massive set of rapidly rotating coils would be difficult to construct. In addition, it would be necessary to carry the large amounts of current generated through rotating contacts. It is more convenient to allow the electromagnet producing the magnetic field, which requires relatively little current, and is much less massive than the set of coils, to rotate.

In view of the foregoing, practical alternators are constructed with a stationary armature or stator and a rotating field. The rotor consists of an even number of poles of alternating polarity. On each pole, there is a field coil. The field coils are connected together to form a field winding. A small dc generator, called an *exciter*, feeds the dc current to the field winding. This results in a magnetic field flux through the stator as shown schematically in Fig. 9.1(b).

The stator or armature winding, in which the electromotive forces (EMFs) are generated, are placed in equidistant slots in the stator surface. This winding consists of coils which are placed so that the coil sides are one pole division apart (in Fig. 9.1 this is 90°).

As the rotor spins, the flux sweeps by the armature winding and the armature (stator) iron will experience a changing flux. If the stator were constructed from one solid iron block, there would be a current flow through the stator. This would result in excessive heating and core losses. The stator core is therefore constructed from individually insulated laminated iron sheets to prevent the current flow.

Alternator current frequency. For the dashed flux path shown in Fig. 9.1(b), the total magnetic flux linked to one of the armature coils is zero. If the rotor turns 45°, the linked coil flux will reach a maximum value. The continued rotation for another 45° again reduces the flux to zero. Another 45° rotation will produce a maximum coil flux with opposite polarity. Just as in the simple alternator, this results in a periodic EMF induced in the coil. In a generator containing n_p poles, $n_p/2$ EMF cycles will be induced per rotation. If the rotor completes ω_r revolutions per minute [$\omega_r/60$ rev/s], the generator frequency f_z, is

$$f_z = \frac{n_p \omega_r}{120} \quad \text{Hz} \tag{9.2}$$

While Fig. 9.1 was useful in describing the basic operational theory and principles of an alternator, the figure depicted what is called a *salient rotor design*. This type of rotor is only used in low-speed systems, such as encountered in hydroelectric power applications (see Section 9.3). Steam-turbine-driven generators are high-speed devices. The most widely used generators have two-pole and four-pole rotors which operate at speeds of 3600 and 1800 rpm. The rotor coils of a salient generator would not survive the large centrifugal stresses developed by the high-speed generators. Steam-driven generators are designed with a dc rotor winding placed in slots in much the same manner as the ac stator armature winding. This type of rotor design, called a *nonsalient rotor*, is used in all large central station generators.

Three-phase generator design. Electric motors driven by polyphase current show much less torque pulsation than those driven by single-phase current. This has led to the almost universal use of three-phase alternators since a three-phase system is the simplest possible symmetric polyphase system.

A three-phase alternator, such as shown in the simple schematic of Fig. 9.2(a), contains three separate stator windings or sets of coils. If the rotor has n_p poles, the

Sec. 9.1   Principles of Alternator Design and Operation   413

stator coils of phase A are displaced $(120/n_p)°$ from those of phase B, which are in turn displaced $(120/n_p)°$ from those of phase C. The geometric arrangement for a two-pole machine is shown in Fig. 9.2(a).

The windings for the three phases may be connected in a *delta* (Δ) or *wye* (Y) arrangement, as shown in Fig. 9.2(b). In either case, three sinusoidally varying voltages, V_A, V_B, and V_C, are produced on the bus-bars leaving the generator. These

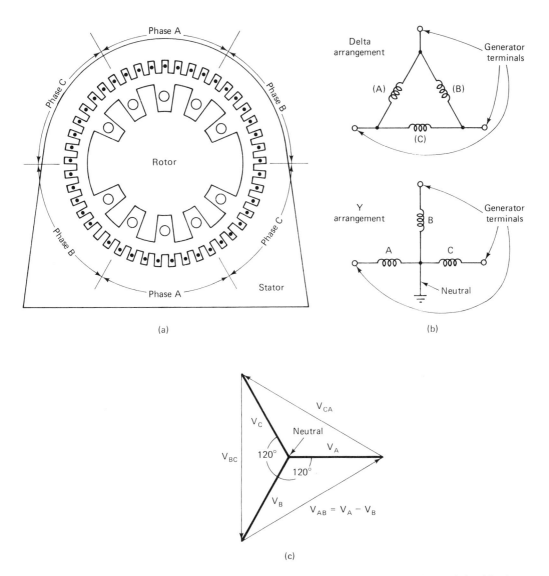

Figure 9.2 (a) Alternator configuration; (b) possible winding connections; (c) voltage relationships in a Y-connected generator.

voltages all have equal rms values, $|V|$, but different phase angles. With a Y connection, the phase voltages are measured against the fourth or ground terminal, g. The voltages shown in Fig. 9.2(c) make up a symmetrical three-phase set. If V_A is assumed to be the reference voltage, then[†]

$$V_A = |V|$$
$$V_B = |V|e^{-120°j} \qquad (9.3)$$
$$V_C = |V|e^{-240°j}$$

These voltages are called *phase* or *phase-to-ground voltages*. The voltages that exist between terminals, V_{AB}, V_{BC}, and V_{CA}, are called *line* or *phase-to-phase voltages*.

$$V_{AB} = V_A - V_B$$
$$V_{BC} = V_B - V_C \qquad (9.4)$$
$$V_{CA} = V_C - V_A$$

These voltage phasors are shown in Fig. 9.2(c). The phasors can also be shown equal to

$$V_{AB} = \sqrt{3}|V|e^{+30°j}$$
$$V_{BC} = \sqrt{3}|V|e^{-90°j} \qquad (9.5)$$
$$V_{CA} = \sqrt{3}|V|e^{-210°j}$$

The rms value of the line voltage equals the product of the $\sqrt{3}$ and the rms value of the phase voltage.

The terminal voltage is less than the EMF generated, E_g, when current is drawn. When the generator is open-circuited (unloaded), the terminal voltage equals the EMF or $V = E_g$. When the generator is loaded, the current, I_a, flows through the stator resulting in a voltage drop, $I_a Z_s$, through the winding impedance, Z_s. The terminal voltage would be $V = E_g - I_a Z_s$.

The Synchronous Generator

Most power plant generators are part of the national power grid, which operates on a common frequency of 60 ± 0.05 Hz. Before the output of a generator can be connected to this grid, it must be synchronized with the grid. Synchronization means:

1. The machine frequency must match the grid frequency.
2. The phase sequence of the machine must match the network.
3. The generator output voltage must be made to match the system line voltage.
4. The generator EMF must have the same phase as the system voltage.

[†] Note that $j = \sqrt{-1}$ and quantities in the format $Ae^{j\alpha}$ are exponential forms of the complex number $A(\cos\alpha - j\sin\alpha)$.

Sec. 9.1   Principles of Alternator Design and Operation

When all these conditions are satisfied, the breaker can be closed, bringing the generator on-line.

Phasor diagram and power output. Although the grid voltage, V_g, remains constant, the generator EMF, E_g, will change in phase and magnitude in response to load and excitation changes. As the generator is loaded, an increase in turbine torque tends to accelerate the generator away from the network. Since it is locked into the network, the generator cannot pull away, but it advances an angle δ_g. As the EMF wave follows the rotor, E_g is advanced in phase.

If the field excitation is varied, a voltage difference, ΔV, is produced where

$$\Delta V = E_g - V_g \quad \text{volts/phase} \tag{9.6}$$

and V_g represents the grid voltage. The current due to ΔV is

$$I_a = \frac{\Delta V}{Z_s} \quad \text{amperes/phase} \tag{9.7}$$

where $Z_s \equiv$ synchronous impedance of the stator winding. The Z_s term has a real and a reactive component, which are

$$Z_s = R_s + j\omega L_s \quad \text{ohms/phase} \tag{9.8}$$

where R_s = stator resistance, ohms
L_s = stator inductance, henrys
ω = rotational speed, rad/s

For power plant generators, $\omega L_s \gg R_s$ and Eq. (9.8) becomes

$$Z_s \simeq j\omega L_s \equiv jX_s \quad \text{ohms/phase} \tag{9.9}$$

where X_s is the synchronous reactance.

The simple schematic shown in Fig. 9.3(a) depicts the equivalent generator circuit. The generator is equivalent to the EMF E_g in the circuit shown. The phasor diagram corresponding to this circuit is shown in Fig. 9.3(b). The angle γ_{cf} represents the lag between current and EMF in each phase of the generator. The angle between voltage and EMF, δ_{cv}, is the power angle and is a measure of the real three-phase power delivered by the generator.

It will be recalled that ac power is a complex quantity having a scalar part, P_{real}, and an imaginary portion, jQ_{rp}. The quantity Q_{rp} is called the *reactive power* and represents the peak value of the power component traveling in the line. The power component traveling back and forth in the line has an average value of zero and hence does not do useful work. When the angle between current and voltage is ψ, the real power P_{real} and the reactive power Q_{rp} are given by

$$P_{real} = |V||I_a|\cos\psi \tag{9.10}$$

$$Q_{rp} = |V||I_a|\sin\psi \tag{9.11}$$

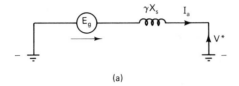

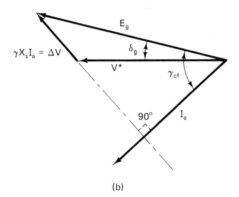

Figure 9.3 (a) Equivalent generator circuit; (b) phasor diagram for an equivalent generator circuit.

From the vector diagram of Fig. 9.3(b), we have

$$|E_g|\cos\delta_{cv} = |V| + X_s|I_a|\sin\psi \tag{9.12}$$

$$|E_g|\sin\delta_{ev} = X_s|I_a|\cos\psi \tag{9.13}$$

or

$$|I_a|\cos\psi = \frac{|E_g|}{X_s}\sin\delta_{cv} \tag{9.14}$$

$$|I_a|\sin\psi = \frac{|E|\cos\delta_{cv} - |V|}{X_s} \tag{9.15}$$

By using the values for $|I_a|\cos\phi$ and $|I_a|\sin\phi$ in the equation for real and reactive power, we obtain

$$P_{real} = \frac{|E_g||V|}{X_s}\sin\delta_{cv} \quad \text{watts/phase} \tag{9.16}$$

$$Q_{rp} = \frac{|V||E_g|\cos\delta_{cv} - |V|^2}{X_s} \tag{9.17}$$

$$|I_a| = \frac{\sqrt{P_{real}^2 + Q_{rp}^2}}{|V|} \quad \text{amperes/phase} \tag{9.18}$$

Example 9.1

A generator is producing an EMF that is 20% above the phase voltage of 8.66 kV. The steam turbine torque output is set at a value for the generator to deliver 12 Mw

Sec. 9.1 Principles of Alternator Design and Operation

(three-phase) to the grid. If the synchronous reactance is 11 Ω per phase, find δ_{cv}, Q_{rp}, and $|I_a|$.

Solution

$$|E_g| = 1.2|V| = 10.39 \text{ kV/phase}$$

Then

$$P_{real} = \frac{P_{3\phi}}{3} = \frac{12 \times 10^6}{3} = 4 \times 10^6 \text{ W}$$

and by Eq. (9.16),

$$\sin\delta = \frac{X_s P_{real}}{|E||V|} = \frac{(11)(4 \times 10^6)}{(10.39 \times 10^3)(8.66 \times 10^3)} \qquad \delta_{cv} = 29.3°$$

From Eq. (9.17),

$$Q_{rp} = \frac{(8.66) \times 10^3 (10.39) \times 10^3 (\cos 29.3) - (8.66 \times 10^3)^2}{11} = 0.32 \times 10^6 \frac{\text{volt-amperes}}{\text{phase}}$$

and then $|I_a|$ is obtained from Eq. (9.18), as

$$|I_a| = \frac{\sqrt{(4 \times 10^6)^2 + (0.32 \times 10^6)^2}}{8.66} = 0.46 \text{ kA/phase}$$

We see from the foregoing example that when $|E_g|$ exceeds $|V|$, the generator delivers reactive power to the system (machine acts as a capacitative load). If $|V| > |E_g|$, the machine will draw reactive power from the network, thus acting as an inductance load.

It should also be noted from Eq. (9.16) that the real power reaches a maximum when $\delta_{cv} = 90°$. If the torque is then increased further, a decrease in electrical power results. The generator pulls out of synchronism at that point and the rotor slips until the excess torque is removed. Excessive current will flow in the stator at these times since the generator EMF and voltage phasors are in opposite phase. This behavior leads to substantial overheating of the stator.

Generator power and generator design. To obtain the power production in terms of the design variables, we consider a thin strip of stator surface of length L_w and tangential width dx. As shown in Fig. 9.4, the strip is subjected to a magnetic flux density, $\mathbf{B}_m$, varying with time. If the current in the strip is dI, then the force $d\mathbf{F}$ on the strip will be

$$d\mathbf{F} = \mathbf{B}_m dI_a L_w \qquad (9.19)$$

It is convenient to express the current dI_a in terms of the surface current density, a_s. That is,

$$dI_a = a_s dx \qquad (9.20)$$

where a_s is in amperes/meter. The magnetic flux $\mathbf{B}_m$ varies with time and position

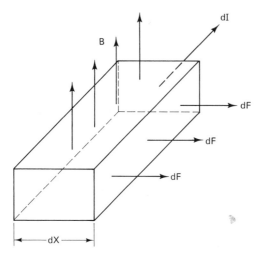

Figure 9.4 Current, flux, and force acting on a stator element.

as

$$\mathbf{B}_m = \mathbf{B}_{m(\max)} \cos\left(\frac{n_p x}{D_s} - \omega t\right) \qquad (9.21)$$

where t = time
n_p = number of poles
ω = angular rotation speed
D_s = diameter of stator

The current wave will follow the flux wave but lag by the angle γ_{cf}. Hence

$$a_s = (a_s)_{\max} \cos\left(\frac{n_p x}{D_s} - \omega t - \gamma_{cf}\right) \qquad (9.22)$$

and therefore

$$dF_s = L_w (\mathbf{B}_m)_{\max} (a_s)_{\max} \cos\left(\frac{n_p x}{D_s} - \omega t\right) \cos\left(\frac{n_p x}{D_s} - \omega t - \gamma_{cf}\right) dx \qquad (9.23)$$

The total force on the stator is obtained by integrating over the entire circumference

$$F_s = \int_{x=0}^{x=\pi D_s} d\mathbf{F}\, dx = \tfrac{1}{2} \pi D_s L_w \mathbf{B}_{m(\max)} (a_s)_{\max} \cos \gamma_{cv} \qquad (9.24)$$

Note that although the force on any one strip is a function of time, the force over the entire stator is invariant with time. The three-phase power, $P_{3\phi}$, is simply the product of the force and tangential velocity:

$$P_{3\phi} = F_s \frac{\pi D_s \omega_r}{60} = \frac{\pi^2}{120} \omega_r L_w D_s (\mathbf{B}_m)_{\max} (a_s)_{\max} \cos \gamma_{cf} \quad \text{watts} \qquad (9.25)$$

Since $\pi L_w D_s^2 / 4$ equals the rotor volume v_r, the expression for generator power can

Sec. 9.1   Principles of Alternator Design and Operation   419

be simplified to

$$P_{3\phi} = \frac{\pi}{30} \omega_r v_r (\mathbf{B}_m)_{\max} (a_s)_{\max} \cos \gamma_{cf} \qquad (9.26)$$

It may be seen that the power will increase directly in proportion to speed, rotor volume, magnetic flux density, and current density in the stator surface. Since the flux density of the iron core of the rotor reaches a saturation value and since surface current density is limited by the ability to cool the current conductors, high-power outputs require large rotor volumes and high generator speeds.

Generator control.   There are two control actions available for control of a synchronous generator: control of the field current and control of the prime-mover torque. By varying the field current, we change the flux $\mathbf{B}_m$ and hence the EMF generated, E_g. However, since a generator is almost always connected to a very large system, the terminal voltage, V, is not significantly changed. Therefore, we can control the EMF generated by any particular generator, without changing the system voltage, by varying the field current of the particular generator. Normally, the generator EMF is compared to a standard voltage and the error signal obtained is used to adjust the field current.

Note that variation of the field current cannot change the real power delivered by the generator as long as the prime-mover torque and speed are unchanged. Variation in E_g therefore causes a corresponding change in δ_{cv} [see Eq. (9.16)] so that real power stays constant. As shown earlier, the reactive power will change as E_g changes.

In Chapter 8, turbine control in response to load changes was discussed. It was pointed out that load changes were accommodated by changing the position of the speed controller (change speed setting) so as to admit additional steam to the turbine. This changes the speed–load characteristics of the turbine and changes the torque on the generator.

In the case of response to increasing load, the turbine generator initially responds by decelerating to a slightly lower speed. The speed controller then causes an increase in the vapor flow to the turbine. The vapor flow to the turbine is adjusted so that the new equilibrium speed is the desired synchronous speed (original speed before load perturbation). Since more power is now being generated at the same speed, γ_{cf}, the angle between E_g and I_a, more closely approaches zero ($\cos \gamma_{cf} \to 1$). At the same time, the power angle, δ_{cv}, increases.

If the load on the generator is reduced, the reverse behavior is seen. The power angle δ_{cv} will then decrease. If δ_{cv} is ever less than zero, the force and torque shift polarity and the machine now acts as a motor. (A synchronous motor is nothing more than alternator in which the stator windings are supplied with current. Such devices have a constant speed–load characteristic and are often used as the drive units in dc motor generator sets.) When the machine is being used as a generator, motoring must be avoided.

9.2 GENERATOR CONFIGURATIONS

High-Speed Machines

Alternators driven by steam or gas turbines are driven at high speeds, generally 1800 or 3600 rpm in the United States (as of 1982). In some European countries where 50-Hz current is used (e.g., the USSR), 1500- and 3000-rpm units are common. To produce the required 60-Hz current, Eq. (9.2) tells us that 1800-rpm units will be four-pole machines, while 3600-rpm units will be two-pole machines.

The advantages of using a high-speed machine are obvious if we examine Eq. (9.26). Total power output is proportional to speed and hence a much smaller machine can be used for production of a given amount of power when a high-speed unit is used. The size of the device is described by the rotor volume, v_r. A high-speed machine has a lower v_r.

All large modern, high-speed units are of the horizontal type. A typical unit is shown in Fig. 9.5. Units operating at 1800 rpm have been built in sizes up to 1600

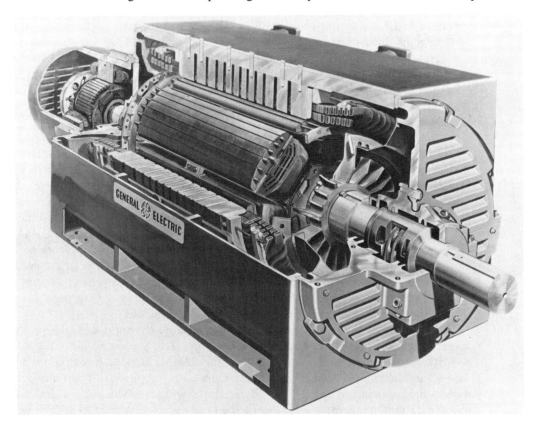

Figure 9.5 Synchronous generator. (Courtesy of General Electric Company, Schenectady, N.Y.)

Sec. 9.2 Generator Configurations

MVA† (~ 1600 MW). It is more difficult to eliminate vibration and other mechanical problems in the 3600-rpm units and maximum sizes have been somewhat lower (~ 1150 MVA). Energy-conversion efficiencies of large units are generally in excess of 98.5%.

The actual design of an alternator is, of course, more complex than indicated by the previous discussion. In practice, elimination of higher harmonics can be gained by allowing some overlap in the phase belts of the stator winding. In addition, a damper (or amortisseur) winding is provided on the rotor. These windings, which consist of short-circuited copper bars, carry no current during normal operation. If, during a transient, the rotor changes position relative to the stator current wave (γ_{cf} changes), currents are induced in the damper winding. These currents have a direction that counteracts the position change, and hence the currents have a damping effect on rotor swings.

Although laminated cores and high-resistivity silicon steel limit stator eddy-current losses, heating effects due to these currents are still appreciable. In addition, hysteresis losses, due to reorientation of the magnetic moments in the steel as ac flux changes, cause significant heating in both stator and rotor. The combined eddy-current and hysteresis losses are called *iron* or *core losses*. In addition, there are substantial $I_a^2 R_s$ losses in the rotor and stator windings. The heat generated by these losses must be removed. The rotor windings are generally cooled by hydrogen, which fills the atmosphere inside the generator. Cold hydrogen is scooped up in the gas gap between rotor and stator and then driven diagonally in and out of the rotor copper to remove the heat. The stator core is also cooled by the hydrogen, some of which flows in and out of the stator. Hydrogen coolers located in the upper portion of the machine remove the heat transferred to the hydrogen.

The $I_a^2 R_s$ losses in the stator windings of large modern machines are removed by liquid cooling. Half or more of the copper bars making up the stator windings are hollow and liquid is circulated directly through the hollow region. Early designs used oil cooling, but current designs use deionized water.

Low-Speed Machines (Hydroelectric Generators)

As our earlier analysis showed [see Eq. (9.26)], low-speed alternators will be considerably larger (larger v_r) than high-speed machines. At a typical water turbine speed of 360 rpm, size limitations generally restrict generator capacity to 120 MVA or less. At 360 rpm, 20 poles would be required for 60-Hz current generation.

Alternators at large dam sites are typically vertical and mounted directly over the radial water turbine. However, horizontal units connected to a Pelton turbine are also used. In pumped storage projects, horizontal units connected directly to horizontally mounted axial-flow turbines are common. Direct liquid cooling of the

---

†Machines should be rated in terms of the product of voltage and current since heat generation, which limits operation, is proportional to the product and does not depend on angular displacement between I and V, as does power.

stator windings is generally not required and indirect cooling via the hydrogen also used for rotor and stator cooling is generally sufficient.

Advanced Superconducting Generators

It is expected that during the 1980s experimental operation of a large (300-MVA) superconducting generator will begin. "Superconductivity" refers to the complete disappearance of electric resistance in some metals (e.g., aluminum, niobium–tantalum alloys) when these are cooled to close to absolute zero by liquid helium. In the superconducting generator, the rotor windings are cooled to near absolute zero, thus allowing very high current densities without significant $I_a^2 R_s$ loss. The very strong magnetic fields made possible by this allow a superconducting generator to be much smaller and more efficient than a conventional unit. It is also expected that such generators will be more resiliant to load variations than are present units. The reliability and economy of these units remain to be proven.

9.3 POWER PLANT ELECTRICAL SYSTEMS

Each generating station must be connected to the electrical grid so that it can feed power to the system and have power available for startup. In addition, the necessary transformer and switchgear for station equipment (motors for pumps, fans, crushers, pulverizers, conveyors, etc.) must be available.

Since current is generated at a voltage less than that used for main-line transmission (generation voltages are on the order of $\sim 350 \times 10^3$ V, while main-line transmission voltages are of the order of 750×10^3 V), a step-up transformer must be provided between the generator and transmission line. Further, since station equipment runs at substantially lower voltages than are generated, step-down transformers must be provided to meet station requirements. To assure power to operate equipment when the generator is down and to start up the plant, one or more separate tie lines feeding the plant from the grid are needed. Step-down transformers from this line are also needed.

A typical layout of the electrical system at a coal-fired plant is shown in Fig. 9.6. The first step-down transformer provides current at about 6.6 kV for large, high-voltage motors. Current at 380 V is supplied for most other large equipment.

The switchgear for the auxiliary systems is divided into a unit-connected bus and a general-purpose bus. Startup and equipment operation during shutdown of the power station is accomplished through the startup transformer. The unit-connected auxiliary transformer and the startup transformer have the same rating which is sufficient for the full unit and the general-purpose auxiliary supply. An emergency auxiliary generator is required to ensure adequate power during startup and shutdown even if all off-site power is lost. As the plant size increases, the auxiliary equipment ratings increase. In-plant system voltages may also increase above those shown in Fig. 9.6.

Sec. 9.3   Power Plant Electrical Systems

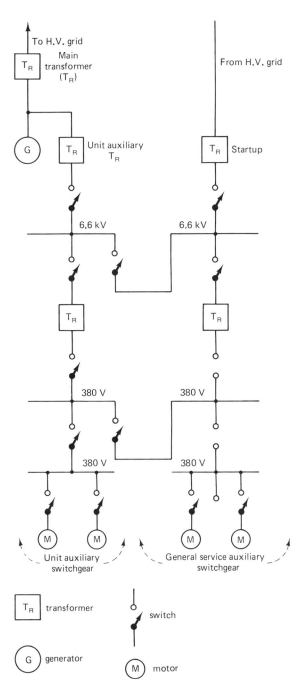

Figure 9.6 Typical electrical systems arrangement at a fossil-fueled power station.

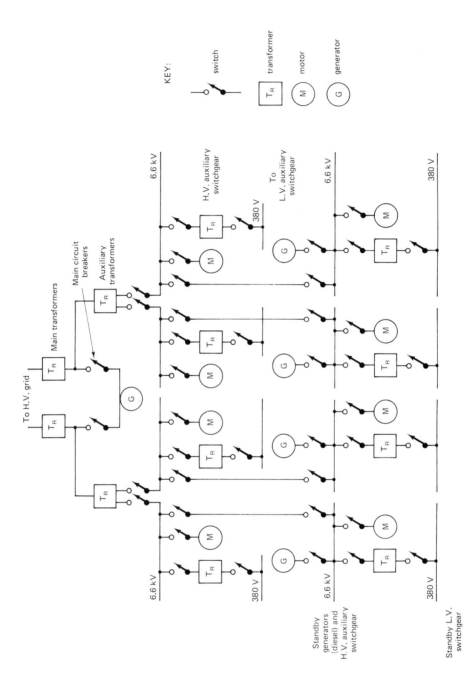

Figure 9.7 Typical electrical systems arrangement at a BWR power station.

Sec. 9.3 Power Plant Electrical Systems

Electrical Systems for Nuclear Power Stations

The electrical equipment and plant layout for a typical BWR system is shown in Fig. 9.7. Quick-starting standby diesel generator units are required to ensure emergency core cooling and residual heat removal after shutdown in the event of loss of external power supply. Regulations also require two independent network feeds for normal startup and shutdown supply. Nuclear power stations also have two generator circuit breakers, which eliminate the need for separate startup transformers. By providing two half-load main transformers and dividing the station electrical supply between two three-winding transformers, the power station can operate at a reduced power output even if a line or generator transformer fails.

Power Consumption by Electrical Auxiliaries

The power requirement of the auxiliary equipment just described increases as the station capacity increases. Auxiliary power consumed at the site can be given in terms of percent of generator output, horsepower, or kilowatts required at rated load. Also, the auxiliary power consumption increases dramatically if sulfur dioxide removal equipment is required at coal-burning stations. Table 9.1 lists the power consumed by auxiliary equipment at the site, as a percent of generator output. The auxiliary power requirements for a typical 1000-MWe BWR plant are listed in Table 9.2.

TABLE 9.1 AUXILIARY POWER CONSUMED AT THE SITE AS A PERCENT OF GENERATOR OUTPUT

	Low estimate (%)	High estimate (%)
Oil-fired station	5	6
Coal-fired station	6	7
Coal with SO_2 removal	7	12
Nuclear LWR station	4	7

Source: Data from U.S. Nuclear Regulatory Commission Report NUREG 0248, 1978.

TABLE 9.2 MAJOR AUXILIARY POWER REQUIREMENTS FOR A 1000-MWE BWR

Component	Power required (kW)
Reactor circulating pumps	12,000
Feedwater pumps	25,000
Condensate pumps	5,000
Cooling water pumps	13,000
Miscellaneous electrical equipment	10,000

Source: Data from U.S. Nuclear Regulatory Commission Report NUREG 0248, 1978.

9.4 THE ELECTRIC POWER NETWORK

Structure

The economics of electric power generation dictate that electricity must be generated in large quantities at central power stations. The electric energy must then be transmitted to every customer via the electric power network.

The power network can be divided into categories in accordance with the voltage at which it operates. At the highest voltage (voltages of 300 to 750 kV) is the *transmission* or *power grid*. This is the national grid over which large blocks of power are transmitted. These lines interconnect large generating stations within a given utility system as well as interconnecting utility systems.

Power is withdrawn from the transmission grid at a number of bulk power substations. Here the voltage is stepped down and the power is fed into subtransmission systems. At a large number of distribution substations, the voltage is stepped down and the power fed to the distribution system.

In operating a power system that is part of the national grid, a number of objectives must be met. These include:

1. Balancing the real power generated with that consumed
2. Control of the system frequency
3. Maintaining a reactive power balance
4. Maintaining the desired voltage
5. Generating the required power in the most economical manner
6. Routing the power over the grid in the most economical manner
7. Operating the system in a manner that minimizes system disturbances

Power Flow in Transmission Lines

For the simple lossless transmission line shown in Fig. 9.8 (actual line losses are low but not zero), the real and reactive power transmitted along the line may be shown to be

$$P_1 = P_2 = P_{real} = \frac{|V_1||V_2|\sin\theta_{21}}{X_s} \quad \text{watts} \quad (9.27)$$

$$Q_1 = \frac{|V_1|^2 - |V_1||V_2|\cos\theta_{21}}{X_s} \quad \text{watts} \quad (9.28)$$

$$Q_2 = \frac{|V_1||V_2|\cos\theta_{21} - |V_2|^2}{X_s} \quad \text{VA reactive} \quad (9.29)$$

where θ_{21} is the phase angle between V_1 and V_2 and X_s is the synchronous reactance.

Sec. 9.4 The Electric Power Network

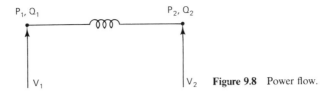

Figure 9.8 Power flow.

With voltages fixed in magnitude, we see from the foregoing that real power transmitted along the line depends on the phase angle θ_{21}. For positive values of θ_{21} (V_1 leading V_2), the real power flows from bus 1 to bus 2. If θ_{21} becomes negative, the real power flow is reversed. Note the difference between ac and dc power flow. In dc power flow the voltage magnitudes determine the direction of power flow, while in ac systems the direction is not determined by the voltage magnitude but by the voltage phase angle.

The reactive power flow depends on both the magnitude and the phase angle of the voltages. However, in normal operation $\cos\theta_{21}$ is fairly close to unity. We can therefore reverse the direction of reactive power flow by changing the magnitude of one of the voltages. The reactive line power flow tends to be in the direction of the lowest voltage.

Assume that in Fig. 9.8 a generator is located in position 1 and that a large load is located at position 2. Most loads are inductive loads, and all aggregate loads are inductive loads. There will then be a flow of reactive power from position 1 to position 2. This will cause V_2 to be lower than V_1.

Control of Line Voltage and Reactive Power Flow

In our previous discussion of generator control, it was noted that at each generator, the voltage is compared with a standard voltage. The error signal so generated is used to control the field current and keep the voltage constant. Application of this control technique to all the generators in a system allows the system voltage profile to be controlled.

If the voltage level at any bus falls below that desired, reactive power must be injected at or near that bus. If a generator is at or near the bus in question, reactive power may be added by the generator. It will be recalled that an overexcited generator, $|E_g| > |V|$, will add reactive power to the system. Thus by appropriate adjustment of generator voltage, significant control of the system voltage profile may be achieved.

In many cases, generators are not located near the large inductive load centers. It is therefore common practice to install large shunt capacitors on the lines to inject reactive power at the desired locations. During the night, when most of the inductive load is lost, the presence of these shunt capacitors leads to line voltage increases. It may then be necessary to run some generators as underexcited machines ($|E_g| < |V|$).

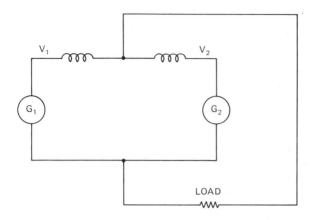

Figure 9.9 Parallel operation of generators.

Maintenance of Synchronism

Consider Fig. 9.9, in which generators are now located at positions 1 and 2 and supply a common load. Suppose that V_1 is leading V_2, then, as in Fig. 9.8, real power will flow from generator 1 to generator 2. If generator 1 were to speed up temporarily, V_1 will advance farther with respect to V_2 (θ_{21} will increase). Additional power will then flow from generator 1 to generator 2. That is, the power load on generator 1 will be increased. The increased power load will then tend to slow generator 1 to the original speed. If, on the other hand, generator 1 were initially to slow down, θ_{21} will be reduced and the load on generator 1 would be reduced. This would cause the generator to speed up slightly. Thus, if alternators which are operating in parallel tend to pull out of step, the power flow between the machines tends to accelerate the lagging alternator and retard the leading alternator. This power flow, called the *synchronizing power*, acts to prevent the alternators from pulling out of synchronism.

Automatic Load-Frequency Control

The description of turbine control provided in Chapter 8 noted that all the generators in an electric system must operate at a single common frequency. An increase in load on the system causes all the turbines to slow down slightly, while a decrease in load causes all turbines to speed up slightly. Thus the system frequency tends to change with system load. It was pointed out that by changing the speed–load characteristics of one of the turbosets (i.e., opening or closing the turbine steam admission valves), the system can be brought to a new equilibrium at the desired frequency and new power level.

It is usually desired to control the system frequency to 60.0 ± 0.05 Hz. This is generally accomplished automatically by the *automatic load-frequency control* (ALFC) system, whose operation is illustrated schematically in Fig. 9.10. A frequency

Sec. 9.4 The Electric Power Network 429

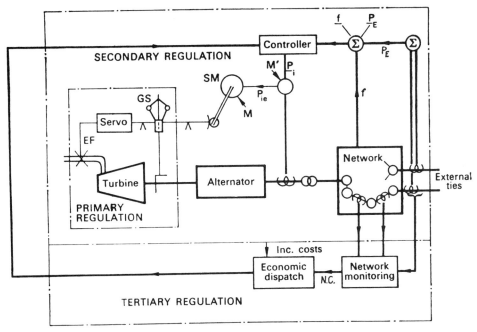

Figure 9.10 Dispatching of power from a central station. (Reprinted with permission from U. K. Knight, *Power Systems Engineering and Mathematics*, Pergamon Press, Elmsford, N.Y., 1972.)

sensor-comparator senses the system frequency and compares it with a reference frequency at 60.00 Hz to generate an error signal. The error signal is amplified and sent to the speed controller of the turbine that is to accept the load change. The controller's setting is adjusted and this changes the steam admitted to the turbine. The direct regulation of the turbine speed at a given speed control setting is called *primary regulation*. The adjustment of this setting by automatic frequency control is called *secondary regulation*.

The secondary regulation based on system frequency deviation is modified by a signal based on tie-line power flows. If a large generator were suddenly lost on an interconnected system, the entire system would respond. If the utility operating the generator that was lost generated approximately 20% of the interconnection power, only 20% of the lost power would be picked up instantaneously by the local utility and 80% by the interconnection. It is generally not desirable to let this situation persist for very long. The local utility, to the extent that it is able, attempts to readjust its generating capacity to meet its local requirements and return the tie-line power flow to its normal value. The signal indicating generation power charge, ΔP, within a given local utility area is often written as

$$\Delta P = k_{ag}(f_{\text{ref}} - f_{\text{me}}) + (P_{\text{ref}} - P_{\text{me}}) \tag{9.30}$$

where f_{ref}, f_{me} = reference and measured frequencies, respectively
P_{ref}, P_{me} = reference and measured tie-line power transfers, respectively
k_{ag} = constant (generally of the order of 3% of area generating capacity per hertz)

Economic Power Dispatch

The determination of which turboset within a system shall accept the next block of load is called *tertiary regulation*. With an automatic load-frequency control system, this is accomplished by determining the settings of the secondary regulation system in accordance with incremental power generation costs. For each significant load change, a computer program is solved to find the most economical system configuration considering the capacity and availability of generators, limitations imposed by system transmission capability and tie-line transfer requirements.

In determining the dispatch schedule, two *merit orders* are generally distinguished. In the first of these, the incremental costs, or *heat rates*, are inclusive of fixed operating charges. In the second merit order, incremental costs are computed only for running plants and exclude fixed charges. The first merit order is used in scheduling decisions, that is, the choice of which plant is to be run. The second merit order is used in dispatching decisions, that is, allocation of power production among plants on the line.

The determination of value of P_{ref}, the desired interconnection tie-line power in Eq. (9.30), is also a part of the economic dispatch analysis. Periodic recomputations of this value are made based on availability of generator capacity, load requirements, and relative costs of purchasing power from outside the system versus generating it internally.

9.5 MAGNETOHYDRODYNAMICS

Although essentially all large central station power plants generate electricity by means of synchronous alternators, other generation methods are being actively studied. Magnetohydrodynamic power generation is one of the concepts being actively pursued in both the United States and the USSR. Successful commercialization of magnetohydrodynamic generators could lead to generating stations with considerably improved thermodynamic efficiencies.

Basic Concepts

Ionized gas is produced when some of the electrons surrounding an atom are freed from their orbits, leaving the atom positively charged. When a flowing ionized gas is passed through a strong magnetic field which is perpendicular to the flow direction, an electric potential is created across the channel containing the ionized gas or *plasma*. The electric field potential will be perpendicular to both the flow direction

Sec. 9.5 Magnetohydrodynamics

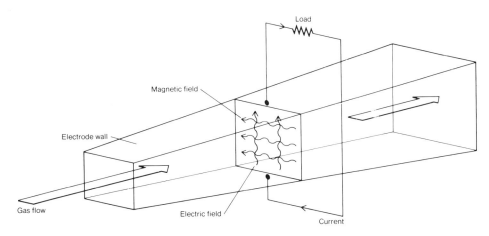

Figure 9.11 Principles of an MHD generator. (From the *EPRI Journal*, April 1980.)

and the magnetic field (see Fig. 9.11). If the upper and lower channel walls are insulated from each other, a significant voltage difference can be maintained. Charge accumulation on the upper and lower plates can be prevented, and useful work obtained, by connecting an external load across the plates.

In a MHD generator, hot combustion gases moving at a high velocity through the MHD channel (Fig. 9.12) are subjected to a strong magnetic field at right angles

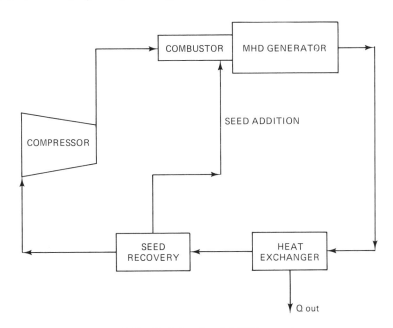

Figure 9.12 Ideal closed-cycle MDH generator.

to the flow direction. Although all gases are highly ionized at sufficiently high temperatures, combustion product gases do not ionize appreciably at combustion temperatures (2500 to 3000°C). To produce the desired ionization it is necessary to add a "seed" material to the generator. Salts of cesium or potassium, which have significant vapor pressures at combustion temperatures, and which have substantially lower ionization energies (gas will ionize at lower temperature) than combustion gases, are used for this purpose. Because of the high cost of cesium salts, a potassium salt, particularly potassium carbonate, is the preferred seed material.

The MHD Cycle

In the ideal cycle shown in Fig. 9.12, the MHD generator can be considered in much the same fashion as a Brayton cycle gas turbine. The compressed gas is expanded adiabatically across the MHD channel. The gas stream is decelerated by an electromagnetic force rather than the turbine blades. Electrical work is done on the external load rather than producing mechanical work as in a turbine.

MHD generators can be operated only at high temperatures at which a high degree of gas ionization, and hence high gas conductivity, can be maintained. Hence the gases leaving an MHD generator are still at very high temperatures. It would be uneconomical to reject this heat to the atmosphere, and thus in any real MHD cycle the heat would be used to generate steam to drive a steam turbine. The MHD generator is a "topping cycle" just as the gas turbine is in combined Brayton–Rankine cycle plants. The MHD generator would have the same ideal thermodynamic efficiency as a Brayton cycle turbine working over the same temperature range.

Combined Brayton–Rankine cycle and MHD plants may be considered as competing technologies. Both schemes provide improved thermodynamic efficiencies. Since MHD generators operate at higher temperatures than Brayton cycle turbines (large quantities of excess air need not be added), MHD generation as a topping cycle leads to the highest overall cycle efficiency. However, this is achieved at the expense of considerable complexity and the overall economics may not be improved.

MHD Generators

We have seen that in conventional electric power generation, a conductor is moved through a magnetic field such that

$$\mathbf{E}_i = \mathbf{u}_c \times \mathbf{B}_m \quad \text{V/m} \tag{9.31}$$

where $\mathbf{u}_c$ = velocity of conductor
$\mathbf{B}_m$ = magnetic field

A voltage is developed which results in a current when connected to a load resistance, R. Together with the conductor current, I_a, the magnetic field produces

Sec. 9.5 Magnetohydrodynamics

a force, $\mathbf{F}_l$, in the conductor:

$$\mathbf{F}_l = \mathbf{I}_a \times \mathbf{B}_m \qquad \text{newtons/meter of length} \qquad (9.32)$$

The resultant force, $\mathbf{F}_l$, is in a direction opposite to the conductor movement. The work done on the conductor is equivalent to the energy dissipated by the resistance.

An MHD generator employs the same basic principle as a conventional generator, except that the high-velocity ionized gas serves as the conductor, thereby eliminating the necessity of rotating shafts with mechanical connections. In the simple MHD generator of Fig. 9.11, the gas flow is assumed to be along the x axis, with a velocity of u_0. The magnetic field is assumed to be parallel to the z axis. The induced electric field, $\mathbf{E}_i$, is found from the vector cross product

$$\mathbf{E}_i = \mathbf{u}_0 \times \mathbf{B}_m \qquad (9.33)$$

or for the simple case shown in Fig. 9.11,

$$E_i = u_0 B_m$$

where u_0 and B_m are scalars. The field just described causes the positive ions to move upward and the electrons to move downward. A current will flow when an external load resistance is connected across the MHD generator.

Ideal MHD generator operation. The transverse magnetic force shown in Fig. 9.11 decelerates the charged particles in the hot gas stream. The electrons are induced to spin in a circular orbit about the transverse axis in the magnetic field. This negative field acts as an attractive or Coulomb force to slightly decelerate the positive ions. The positive ions, because of their large mass compared to an electron, have very low velocity. The net result is a dense area or screen of plasma, containing both positive ions and free electrons. This screen serves to decelerate the neutral particles through direct elastic collisions. The gas stream exerts a force on this screen and pushes the electrons through the applied magnetic field. This reduces the kinetic energy of the gas stream and produces an equivalent direct current electrical power in the external load circuit. The electron movement is also shown in Fig. 9.11. The electrons move to the electrode wall (anode), through the external load, back to the cathode, and back into the gas stream.

For steady-state operation and a one-dimensional analysis of the MHD system, the current density, J_c, can be expressed as

$$J_c = \sigma_{ec}(u_0 B_m - E_i) \qquad (9.34)$$

where the new terms introduced are

$J_c \equiv$ current density, current per unit area

$\sigma_{ec} \equiv$ electrical conductivity of the ionized gas

This is one of the Maxwell equations in scalar form and is just a form of Ohm's law with a magnetic field. For the simple system shown in Fig. 9.11, the power output per unit volume, P_v, would be

$$P_v = E_i J_c - \sigma_{ec} E_i (u_0 B_m - E_i) \tag{9.35}$$

or

$$P_v = \frac{J_c}{\sigma_{ec}} (\sigma_{ec} u_0 B_m - J_c) \tag{9.36}$$

The *electrical* efficiency of the system would be

$$\eta_{\text{ele}} = \frac{P_v}{P_v + J_c^2/\sigma_{ec}} \tag{9.37}$$

where J_c^2/σ_{ec} is the joule heating per unit volume. By combining Eqs. (9.36) and (9.37), the electrical efficiency of the system can be written as

$$\eta_{\text{ele}} = \frac{E_i}{u_0 B_m} \tag{9.38}$$

Since E_i is the voltage gradient (V/m), the output voltage, V_{out}, is simply

$$V_{\text{out}} = E_i b \tag{9.39}$$

where b is the channel width.

MHD generator losses: the Hall effect. In addition to normal joule heating losses, MHD generators have other unique losses. One such loss is due to the *Hall effect*. The Hall effect loss is a result of the Lorentz force acting on the electrons in the interelectrode gap. The *Lorentz force* is defined as

$$\mathbf{F}_L = q_c (\mathbf{E}_i + \mathbf{u}_0 \times \mathbf{B}_m) \tag{9.40}$$

where $\mathbf{F}_L \equiv$ Lorentz force
$q_c \equiv$ charge, coulombs (C)
$\mathbf{E}_i \equiv$ electric field, V/m
$\mathbf{B}_m \equiv$ magnetic field
$\mathbf{u}_0 \equiv$ velocity of gas

The Lorentz force makes the electrons travel in a circular orbit. Physically, this implies that all the electrons attempt to migrate to a particular end of the collecting electrode. This results in very large currents with correspondingly large resistance losses at that end. As will be discussed later, one solution to this phenomenon is to design an insulated, segmented collecting electrode.

In Fig. 9.13, $\mathbf{B}_m$ is directed upward out of the page. The vector cross product, $\mathbf{u}_0 \times \mathbf{B}_m$, results in the induced field, $\mathbf{E}_i$. If the applied electric field is designated as $\underline{\mathbf{E}}$, then the field $\mathbf{E}'$ relative to the gas is the vector sum

$$\mathbf{E}' = \underline{\mathbf{E}} + \mathbf{u}_0 \times \mathbf{B}_m \tag{9.41}$$

The field expressed by Eq. (9.41) is the EMF field experienced by the charged particles in the ionized gas. The current flows in the direction of $\mathbf{E}'$, which gives an

Sec. 9.5 Magnetohydrodynamics

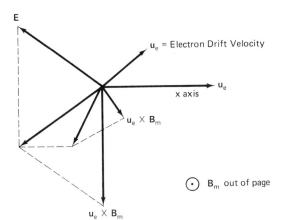

Figure 9.13 Vector diagram to illustrate the Hall effect.

additional EMF due to the Hall effect. Electrons with a drift velocity, $\mathbf{u}_e$, experience forces due to both the transverse induced field, $\mathbf{u}_e \times \mathbf{B}_m$, and $\mathbf{E}'$. This interaction gives rise to the current density, $\mathbf{J}$, directed opposite to $\mathbf{u}_e$. The electric field $\mathbf{E}''$ relative to the drift velocity is the vector sum

$$\mathbf{E}'' = \mathbf{E}' + (\mathbf{u}_e \times \mathbf{B}_m) \tag{9.42}$$

and we obtain the current density

$$\mathbf{J}_c = \sigma_{ec}\mathbf{E}'' = \sigma_{ec}\mathbf{E}' + \sigma_{ec}(\mathbf{u}_e \times \mathbf{B}_m) \tag{9.43}$$

The gas conductivity, σ_{ec}, is given by

$$\sigma_{ec} = \mu_e N_e \bar{e} \tag{9.44}$$

where $\mu_e \equiv$ electron mobility, m^2/V·s (we obtain μ_e from $\mathbf{u}_e = -\mu_e \mathbf{E}$)
$\bar{e} \equiv$ electron charge, coulombs
$N_e \equiv$ electron density

We also can write that

$$\mathbf{J}_c = -\mathbf{u}_e N_e \bar{e} \tag{9.45}$$

Hence by, combining Eqs. (9.44) and (9.45), we have

$$-\mu_e \mathbf{J}_c = \mathbf{u}_e \sigma_{ec} \tag{9.46}$$

We now rewrite Eq. (9.43) as

$$\mathbf{J}_c = \sigma_{ec}\mathbf{E}' - \mu_e \mathbf{J}_c \times \mathbf{B}_m \tag{9.47}$$

The components of the vector $\mathbf{J}_c$ in the x, y, and z directions are obtained from Eq. (9.47) and are

$$\begin{aligned} J_{cx} &= \sigma_{ec} E'_x - \mu_e J_{cy} B_m \\ J_{cy} &= \sigma_{ec} E'_y + \mu_e J_{cx} B_m \\ J_{cz} &= \sigma_{ec} E'_z = 0 \end{aligned} \tag{9.48}$$

After defining the Hall coefficient, β_H, as $B_m\mu_e$, we solve Eq. (9.48) for J_{cx} and J_{cy}.

$$J_{cx} = \frac{\sigma_{ec}(E'_x - \beta_H E'_y)}{1 + \beta_H^2}$$

$$J_{cy} = \frac{\sigma_{ec}(E'_y - E'_x)}{1 + \beta_H^2} \tag{9.49}$$

The Hall effect and MHD generator performance. The electric and magnetic field interaction is shown in Fig. 9.13. For the system shown, $\mathbf{E}_x = 0$ and $\mathbf{E}'_x = 0$. Let α_H be defined by

$$\alpha_H = \frac{E_y}{u_0 B_m} \tag{9.50}$$

and

$$E'_y = E_y - u_0 B_m \tag{9.51}$$

Using Eq. (9.49) for J_{cx} and J_{cy}, we get

$$J_{cx} = \frac{\sigma_{ec} u_0 B_m \beta_H (1 - \alpha_H)}{1 + \beta_H^2}$$

$$J_{cy} = \frac{-\sigma_{ec} u_0 B_m \beta_H (1 - \alpha_H)}{1 + \beta_H^2} \tag{9.52}$$

The energy removed from the MHD generator per unit area, $\dot{w}$, is the product of the current and potential difference in the y direction.

$$\dot{w} = -E_y J_{cy} = \frac{\sigma_{ec} u_0^2 B_m^2 (1 - \alpha_H) \alpha_H}{1 + \beta_H^2} \tag{9.53}$$

The Hall effect thus reduces MHD generator output by the factor $(1 + \beta_H^2)$. Since a typical value for $\beta_H = 5$, the Hall effect would reduce the output to less than 4% of that predicted by the simple model.

The problem is reduced through the use of a segmented electrode, as shown in Fig. 9.14. For the segmented electrode design,

$$E'_x = \beta_H E'_y \tag{9.54}$$

$$E_x = \beta_H (E_y - u_0 B_m) \tag{9.55}$$

The current components are

$$J_{cx} = 0$$

$$J_{cy} = \sigma_{ec}(E_y - u_0 B_m) \tag{9.56}$$

and the power generated for unit length is

$$\dot{w} = \sigma_{ec} u_0^2 B_m^2 \alpha_H (1 - \alpha_H) \tag{9.57}$$

The penalty factor $(1 + \beta_H^2)$ has been removed and the power output is now that predicted by the simple model.

Sec. 9.5 Magnetohydrodynamics

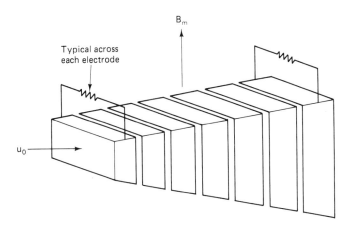

Figure 9.14 MHD generator with segmented electrodes.

MHD Power Plants

As mentioned earlier, a commercial MHD system would be expected to be used as a topping cycle for a conventional steam power station. This would result in significant improvements in the overall efficiency. The combined conversion efficiency for the cycle may be shown to be

$$\eta_c = \eta_{MHD} + \eta_{steam} - \eta_{MHD}\eta_{steam} \tag{9.58}$$

where η_c = combined power plant efficiency
η_{MHD} = efficiency of the MHD plant alone
η_{steam} = efficiency at the steam plant alone

For expected operating conditions, conversion efficiencies may be increased to 50%, compared to ~40% for the best steam plant.

A typical configuration of a proposed MHD generation plant using coal as fuel is shown in Fig. 9.15. The system shown in this figure is an open-cycle system, since the working fluid (the hot gas) is not recirculated. The seed material, for economic and environmental reasons, must be recovered. It should be noted that the potassium ions of the seed will react with the SO_2 in the combustion gases to produce potassium sulfate which is removed as a solid from the exit gases. The necessity for expensive SO_2 scrubbers is thus removed.

Although SO_2 scrubbers are not needed, an MHD plant requires a number of costly auxiliaries not needed in the standard steam plant. In order to obtain the very high temperatures (2200°C or better) needed by the MHD channel, present design calls for the air entering the combustion to be oxygen enriched. An oxygen plant is then required and some of the power produced will have to be used for operating the compressors needed. Further, the plant must be arranged so that the slag can be

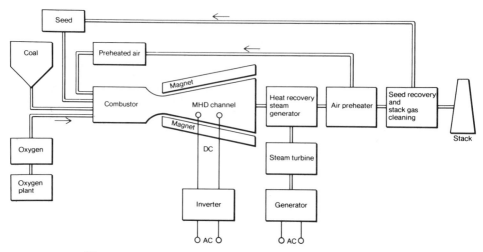

Figure 9.15 MHD power plant. (From the *EPRI Journal*, April 1980.)

separately removed at temperatures above those at which the seed particles will condense. If this is not done, the relatively expensive seed will be lost.

It should also be observed that the MHD channel produces dc current. Inverters (devices producing ac current from dc current) are therefore needed. These add additional expense and complexity to the plant.

The very high magnetic fields required by MHD generators with reasonable current densities would lead to extensive energy losses in conventional iron core magnets operating at ambient temperatures. To eliminate these losses, present designs propose that the MHD channel be surrounded by a superconducting magnet at liquid-helium temperatures (see Section 9.2 for a discussion of superconducting alternators). This poses a difficult design problem since the superconducting magnet must be in close proximity to the high-temperature MHD channel. Note that a portion of the liquid N_2 by-product of the oxygen plant may be used as a buffer between the plant system and liquid helium.

Prospects for Commercialization

There are a number of technical and economic problems that must be solved before MHD power would be considered feasible for utility application. The economics will not become clear until after a few large demonstration devices are built and operated. Generator electrode life must be increased to 1 year at suitable cost. An MHD combustor must be developed for practical operating temperatures and pressures. A slag removal system must be developed. Environmental controls for NO_x emissions must be developed or combustion must be carried on in a way that limits NO_x to acceptable levels.[†] Commercial devices, which are years away, will also face stiff competition from other energy sources and cycles.

[†] NO_x production is increased as the combustion temperature is increased.

9.6 FUEL CELLS

Basic Principles

If successfully commercialized, fuel cells could provide improved energy-conversion efficiencies as well as a means for efficiently reconverting stored chemical energy. The fuel cell is an electrochemical system which converts the chemical energy of a conventional fuel directly into low-voltage, direct-current electrical energy. Since these systems do not rely on thermal energy conversion, they are not bounded by Carnot efficiency limitations. A fuel cell can be viewed as a primary battery in which the fuel and oxidizer are stored externally. The fuel cell is an apparatus that prevents fuel and oxidizer molecules from mixing but permits the transfer of electrons through a metal path. A highly simplified schematic of a fuel cell is shown in Fig. 9.16.

Gaseous fuels require porous electrodes which allow the desired electrochemical reactions to take place at the electrode–fluid interfaces. In the hydrogen–oxygen cell shown in Fig. 9.16, hydrogen is supplied through one of the porous electrodes while oxygen is supplied through the other. At the anode electrode–electrolyte interface, the hydrogen dissociates into H^+ ions flow through the electrolyte to the cathode. At the cathode electrode–electrolyte interface, the hydrogen ions, together with the electrons that have traveled through the load, combine with oxygen to form water.

If a H_2–O_2 fuel contains an acidic electrolyte, the chemical reactions occurring are

$$2H_2 \rightarrow 4e^- + 4H^+ \tag{9.59}$$

$$4e^- + 4H^+ + O_2 \rightarrow 2H_2O \tag{9.60}$$

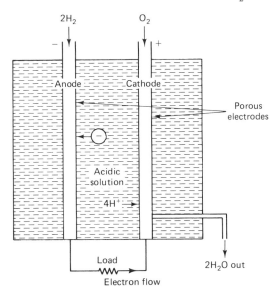

Figure 9.16 Basic elements of a fuel cell.

TABLE 9.3 ENTHALPIES AND FREE ENERGIES OF FORMATION AT STANDARD CONDITIONS[a]

Material	ΔG_{fo} (MJ/kg mol)	ΔH_{fo} (MJ/kg mol)
$CO(g)$	−137.4	−110.6
$CO_2(g)$	−394.7	−393.8
$CO_2(aq)$	−380.5	−413.2
$CH_4(g)$	−50.83	−74.9
$C_8H_{18}(l)$	7.41	−250.1
$H_2O(l)$	−237.1	−286.1
$H_2O(g)$	−228.8	−242.0
Cl	−131.3	−167.6
H^+	0	0
Na^+	−262.0	−239.8
OH^-	−157.4	−230.1

[a] Values determined at 298°K (20°C) and 1 atm.
Source: Data from *National Bureau of Standards Circular 500*, 1952.

If an alkaline electrolyte is used, the reactions are

$$2H_2 + 4(OH)^- \rightarrow 4H_2O + 4e^- \quad (9.61)$$

$$2H_2O + O_2 + 4e^- \rightarrow 4(OH)^- \quad (9.62)$$

Fuel cells can also employ a solid electrolyte which serves as an ion-exchange membrane. These cells are more expensive to construct and operate and not expected to be used for utility applications.

Fuel Cell Analysis

The energy released in the fuel cell reaction is equal to the change in the enthalpy of formation, ΔH_{fo}.

$$\Delta H_{fo} = \sum (\Delta H_{fo})_{reactants} - \sum (\Delta H_{fo})_{products}$$

Values for ΔH_{fo} can be readily found in the literature. Some typical values are listed in Table 9.3. For example, in the fuel cell reaction described previously,

$$4H^+ + O_2 \rightarrow H_2O \quad (9.63)$$

$$\Delta H_{fo} = (\Delta H_{fo})_H + (\Delta H_{fo})_{O_2} - (\Delta H_{fo})_{H_2O} \quad (9.64)$$

$$\Delta H_{fo} = 0 + 0 - (-286 \times 10^3) = +286 \times 10^3 \frac{kJ}{kg\ mol\ H_2O}$$

Not all of the heat of formation[†] above can be utilized to produce electrical energy, as some of the energy is lost in the process through the change in entropy of

[†] The heat of formation of any element is zero.

Sec. 9.6 Fuel Cells

the system. The minimum amount of energy loss is equal to

$$E_{loss} = \int T\,dS \qquad (9.65)$$

Since the fuel cell is essentially an isothermal process,

$$E_{loss} \simeq T\Delta S \qquad (9.66)$$

Under these conditions, the maximum electrical energy produced by the fuel cell per kilogram of fuel would be

$$E_{max} = \Delta H_{fo} - E_{loss} \qquad (9.67)$$

From our definition of free-energy change, ΔG [see Eq. (4.30)], we have

$$E_{max} = \Delta G$$

From the discussion of Chapter 4, we know that the change in free energy during a reversible chemical reaction can be obtained from the standard free energies of formation, ΔG_{fo}, of the reactants and products. That is, at standard conditions

$$\Delta G = \sum (\Delta G_{fo})_{reactants} - \sum (\Delta G_{fo})_{products} \qquad (9.68)$$

The Gibbs free energy is dependent on the temperature and pressure at which the reaction occurs. Values of Gibbs free energy of formation are also included in Table 9.3 for typical compounds at standard temperature and pressure conditions. For nonstandard conditions and a typical reaction of the form

$$\alpha A + \beta B \rightleftharpoons \gamma C + \delta D \qquad (9.69)$$

we have from Chapter 4 [see Eq. (4.42)] that if products and reactants are gases, the free-energy charge is

$$\Delta G = \Delta G_{fo} + RT \ln \frac{p_C'^{\gamma} p_D'^{\delta}}{p_A'^{\alpha} p_B'^{\beta}} \qquad (9.70)$$

where $\Delta G \equiv$ Gibbs free-energy change at nonstandard conditions
$R \equiv$ ideal gas constant
$T \equiv$ absolute temperature of reaction
$p_i' \equiv$ partial pressure of reactant or product i

For a reaction to be useful in a fuel cell, it must be exothermic (ΔG negative).

We may use the free energy of formation to compute the maximum voltage, ΔV, which can be produced by a cell in which the reaction of Eq. (9.69) occurs. Since a charge of 1 C accelerated through a 1-V potential gains 1 J of energy, we have

$$-\Delta G = \Delta V n_e F_a \qquad (9.71)$$

or

$$\Delta V = \frac{-\Delta G}{n_e F_a}$$

where ΔV = voltage output of the fuel cell, V
n_e = number of moles of electrons released per γ moles of product C produced [see chemical reaction of Eq. (9.69)]
F_a = Faraday's constant = number of coulombs released per equivalent weight = 96,494 C/g equiv [if ΔG were expressed in calories, F_a = 23,062 cal/V·g equiv)]

Note that the minus sign arises since a positive voltage is generated by an exothermic reaction.

By making use of Eq. (9.71), we may rewrite Eq. (9.70) as

$$\Delta V = \Delta V_o + \frac{RT}{n_e F_a} \ln \frac{p_C^{\prime \gamma} p_D^{\prime \delta}}{p_A^{\prime \alpha} p_B^{\prime \beta}} \qquad (9.72)$$

where ΔV_o is the ideal voltage output of a cell operating at standard conditions = $-\Delta G_{fo}^\circ / n_e F_a$.

Equation (9.72) is a special case of the *Nernst equation*. As predicted by this equation, increasing the fuel cell temperature will reduce the output voltage.

The maximum conversion efficiency, η_{max}, of a cell is given by

$$\eta_{max} = \frac{\Delta H_{fo} - E_{loss}}{\Delta H_{fo}} = \frac{\Delta G}{\Delta H_{fo}} \qquad (9.73)$$

The actual conversion efficiency is dependent on the actual energy output. That is,

$$\eta_{act} = \frac{n_e F_a V_a}{\Delta H_{fo}} \qquad (9.74)$$

where V_a is the actual voltage by cell. Actual fuel cell efficiencies are in the range 60 to 70%.

Fuel Cell Status

Most of the presently available fuel cells operate at low temperature, below 250°C, and use hydrogen and oxygen as the reactants. A single cell can generate 1 V and about 100 to 200 W per square foot of electrode. Individual cells can be series connected to increase the voltage and connected in parallel to increase the power output. A complete fuel cell system consists of fuel and oxidant storage tanks, a large number of fuel cells in a series–parallel arrangement and a dc-to-ac conversion system (inverters).

Sec. 9.7   Thermionic Generators

Fuel cells have been used for specific applications that require low-power-level devices. The current-generation hydrogen oxygen cell produces expensive electric power. Utility applications require inexpensive high-power-output cells for peak-load generation. Improved low-temperature H_2–O_2 fuel cells might be used for peak power generators. During off-peak hours, excess power could be used to electrolyze water, producing H_2 and O_2. The gases could be stored and recombined during peak hours (see Section 10.1).

Current research is directed toward the development of fuel cells that would operate with inexpensive hydrocarbon fuels and air. Most of these studies are based on fuel cells operating at high temperature with a molten salt electrolyte or a solid electrolyte. Development of high-temperature fuel cells with long operating lifetimes appears to be a number of years away.

9.7 THERMIONIC GENERATORS

Basic Principles

The basic thermionic cycle is very simple. A metal electrode, which is called the *emitter*, is heated until it is hot enough to release electrons from its surface. The electrons cross a small gap and collect on a colder metal electrode, the *collector*. To minimize energy losses as electrons cross the gap, the space between electrodes is maintained at either a high vacuum or filled with a highly conducting plasma. The electrons enter the collector and return through an external load to the emitter, thereby producing electrical power. In each of these simple steps, there are complications that tend to retard the electron flow. Figure 9.17 illustrates this basic concept.

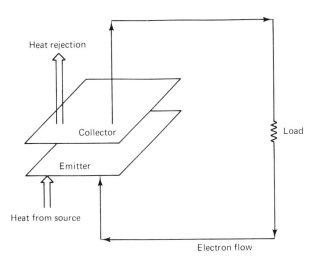

Figure 9.17 Basic thermionic generator.

At absolute zero, the kinetic energy of the free electrons would occupy quantum states, or energy levels from zero up to some maximum value defined by the Fermi level, ε_f. These energy levels are separated from each other by very small, discrete amounts of energy. Each quantum level contains a limited number of free electrons, in much the same manner as an electron orbit contains a limited number of electrons. At temperatures above absolute zero, some electrons may have energies higher than the Fermi level. The electrons may be pictured as vibrating about the Fermi energy level with an amplitude of vibration that is dependent on the temperature.

The energy that must be supplied to overcome the weak attractive force on the outermost orbital electrons is called the *surface work* or *work function*, ϕ_{wf}. The units are electron volts per electron or simply eV. It is understood that the work must be supplied to raise the electron to the energy level $\phi_{wf} + \varepsilon_f$ is simply ϕ_{wf}. It is assumed that the electron that escapes from the surface has already acquired the upper level of the Fermi energy distribution, ε_f.

When heat is supplied to the emitter, some of the high-energy free electrons at the Fermi level will obtain the necessary energy to escape the emitter surface. Thus they possess energy equal to the emitter work function, ϕ_{wf}, plus some additional kinetic energy. These electrons move through the gap to the collector, strike the collector, and give up their kinetic energy plus the energy equal to the collector work function. It requires an amount of energy equal to the collector work function to have an electron pass through the collector surface. This energy must be rejected as heat from the low-temperature collector.

The electron energy is reduced to the Fermi energy level of the collector, but this energy state is higher than that of the electron at the Fermi energy level of the emitter. This higher energy potential is the reason the electron will pass through an external load to the emitter.

Based on statistical mechanics, the maximum electron current per unit area that an emitting surface can provide is given by the *Richardson–Dushman equation*:

$$J_c = C_1 T^2 \exp\left(\frac{-\phi_{wf}}{k_B T}\right) \qquad (9.75)$$

where $J_c \equiv$ current density, A/m^2
$T \equiv$ absolute temperature, °K
$\phi_{wf} \equiv$ work function, eV
$k_B \equiv$ Boltzmann's constant, eV/°K
$C_1 \equiv$ emission constant

The emission constant and the work function are both material properties.

The electrons that move across the gap between the emitter and collector must also overcome a space-charge barrier, δ_{sc}. This space charge is the result of two forces acting on the electrons that are trying to cross the gap: (1) the emitter is left with a positive charge and tends to pull the electrons back, and (2) electrons already

Sec. 9.7 Thermionic Generators

in the gap exert a force on the electrons trying to cross. The effect of this space-charge barrier is to reduce the voltage output of the thermionic generator.

The net voltage output of the generator, V_{tg}, is

$$V_{tg} = [(\phi_e - \phi_c) + (\phi_k - \delta_{sc})] \tag{9.76}$$

where $\phi_e \equiv$ emitter work function
$\phi_c \equiv$ collector work function
$\phi_k \equiv$ excess kinetic energy of the electron
$\delta_{sc} \equiv$ space-charge barrier

The work functions ϕ_c and ϕ_e are properties of the collector and emitter and are independent of the current density in the device. However, the space-charge terms, $(\phi_k - \delta_{sc})$, vary with current density. A simple estimation of the output voltage can be made for the special case in which $\phi_k - \delta_{sc} \simeq 0$. For this case, the voltage output is approximately equal to the difference in work functions between the emitter and collector. This situation is almost achieved in practice for thermionic devices that utilize a conducting gas (low-pressure ionized gas) or a plasma in the interelectrode gap instead of a vacuum.

The use of an ionized gas or vapor in the interelectrode gap significantly reduces the charge barrier problem. Most practical devices have employed cesium vapor between the emitter and collector. However, to achieve a high degree of ionization of the cesium, the emitter temperature must be in the range 1500 to 1600°C (2730 to 2910°F). Other research is directed toward the use of an ionized inert gas to minimize electrode corrosion problems.

It is difficult to predict the thermal efficiency of a thermionic generator. Many of the losses depend on the actual materials used in the device and the operating conditions.

Thermionic Systems for Utility Applications

There are a number of concepts being studied to utilize thermionic generators for utility applications. A thermionic converter could improve the overall efficiency of a fossil-fueled power plant by first using the high-temperature combustion gas before the steam production cycle. As previously indicated, the use of the hot combustion gases to produce extra energy before the steam cycle is called a *topping cycle*. The total quantity of heat utilized is increased, as is the overall plant efficiency.

A second interesting application is in a topping cycle combining an MHD generator with a thermionic generator. The waste heat from an MHD generator is often as high as 1900°C. The thermionic device could utilize this heat prior to its use in a conventional steam cycle. Estimates show that an MHD topping cycle plus a conventional steam cycle could achieve an overall efficiency of 50%. With the addition of a thermionic device between the MHD system and steam generator, estimates indicate that the overall cycle efficiency could be increased to 55%.

Practical application of thermionic converters to large-scale power applications remains some time in the future. High-temperature emitters, of such materials as tungsten on rhenium, are expensive. Ceramic shields must be provided to protect these emitters from the corrosive combustion gases. Collectors, often made of molybdenum coated with cesium to obtain a low work function, are somewhat less expensive but still costly. In addition, current thermionic converter efficiencies are generally less than 15% and lifetimes are still considerably shorter than desired.

SYMBOLS

a_s	surface current density, A/m
b	width of the MHD device channel, m
$\mathbf{B}_m$	magnetic field flux density, T or Wb/m²
$|B_m|$	magnitude of the magnetic field, T or Wb/m²
C_1	emission constant of thermonic device
D_s	diameter of stator, m
$\bar{e}$	electron charge, C
$\underline{\mathbf{E}}$	applied electric field, V/m
E_g	generator EMF, V
$\mathbf{E}_i$	electric field, induced, V/m
E_{loss}	voltage losses in a fuel cell, V
$E_{\max}$	maximum electric energy produced by fuel cell, J/s
$\bar{\mathbf{E}}'$	electric field relative to MHD gas, V/m
$\bar{\mathbf{E}}''$	electric field relative to drift velocity, V/m
f_{me}	measured generator frequency, Hz
f_{ref}	generator reference frequency, Hz
f_z	generator frequency, Hz
$\mathbf{F}, \mathbf{F}_l$	force on conductor, N/m
F_a	Faraday's constant
$\mathbf{F}_L$	Lorentz force, N
F_s	stator force, N
ΔG	free-energy change, J
ΔG_{fo}	standard free energy of formation, J
ΔH	enthalpy change, J
ΔH_{fo}	standard enthalpy of formation, J
I_a	current, A
J_c	current density, A/m²
k_{ag}	generator constant

Chap. 9 Symbols

k_B	Boltzmann's constant
k_{ps}	power system constant
L_s	stator inductance, H
L_w	length of stator winding strip, m
n_e	number of moles of electrons
n_p	number of poles in generator
N_e	electron density, electrons/m^3
p'	partial pressure of each component, Pa
P_1, P_2	real power in lines 1 and 2
$P_{3\phi}$	three-phase power, W
P_{me}	tie-line power transfer: measured, W
P_{real}	real component of ac power, W
P_{ref}	tie-line power transfer: reference, W
P_v	power output per unit volume, W/m^3
ΔP	generation power change, W
q_c	charge, C
Q_1, Q_2	reactive power in lines 1 and 2, W
Q_{rp}	reactive power, W
R	ideal gas constant
R_s	stator resistance, Ω
S	entropy, J/°K
t	time, s
T	temperature, °C or °K
u	velocity, m/s
$\mathbf{u}_e$	electron drift velocity, m/s
u_0	velocity of gas flow, m/s
$\mathbf{u}_c$	velocity of conductor, m/s
v_r	volume of the rotor, m^3
V_a	actual fuel cell voltage, V
V_A, V_B, V_C	voltage, V
V_{AB}, V_{BC}, V_{CA}	phasors (voltages between terminals), V
V_g	grid voltage, V
V_{out}	output voltage of the MHD device, V
V_{tg}	net voltage output of thermonic generator, V
ΔV_o	voltage output of fuel cell at standard conditions, V
V_1, V_2	voltage in lines 1 and 2
$|V|$	rms line voltage, V
ΔV	voltage output, V

dV/dx	induced voltage gradient, V/m
$\dot{w}$	energy output of the MHD device, W/m²
x	distance, m
X	reactance
X_s	synchronous reactance
Z_s	synchronous impedance of stator winding, Ω

Greek Symbols

α_H	$E_y/u_e B_m$
β_H	Hall coefficient (B_m/μ_e)
γ_{cf}	lag angle between current wave and flux wave, degrees
ε_f	Fermi level
δ_{cv}	angle between the current and voltage, degrees
δ_g	generator advance angle, degrees
δ_{sc}	space charge barrier, V
η_{act}	actual efficiency of the fuel cell
η_c	combined power plant efficiency
η_{max}	maximum conversion efficiency of a fuel cell
η_{MHD}	efficiency of the MHD device
η_{steam}	efficiency of the steam power station
θ_{21}	phase angle between voltages produced at two different bus bars (i.e., $V_2 - V_1$)
μ_e	electron mobility, m²/V·s
ϕ_c	collector work function, V
ϕ_e	emitter work function, V
ϕ_k	excess KE of the electron, V
ϕ_{wf}	work function, V
σ_{ec}	electrical conductivity of ionized gas, A/V·m
ψ	angle between current and voltage, degrees
ω	angular velocity, rad/s
ω_r	angular velocity, rpm

Subscripts

0	initial condition or position
1, 2, ...	first condition or position, second condition or position, etc.
x	x component or direction

y	y component or direction
z	z component or direction

PROBLEMS

9.1. A low-speed hydraulic turbine drives a generator at 360 rpm. How many poles must the rotor have to produce 60-Hz alternating current?

9.2. Determine the power at which the generator described in illustrative Example 9.1 (Section 9.1) will pull out of synchronism. Assume that the phase voltages remain constant.

9.3. A three-phase synchronous generator supplies current to a 14-kV line. The magnitudes of generator EMF and bus voltage are equal. The synchronous reactance is 6 Ω per phase. If the alternator delivers 20 MW to the grid, determine the power angle (δ_{cv}), phase current magnitude and direction, and magnitude of the reactive power.

9.4. An alternator contains a rotor that is 1 ft in diameter and 5 ft long. The rotor coils are supplied with 60 A (rms) per phase and there are two conductors per rotor slot [we then have 120 A (rms) per slot]. The rotor has a peak magnetic flux of 1.5 T and rotates at 1800 rpm. Determine the total power output if the current lags the flux wave by 10°.

9.5. The total kinetic energy of all the spinning turbines and rotors in a given system amounts to 500×10^6 J. Assume that the system is operating at constant frequency and in power balance. If there is a sudden power decrease of 2.5 MW and the control system fails to change turbine power, at what rate (in Hz/s) will the system accelerate?

9.6. If resistive losses in the stator account for 80% of the electrical losses in the generator described in Problem 9.3, what is the generator efficiency?

9.7. Fifty megawatts of real power per phase is transmitted between two locations along a 120-kV power line. If the power line has a synchronous reactance of 20 Ω per phase, what is the phase angle between the voltages at the two locations? What fraction of the transmitted power will be lost if the line has a resistance of 4.5 Ω per phase?

9.8. A section of an MHD generator with segmented electrodes has the following characteristics:

Applied voltage:	1200 V
Width of each electrode:	0.4 m
Mean gas velocity:	2300 m/s
Magnetic flux density:	2 T
Gas conductivity:	45 $(\Omega \cdot m)^{-1}$
Distance between electrodes:	0.2 m

Determine

(a) The power generated per unit length of generator assuming that $\beta_H = 5.1$.

(b) The electrical efficiency of the generator.

(c) The pressure drop assuming that the gas has the properties of air at 1000°C and 15 bar.

9.9. A new fuel cell is designed to burn CO with atmospheric air at 1 bar and 50°C. Calculate:

(a) The maximum (theoretical) voltage that the cell can develop. Assume that the partial pressure of CO_2 in the exit stream is 0.05 bar.

(b) The theoretical power output per kilogram of CO burned.

(c) The ratio of the useful energy produced by a cell, having an efficiency of 80% theoretical, to that produced by combustion of the same quantity of fuel in a thermal power plant with an overall efficiency of 37%.

9.10. An emitter of a thermoionic device consists of a 0.25-cm-diameter tungsten rod which is 0.8 m long. The barium oxide collector has a work function of 1.2 V. Calculate the maximum theoretical power from such a unit. You may assume that the emission constant is 100 A/cm^2, the work function for tungsten is 4.5 V, and that $\phi_k - \delta \simeq 0$.

9.11. Show that the cycle efficiency predicted by Eq. (9.58) for a combined MHD–stream cycle is correct when the MHD exhaust gases are used for steam production without further heat addition.

BIBLIOGRAPHY

ANGRIST, S. W., *Direct Energy Conversion*, 2nd ed. Boston: Allyn and Bacon, 1971.

ELGRED, O., *Basic Electric Power Engineering*. Reading, Mass.: Addison-Wesley, 1977.

KNIGHT, V. K., *Power Systems Engineering and Mathematics*. Elmsford, N.Y.: Pergamon Press, 1972.

MCPHERSON, G., *An Introduction to Electrical Machines and Transformers*. New York: Wiley, 1981.

NASER, S. A., and L. E. UNNEWEHR, *Electromechanics and Electric Machines*, 2nd ed. New York: Wiley, 1982.

10

Energy Alternatives

10.1 ENERGY STORAGE

Need for Large-Scale Energy Storage

Oil and natural gas were formerly the preferred fuels for generating intermediate- and peak-load electric power. The fuel cost was relatively low and the energy content relatively high. The capital cost for building an oil or gas power plant was also low. These fuels are easy to transport and store. Electric power could be produced by these units at less cost than from other systems available at that time. Public law and escalating costs are forcing utilities to reduce their use of oil and natural gas. As oil and natural gas units are phased out, energy storage is being examined as a way to meet peak-load demand.

It is expensive to use a large coal or nuclear plant for load following during peak-demand periods. As discussed in Chapter 1, the load–duration curve can be used to evaluate the number of hours per year that are in the peak supply area. By meeting this peak demand through energy storage, a utility can save the cost of building a large generating station or avoid the high costs incurred by generating intermediate and peak power through the use of older, less efficient units. Figure 10.1 shows the increase in cost associated with peak-load generation.

An ideal energy storage system would utilize electric power from base-load coal or nuclear plants at off-peak hours (12 midnight to early morning) to store energy for peak-hour use. The ideal storage system thus makes better use of existing power stations by increasing the capacity factor while diminishing the need for new power plant construction.

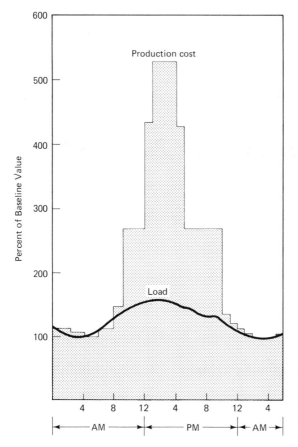

Figure 10.1 Electricity production cost increases as a percent of the base-load cost. (From the *EPRI Journal*, Oct. 1981.)

Energy Storage Technologies

Electricity as such cannot be stored and therefore electrical energy must generally be converted to a form of energy that can be stored. An energy storage system consists of three main subsystems: a conversion subsystem, an energy storage system, and a system to convert the stored energy back into electric power. In the practical case of pumped hydroelectric power, the pump converts the off-peak electrical energy into potential energy of the stored water. The high reservoir is the energy storage system. A hydraulic turbine then converts the stored energy back into electrical power when needed.

The pumped hydroelectric unit just described is the only large-scale storage system presently in use. Table 10.1 lists the technologies now being developed for large-scale energy storage. For economic reasons, current research and development emphasis is on underground pumped hydro, compressed air storage, thermal energy storage, and advanced batteries.

TABLE 10.1 TECHNOLOGIES FOR LARGE-SCALE ENERGY STORAGE

Potential energy storage
  Conventional pumped hydro
  Underground pumped hydro
  Compressed gas
Kinetic energy storage
  Flywheels
Thermal energy storage (TES)
  Sensible heat storage
    Rocks, iron, and other solids
    Water or other fluids
  Latent heat due to phase change
    Sodium sulfate decahydrate (Glauber's salt) ($Na_2SO_4 \cdot 10H_2O$)
    Sodium thiosulfate pentahydrate ($Na_2S_2O_2 \cdot 5H_2O$)
    Water/ice systems
Electrical energy storage
  Electrostatic fields
  Inductive field
Chemical energy storage
  Batteries
  Fuel cells
  Chemical heat pipe
  Chemical reactants (i.e., hydrogen)

Pumped hydro storage, both conventional and underground, was discussed previously. The energy storage concepts discussed in this section are those likely to be developed for utility application in the near-term future: that is, compressed air storage, thermal energy storage, and advanced battery applications. Information on the other technologies listed in Table 10.1 can be found in the Bibliography at the end of this chapter.

Compressed air energy storage. Compressed air energy storage uses off-peak-load power generation to compress air for peak period use in gas-turbine generators. In off-peak hours the generator is disconnected and the turbine runs the air compressor only. The compressed air would be stored in larger underground caverns, salt domes, or abandoned mines. During peak hours, the compressed air is directed back to a gas turbine that drives a generator. During these hours, the compressor is disconnected and all the turbine power goes to the generator.

An even more economical system is one that uses off-peak electric power from coal or nuclear power stations to compress the air to be stored. As we observed in Chapter 9, a synchronous generator may also act as a synchronous motor when power is supplied to it. A synchronous motor–generator may therefore be used to drive the air compressor in the storage portion of the cycle and to generate current during the power portion of the cycle.

A typical system consists of air compressors, electric motor–generator, gas turbine, and a recuperator. A possible power plant arrangement is shown in Fig. 10.2. A typical compressed-air storage power plant would function as follows. In the peak power generation mode, the compressed air, at ~ 85 bar, is released from the storage cavern. It is throttled to about 40 bar and passed through a recuperator, where the gas is heated by turbine exhaust. The air is ducted to an expansion turbine which drives a small generator. As the heated, compressed gas leaves the expansion turbines, it is heated in the gas turbine combustion chamber. The hot, high-pressure gas is expanded through the gas turbine to drive the motor–generator acting as a generator.

During the storage portion of the cycle, the motor–generator, acting as a motor, drives the air compressor with the gas turbine decoupled. The compressed air is stored in the underground caverns beneath the plant. The compressed air storage system utilizing electric power for air compression utilizes only about half the

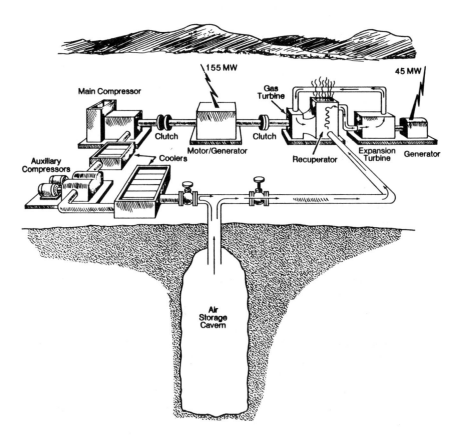

Figure 10.2 Compressed air storage power plant for generating peak-load electricity. (Reprinted with permission from *Power Engineering*, Dec. 1977.)

Sec. 10.1   Energy Storage   455

high-pressure gas turbine fuel per unit energy output that a standard gas turbine system would use.

Compressed-air systems can often be more economical than pumped hydro systems for small capacities. Present system designs range from 200 to 500 MW. A typical system could produce this power level for 2 hours.

Thermal energy storage.   Thermal energy storage systems for peak power applications are those systems in which the stored thermal energy can be used in a heat engine. The heat can be stored in the form of sensible heat of materials or in the latent heat of a phase change.

There are two principal system concepts that are being considered. The first concept, shown schematically in Fig. 10.3, utilizes a separate peaking turbine which would be added to an existing power station. The steam conditions at the throttle of the peaking turbine depend on the thermal storage system and the detailed economic conditions.

The second concept, shown schematically in Fig. 10.4, uses the stored heat to heat the feedwater. This would require a modified turbine design to allow for a wide variation in steam conditions.

Thermal energy storage concepts are divided into two categories: sensible-heat storage or latent-heat storage. In *sensible-heat storage* systems, heat is simply stored by raising the temperature of the solid or liquid. If the specific heat is constant, the energy stored is directly proportional to the temperature increase in the material. Water and the other materials listed in Table 10.2 are likely candidate materials for these systems. The systems operate at low temperatures, normally below 150°C (300°F).

For high-temperature energy storage applications, *latent-heat storage* concepts are being developed. Latent-heat materials store energy by phase change, usually

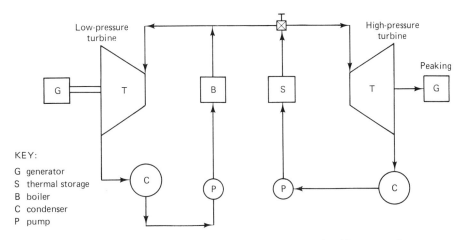

Figure 10.3   Thermal energy storage for use with a peak-load turbine power plant.

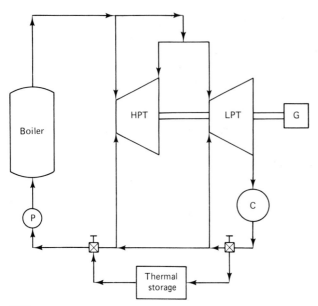

KEY:

P pump
C condenser
HPT high-pressure turbine
LPT low-pressure turbine
G generator

Figure 10.4 Feedwater thermal storage concept.

TABLE 10.2 SENSIBLE-HEAT STORAGE MATERIALS

Material	Chemical formula	Volumetric heat capacity (kJ/m^3 · °C)	Specific heat (kJ/kg · °C)
Water	H_2O	4190	4.2
Isobutyl alcohol	C_4H_9OH	2380	3.0
Amyl alcohol	$C_5H_{11}OH$	2420	2.9
Butyl alcohol	C_4H_9OH	2270	2.8
Ethyl alcohol	CH_3CH_2OH	2230	2.8
Beryllium	Be	5230	2.8
Limestone	$CaCO_3$	2550	0.9
Granite/rock	—	2420	0.8
Sand	SiO_2	1340	0.8
Iron	Fe	3690	0.5

Source: Data from F. Bundy et al., *The Status of Thermal Energy Storage*, Report 76CRD041, General Electric Company, Power Systems Laboratory, Schenectady, N.Y., April 1976.

Sec. 10.1   Energy Storage

TABLE 10.3 PROPERTIES OF LATENT-HEAT STORAGE MATERIALS

Material	Chemical formula	Melting point (°C)	Heat of fusion kJ/kg	Heat of fusion kJ/m³
Water	H_2O	0	335	305,000
Glauber's salt	$Na_2SO_4 \cdot 10H_2O$	32	153	350,000
Sodium hydroxide	NaOH	300	224	467,000
Lithium hydroxide	LiOH	690	3800	3,000,000

Source: Data from D. Golibersuch et al., *Thermal Energy Storage for Utility Applications*, Report 75CRD256, General Electric Company, Schenectady, N.Y., 1975.

from a solid to a liquid. The stored energy is released when the process is reversed. Table 10.3 lists properties of several latent heat storage materials now being considered.

Battery storage. The common battery consists of a set of cells each of which contain two electrodes separated by a solution called an *electrolyte*. Batteries differ from fuel cells in that all the chemical reactants are contained within the battery. When a battery is charged, an endothermic chemical reaction converts the electrical energy into stored chemical energy. The energy recovery, or discharge process, works in the reverse. An exothermic chemical reaction converts the chemical energy into electricity. Since only direct current is produced, an inverter is also required.

The basic cell of a conventional *lead–acid battery*, widely used for automobile service, consists of two lead plates which serve as electrodes and sulfuric acid, H_2SO_4, saturated with $PbSO_4$ as the electrolyte between the plates. Upon charging, an outside power source makes one lead plate the cathode and the other the anode. At the cathode, lead ions are deposited:

$$Pb^{2+} + 2e^- \rightarrow Pb \tag{10.1}$$

At the anode, lead oxide is formed:

$$Pb^{2+} + 6H_2O \rightarrow PbO_2 + 4H_3O + 2e^- \tag{10.2}$$

When the battery is discharged, the foregoing reactions are reversed. During discharge, both anode and cathode dissolve.

The use of batteries for load leveling has the advantage of a quick response system with high modular capability. Currently, the conventional lead–acid storage battery is the only system available for large-scale energy storage applications. However, existing technology does not meet all of the energy storage characteristics required for utility applications. Automotive-type batteries are low in cost but the lifetime is limited to a few hundred deep cycles. Industrial-quality lead–acid batteries have a longer lifetime, perhaps 2000 cycles, but their cost is high.

Research work is under way to develop advanced-design lead–acid batteries with the design characteristics required for utility applications. At present, the lead–acid battery system is not competitive with the technologies described previously for utility applications.

Advanced battery technology may eventually produce batteries that better meet the needs of the utility industry. Two advanced battery vconcepts are being developed toward this application: the zinc chloride battery and the sodium–sulfur battery (called the "beta" battery).

The *zinc chloride battery* utilizes low-cost materials. During the charging and discharging cycles, the aqueous solution of zinc chloride, which serves as the electrolyte, circulates between the graphite electrodes. When a dc current is supplied to charge the battery, zinc metal is electrodeposited on the negatively charged electrode plates and chlorine gas is produced at the positively charged plates. Energy is thus stored in the form of chemical potential energy in the zinc and chlorine formed.

In order to store the chlorine gas safely, the chlorine released is mixed with refrigerated water to form chlorine hydrate, $Cl_2 \cdot 6H_2O$. When it is desired to discharge the battery, the hydrate is decomposed by heating to release the chlorine gas. The chlorine is then mixed with the electrolyte and pumped back to the battery.

During discharge, the electrodeposited zinc dissolves at one electrode and Cl^- ions are formed at the other. All materials are thus returned to solution as zinc chloride. Since the liquid state has no "memory" of the charged condition, long cycle life is expected.

A high-temperature *sodium–sulfur battery* is also being developed for utility applications. The sodium–sulfur battery stores energy in the form of molten sodium and sulfur. The molten sodium serves as the negative terminal, while the sulfur acts as the positive terminal. The operating temperature of this battery is in the range 300 to 350°C.

The unique feature of this battery is the electrolyte, which is a solid ceramic material called *beta-alumina* ($Na_2O \cdot 11Al_2O_3$). The beta alumina serves as an electrolyte by allowing conduction of sodium ions while also acting to separate the sodium and sulfur. The high operating temperature is necessary to develop an appreciable sodium ion conductivity in the electrolyte as well as to keep the sodium and sulfur molten. The properties of the electrolyte also prevent self-discharge of the battery.

A schematic diagram of the basic cell of the sodium–sulfur battery is shown in Fig. 10.5. During discharge, the sodium and sulfur combine electrochemically to produce sodium polysulfide. During the charging cycle this is reversed. In charging, sodium ions migrate from the sulfur electrode, which contains dissolved sodium sulfide, through the beta-alumina to the liquid sodium, where they gain electrons and form sodium metal. At the anode, sulfide ions become sulfur. The reactions that

Sec. 10.2 Geothermal Power

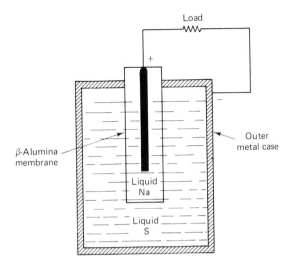

Figure 10.5 Schematic diagram of a sodium–sulfur battery.

occur during charging may be written as

$$2Na^+ + 2e^- \to 2Na \quad \text{(at cathode)}$$
$$S^{2-} \to S + 2e^- \quad \text{(at anode)} \tag{10.3}$$

In present designs, the beta cell is hermetically sealed inside a metal cylinder. One electrode is placed inside the central beta-alumina tube while the metal cylinder serves as the second electrode (see Fig. 10.5). Several design problems remain to be solved prior to commercial use of the beta cell. One of the major problems has been bonding the sodium and sulfur containers to an insulator. A new type of thermocompression metal-to-ceramic seal appears to present a solution. Current cell life is, however, still almost an order of magnitude below that desired.

Summary

The most feasible near-term energy storage system for utility application are pumped hydro systems and compressed air storage for gas turbine use. Thermal energy storage, and possibly one of the advanced battery concepts, may eventually be developed to the point where they are commercially feasible.

10.2 GEOTHERMAL POWER

The interior or core of the earth has a much higher temperature than the surface. The molten, innermost region is thought to be 7000°C. The earth's interior can therefore be viewed as a large thermal reservoir which can be utilized as a source of energy. Many geological formations exist that allow access to this thermal reservoir close to the earth's surface. Generally, useful geothermal formations are along fault lines in the earth's crust.

Geothermal energy resources are divided into three main classifications: hydrothermal, geopressurized water, and hot dry rock. Hydrothermal sources are the only type in commercial use. In a *hydrothermal source*, geothermal energy is available as hot water and/or steam at low to moderate pressures. Steam conditions are typically 400°F and 100 psi. The hot-water or liquid-dominated sites are located primarily on the west coast of the United States, with a few along the east coast. The water temperature ranges between 440 and 160°F. Most commercial utilization of hydrothermal energy is based on the use of hot water. The only exception in the United States is the Geysers field in California, which is a steam-dominated system.

Geopressurized hot water is located primarily along the Texas Gulf Coast. Geopressurized water is hot water, around 350°F, under a pressure of about 1000 atm. The water also contains large quantities of dissolved methane gas. Thus energy may be extracted by virtue of the pressure of the source, the temperature of the source, and from combustion of the associated gas.

Hot dry rock is, as the name implies, an underground rock formation at an elevated temperature of several hundred degrees. Two heat sources contribute to the hot dry rock temperature: igneous-related crustal heat from the interior magma, and conduction of heat from the earth's interior. Hot dry rock reservoirs are scattered throughout the country, especially in the Midwest and West, where there is a lack of underground water to form the hydrothermal reservoir. This resource can be tapped by injecting water down into the rock and transporting the heated water or steam back to the surface for use in power plants, process heat applications, or district heating.

Hydrothermal Power

The classical image of hydrothermal energy is illustrated in Fig. 10.6. Water flows from the surface through a natural crack or fissure in the earth's surface into a porous or crystalline rock formation. As the water migrates through this porous rock, it is heated by the heat flowing radially outward from the convecting magma. If the water is under pressure, it will simply remain as heated water. At lower pressures, the water may be heated to the saturation temperature and become steam. Eventually, the steam or heated water finds another natural fissure and escapes to the surface.

Steam-dominated hydrothermal systems can be used directly in a steam cycle to produce electric power. In the liquid-dominated systems, steam can be produced for steam cycle applications, but more often the hot water is used for space heating or process heat.

Hydrothermal steam systems.
The temperature and pressure of the steam from a geothermal well depend on the rate of steam extraction from the well. A large mass flow of steam will reduce the temperature and pressure. The steam leaves

Sec. 10.2 Geothermal Power

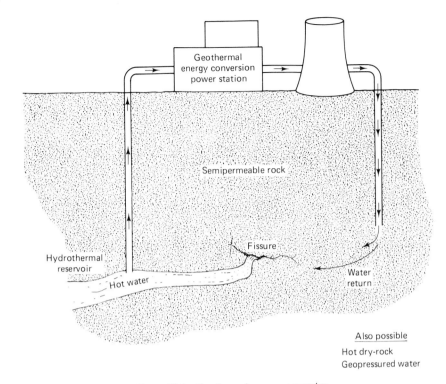

Figure 10.6 Geothermal power conversion.

the wellhead with a mixture of suspended solids and gases such as carbon dioxide, hydrogen sulfide, methane, and ammonia. The solids, together with any suspended liquids, are removed by a centrifugal separator. The purified steam is sent directly to a steam turbine. Because of the low temperature and pressure of the steam (typically 150 to 200°C and 7 to 8 bars), relatively low thermal efficiencies are obtained.

A typical geothermal steam power plant is shown schematically in Fig. 10.7. It will be observed that a direct contact type of condenser follows the turbine rather than the usual shell-and-tube condensers found in nuclear and fossil-fueled plants. Because of the high solids content of the condensate, excessive scale formation is likely in a shell-and-tube condenser and a contact condenser circumvents this problem.

The best practical example of geothermal steam power plants is the Geysers field in California. The highly developed geysers area already generates more than 700 MW of electricity from hot dry steam and is projected to produce up to 2000 MW.

Several steam wells are necessary for each power plant facility. As the production of steam gradually decreases from an old well, a new well must be drilled to replenish the steam supply. The steam supply from a single well can be useful for several years.

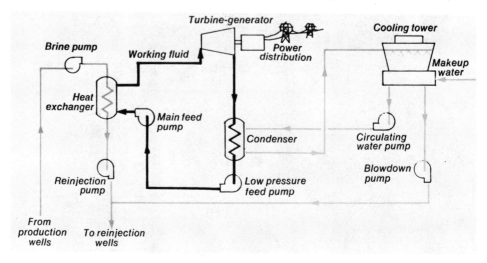

Figure 10.7 Geothermal steam power plant. (Reprinted with permission from *Power*; copyright 1979 McGraw-Hill, Inc., New York.)

Hydrothermal hot-water systems. Liquid-dominated or hot-water hydrothermal systems are more common than the steam-dominated systems. The hot-water systems range in temperatures from 60 to 300°C (140 to 572°F). The higher-temperature water can often be converted to steam at the wellhead by a flashing process. The steam can then be used in a power cycle similar to that used for steam-dominated systems.

If the geothermal water temperature is in the range 100 to 200°C (212 to 392°F), steam production by flashing is not useful. A binary power cycle must be

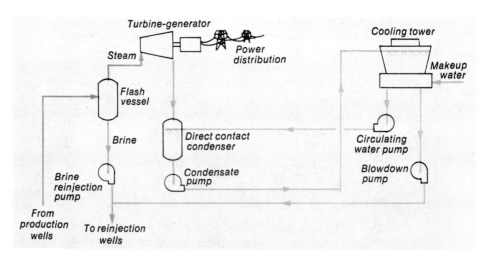

Figure 10.8 Binary power cycle for geothermal hot water. (Reprinted with permission from *Power*; copyright 1979 McGraw-Hill, Inc., New York.)

Sec. 10.2 Geothermal Power

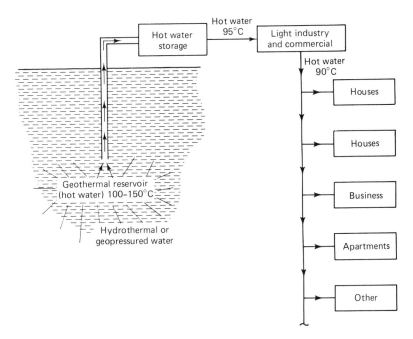

Figure 10.9 Geothermal district heating.

used. Hence the heat from the hot water is transferred to a working fluid with a lower boiling point. The vapor of the working fluid drives the turbine and is exhausted into a condenser, where it is condensed and recycled.

A binary cycle is shown in Fig. 10.8. The working fluids being considered are R22 ($CHClF_2$), R600 (isobutane), R32 (CH_2F_2), R717 (ammonia), and R115 (C_2ClF_5). These provide a range of critical temperatures and pressures and allow adjustment of the cycle to available steam conditions.

In Europe and Iceland, hot water from geothermal sources is utilized in what is called *district heating*. The district heating concept is shown schematically in Fig. 10.9. The hot water is distributed through a piping system from the geothermal field to the point of utilization. The hot water is first used by local industry for process or space heat. It is then distributed by pipes to residential areas for space heat. Some countries also use the hot water for pulp and paper processing, greenhouses, and fish farming.

Hot Dry Rock Geothermal Power

The *hot dry rock* (HDR) geothermal resource is defined as crustal rock that is hotter than 150°C, is at depths less than 10 km, and can be reached with conventional drilling equipment. Based on this definition, the HDR resource base is estimated to

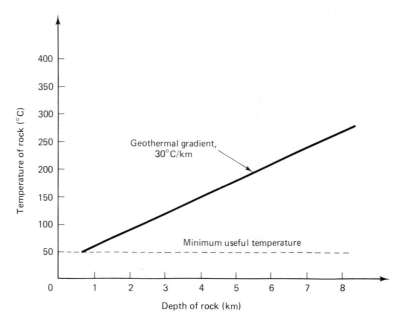

Figure 10.10 Increase in the earth's temperature versus depth.

be 13,000,000 quads.[†] However, there is a real question as to how much of this resource base can be recovered, but even recovery of a small fraction would be significant.

The geothermal gradient (°C/km) is a major factor in the recovery of the geothermal energy. This gradient determines the depth of drilling required to reach a useful temperature. Figure 10.10 is a plot of rock temperature versus drilling depth for a particular site. The maximum drilling depth at this site would depend on the prospective application and detailed economic considerations.

Heat is extracted from HDR formations by introducing an artificial circulation system into the rock formation. Note that the rock is fractured to allow the water to circulate more easily through the geologic formation. Water, introduced into the HDR formation through an injection well, flows through the hot fractured rock and gains heat as it flows toward the extraction pump. The hot water is pumped to the surface, passed through a heat exchanger, and then back down the injection well.

The geothermal heat is transferred to a low-boiling fluid, such as one of the refrigerants, in the heat exchanger. This type of cycle is essentially the same concept as the binary power cycle described by Fig. 10.8. Conversion efficiency ranges only from 8 to 20% because of the low hot-water temperature.

Geothermal power plants using the HDR resource base are not yet in commercial operation. Component development is underway for units with a power output of up to 1 MWe.

[†] A quad is 10^{15} Btu.

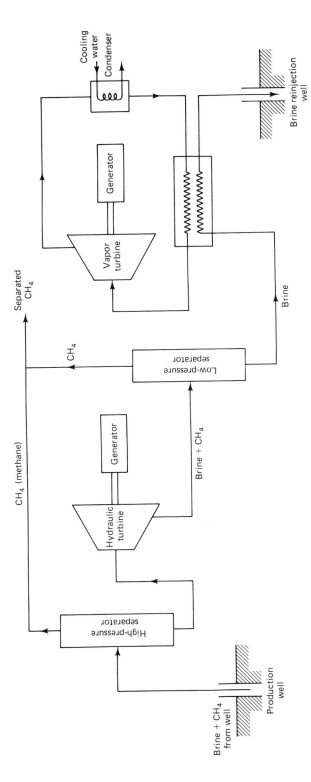

Figure 10.11 Energy from geopressurized water. Three energy sources can be obtained from geopressured water: methane, steam, and hydropower.

Geopressurized Water

A large region of geothermal energy has been located along the Texas Gulf coast. This region consists of hot saline water under a pressure as high as 16,000 psi. The water also contains dissolved salts and a considerable amount of dissolved methane gas.

Plans to develop this resource are based on the concept of extracting the methane from the hot-brine mixture, using the pressure to power a hydraulic turbine, and finally using the heat from the water to power a binary cycle. Figure 10.11 illustrates the scheme for topping the triple energy potential of geopressurized water.

A high-pressure vapor separator is used for initial separation of the bulk of the dissolved methane from the hot-brine solution. This methane is filtered and pumped prior to pipeline use. The liquid, still at a very high pressure, flows through a hydraulic turbine–generator. The low-pressure liquid leaves the turbine, where it is introduced into a low-pressure methane separator, where the remainder of the methane is removed from the liquid. Finally, the sensible heat from the hot water is extracted in a binary power cycle.

Although geopressurized water offers a triple energy benefit, it is not without problems. The system design is complicated by the high content of minerals in the water, which deposit in the pipes and restrict flow. The saline water can contain three times the salt concentration of ocean water. The high-velocity, high-pressure system requirements, coupled with this salt concentration, make material selection difficult. Even stainless steels are attached under these conditions. Erosion, corrosion, and pipe cracking are common.

10.3 SOLAR THERMAL POWER

Solar Energy Fundamentals

The sun and all stars are large fusion reactors. The interiors of the sun and other stars are basically plasmas which have been stripped of their orbital electrons. The hydrogen nuclei have a high velocity (kinetic energy), as the temperature of the sun's interior is of the order of $2 \times 10^6 \,°K$. At these high temperatures fusion of the energetic hydrogen nuclei into helium nuclei will occur.

The fusion of hydrogen nuclei under stellar conditions can take place under two different processes. In the *proton–proton cycle*, direct collisions of protons lead to formation of heavier nuclei, where collisions in turn produce helium nuclei. In the *carbon cycle*, there is a series of steps in which carbon nuclei absorb a succession of protons until they break down into α particles (the helium nucleus) and carbon nuclei. At the temperature of the sun, the proton–proton cycle is most likely to occur and hence provides the majority of the sun's energy.

Sec. 10.3 Solar Thermal Power

The first step in the proton–proton cycle is the combination of two H nuclei (protons) with the formation of a deuteron and accompanied by the emission of a positron:

$$^1H + {}^1H \rightarrow {}^2H + e^+ \tag{10.4}$$

Some deuterons then join with protons to form ^3He nuclei and gamma rays.

$$^1H + {}^2H \rightarrow {}^3He + {}^0_0\gamma \tag{10.5}$$

Finally, two ^3He nuclei react to form a ^4He nucleus and two protons:

$$^3He + {}^3He \rightarrow {}^4He + {}^1H + {}^1H \tag{10.6}$$

Since the mass of the products is less than the mass of the reactants, Einstein's mass–energy conservation principle tells us that a substantial amount of energy is produced by the reactions. The energy is transferred to the outer region of the sun and then radiated into space.

The sun converts 6×10^{11} kg of mass into energy each second. This is equivalent to 1×10^{23} kW. The diameter of the sun is 1.39×10^6 km (6×10^5 mi) and it is located 1.5×10^8 km from the earth (93×10^6 mi). The surface of the sun is at an effective temperature of $5762°K$. This is the temperature at which the sun transmits the electromagnetic energy radiated to the earth.

The foregoing characteristics of the sun and its spatial relationship to the earth produce a nearly uniform intensity of solar radiation outside the earth's surface cover. The solar radiation is referred to as the *solar constant*, S_0. The solar constant is defined as the quantity of energy per unit time passing through a unit area normal to the sun–earth line and outside the effects of the earth's atmosphere. It is a measure of the total energy content of the solar radiation.

Recent measurements have established the average annual value of the solar constant at 1376 W/m² (428 Btu/ft² · h). The wavelength distribution of this radiation is shown in Fig. 10.12. These data were acquired from a satellite located outside the earth's atmosphere. Only a fraction of this energy can be utilized for solar power stations since substations could cover only a very small fraction of the earth and since some of the solar energy never reaches the earth's surface as the earth's atmosphere scatters and selectively absorbs a portion of the incoming electromagnetic radiation.

As seen in Fig. 10.12, most of the solar energy is transmitted in a relatively narrow wavelength (λ) band: $0.2 \ \mu m < \lambda < 2 \ \mu m$ (or in a frequency range of 1.5×10^{14} Hz $< f < 1.5 \times 10^{15}$ Hz). This band is represented by the ultraviolet ($\lambda < 0.38 \ \mu m$), visible ($0.38 \ \mu m < \lambda < 0.78 \ \mu m$), and the infrared ($\lambda > 0.78 \ \mu m$) portions of the spectrum. Only 7% of the energy is contained in the ultraviolet spectrum, with the remaining 93% being 47% in the visible and 46% in the infrared. The total area under the curve shown in Fig. 10.12 represents the rate at which energy is transmitted to a unit area of the earth which is perpendicular to the sun's radiation. This quantity is the *solar constant*, S_0.

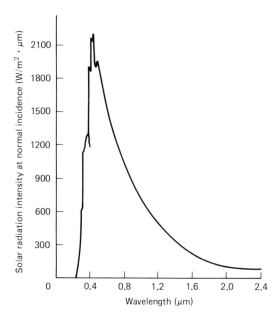

Figure 10.12 Spectral distribution of solar radiation incident on a surface normal to the sun's ray's outside the earth's atmosphere.

The solar radiation actually received at the surface of the earth normal to the sun's rays is less than the value of the solar constant value because of (1) atmospheric scattering by air molecules, water vapor, and dust, and (2) atmospheric absorption by CO_2, H_2O, and ozone. Figure 10.13 illustrates the effect of this absorption on certain wavelengths. This absorption increases as the wavelength increases, such that for wavelengths greater than 2.5 μm, very little energy reaches the surface of the earth. For design purposes, it is assumed that significant quantities of solar energy are received on the surface of the earth only at wavelengths between 0.29 and 2.5 μm.

The quantity of solar energy received at any given point on the surface of the earth depends on its geographic location; the time of the year, the time of day, and local weather conditions. Variations during the year in the distance between the sun and the earth lead to annual variations in the solar constant of $\pm 3.5\%$ (about ± 50 W/m^2). The solar constant is highest during the December–February time period and lowest during the May–August time period.

The solar energy variation due to geographic location is due to the angle that the sun makes with a horizontal plane on the surface of the earth. The solar azimuth and altitude angles are tabulated as functions of latitute, declination, and hour angle by the U.S. Hydrographic Office. Most solar radiation measurements are taken on a horizontal surface. Thus solar energy collectors which are oriented toward the sun must correct the horizontal surface data. Solar energy data are collected by the U.S. Weather Bureau for over 100 geographic locations throughout the United States.

In addition to the foregoing variations in solar radiation received at a particular location, there is also the obvious variations due to the time of day and local

Sec. 10.3 Solar Thermal Power

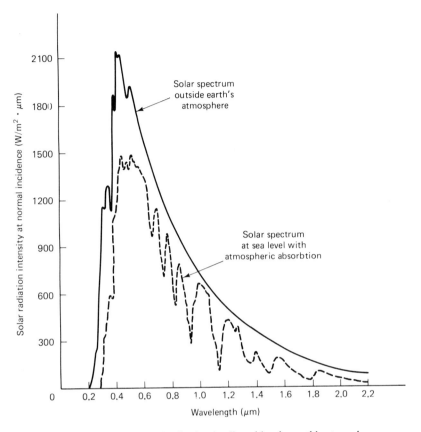

Figure 10.13 Spectral distribution is affected by the earth's atmosphere.

weather conditions. The cosinusoidal curve in Fig. 10.14(a) represents the daily variation in solar radiation on a horizontal surface during a clear day. Figure 10.14(b) illustrates the great reduction in solar radiation due to a cloud cover. Since only a small quantity of solar radiation is transmitted through a cloud cover, the designer must provide sufficient energy storage capability to account for extended periods of cloud cover.

Calculating the Incident Radiation

The angle between the sun's rays and the earth's equatorial plane is called the declination, θ_D. The maximum declination is equal to the tilt of the earth's axis with respect to the axis of the sun, or about 23.5°. The seasonal variation in solar radiation can be accounted for by the declination. The hourly rotation of the earth is defined by an azimuthal angle, ϕ_a. The latitude is determined by the polar angle, θ_p (see Fig. 10.15).

470   Energy Alternatives   Chap. 10

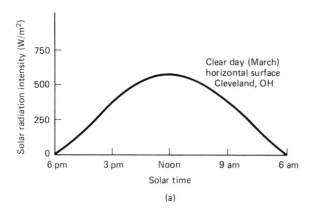

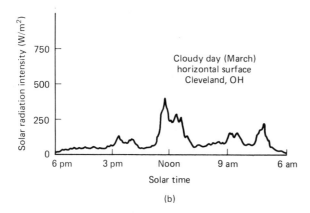

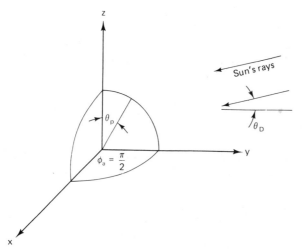

Figure 10.14 Solar radiation on a horizontal surface on (a) a clear day and (b) a cloudy day.

Figure 10.15 Earth–sun coordinate system.

Sec. 10.3 Solar Thermal Power

The incident of solar radiation S_0, is assumed directed toward the earth in the $y - z$ plane of Fig. 10.15 at declination angle, θ_D. Solar noon occurs at $\phi_a = \pi/2$. The geographic and seasonal variations in the sun's intensity can be evaluated from the expression

$$S^1 = S_0 \cos\theta_D \sin\theta_p \sin\phi_a + S_0 \sin\theta_D \cos\theta_p \qquad (10.7)$$

where $S^1 \equiv$ extraterrestrial normal incident solar radiation, W/m^2
$S_0 \equiv$ solar constant, 1376 W/m^2

This equation is valid for positive values of S^1 only since negative values would indicate nighttime conditions. As a very rough first approximation, about half of this incident energy is lost due to the atmospheric factors previously described.

Equation (10.7) can be used to calculate the incident solar radiation at any geographic location for any season. In the United States, the northern hemisphere, a typical latitude would be 40° and $\theta_p = 90 - 40 = 50°$. As an example, consider the two seasonal extremes, the summer solstice ($D_\phi = +23.5°$) and the winter solstice, ($D_\phi = -23.5°$). Using the values above and assuming that approximately 50% of the incident solar radiation reaches the earth's surface, Eq. (10.7) will give the results shown in Fig. 10.16. Note that this analysis is for a horizontal surface.

Another important parameter is the total daily energy input, which is represented by the area under the curve of Fig. 10.16. The average solar incident radiation S^1_{ave} is

$$S^1_{ave} = \frac{S_0}{2\pi} \int_{sunrise}^{sunset} (\cos\theta_D \sin\theta_p \sin\phi_a + \sin\theta_D \cos\theta_p)\, d\phi_a \qquad (10.8)$$

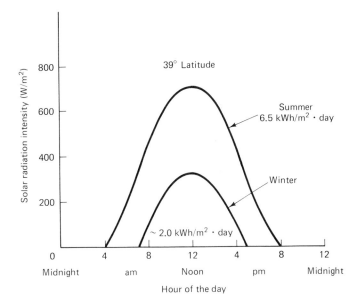

Figure 10.16 Seasonal variation of solar radiation on a horizontal surface.

which integrates to

$$S_{ave}^1 = \frac{S_0}{\pi}\left(\cos\theta_D \sin\theta_p \cos\phi_{sr} + \sin\theta_D \cos\theta_p \phi_{ss}\right) \quad (10.9)$$

where ϕ_{sr} = value of ϕ_a at sunrise
ϕ_{ss} = value of ϕ_a at sunset

For the example case considered, the total solar energy incident on a horizontal surface is only 1.85 kWh/m² per day at the winter solstice. The corresponding value for the summer solstice is 6.35 kWh/m² per day.

These results agree reasonably well with actual U.S. Weather Bureau data. By use of either Eq. (10.7) or (10.9) or actual data, the power plant designer can estimate the solar radiation for any particular location.

Increasing Collection Efficiency

Although the data and most analyses are for horizontal surfaces, actual systems utilize solar collectors that have several degrees of freedom to track the sun in order to maintain a perpendicular surface with respect to the sun's rays. To do this, the collectors' orientation is continually changed throughout the day to face the sun. In the absence of cloud cover, the incident solar radiation remains reasonably constant throughout the day. More important, the variation between summer and winter incident radiation is drastically reduced. The maximum energy collected per day in the summer (at 40° latitude) increases to 10 kWh/m² per day, while the winter energy collection becomes about 6 kWh/m² per day. The difference is due almost entirely to the difference in daylight hours.

Solar Collectors

The production of electric energy through the use of a vapor power cycle requires the concentration of solar energy to achieve high temperatures and reasonable efficiency. The degree of concentration required can be estimated from Table 10.4.

TABLE 10.4 SOLAR CONCENTRATION REQUIREMENTS FOR VARIOUS TEMPERATURES

Concentration factor	Maximum working fluid temperature [°C (°F)]
100	260 (500)
1,000	700 (1292)
10,000	1370 (2500)

Sec. 10.3 Solar Thermal Power

The technology of solar collectors limits the concentration factor, and hence the working fluid temperature, for different collector designs as follows:

Flat-plate collector:	40 to 120°C (104 to 248°F)
Parabolic concentration:	150 to 800°C (302 to 1472°F)
Heliostats:	250 to 1500°C (482 to 2732°F)

Thus solar thermal power plants will require the use of either parabolic concentrating collectors or heliostats.

Parabolic concentrating collectors. Parabolic concentrating collectors use parabolic-shaped mirrors to reflect the solar radiation and concentrate its energy at a single focal area. These collectors track the sun's motion for maximum efficiency. Two types of collectors have evolved, a trough-type and a dish-type collector.

Parabolic Trough Collectors. The parabolic trough collector design is a half-cylinder which rotates on an axis. The device is designed to rotate east to west, such that the parabolic mirror follows the sun and concentrates the sun's rays on a pipe that is parallel to the horizontal axis. This central pipe absorbs the solar radiation and heats the working fluid to the design temperature. Approximately half of the incident solar energy can be converted to heat in the working fluid (collector efficiency $\simeq 50\%$).

Although the collector temperature is high enough to produce steam directly, direct production of steam would require high-pressure piping throughout all the collectors, associated piping system, and heavy movable joints. To avoid this expensive situation, an organic Rankine cycle engine and a heat-transfer loop would be utilized.

Figure 10.17 is a schematic diagram of a solar thermal plant using an organic working fluid in the power cycle. The solar collector loop uses a special working fluid that does not reach the saturation temperature at collector output. Thus the loop can be designed for a relatively low pressure. The second or power loop uses an organic fluid that has a lower saturation temperature. The organic power loop functions in the same manner as a conventional steam power loop. Analytical techniques developed in previous chapters are applicable. The appropriate physical properties would have to be used for each working fluid.

Parabolic Dish Collectors. The parabolic dish collector has the configuration of a partial hemispherical shell. The incident solar radiation is collected and reflected into a single absorber containing a working fluid. This type of collector must track and follow the sun on a multiaxis system. Dish collectors can produce high temperatures in the working fluid.

In some designs, each collector serves as its own heat engine and electric generator. Instead of interconnecting the collector cooling fluids by piping, the

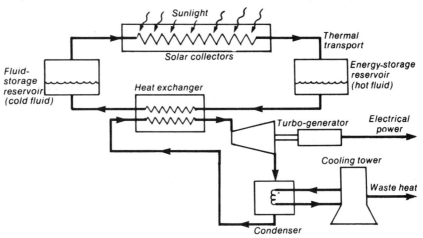

Figure 10.17 Solar thermal power plant for large-scale electric power production. (Reprinted with permission from *Power*; copyright 1977 McGraw-Hill, Inc., New York.)

generator outputs are connected by cable. The interest in parabolic dish collector is due to the potential for higher system efficiency.

Heliostats. Heliostats are highly polished mirror-like devices which are mounted on a multiaxis system. Each heliostate tracks the sun and focuses it onto a central boiler tank mounted on the top of a high tower. An array of heliostats surrounds the central tower.

The working fluid in the boiler tank can be heated to 1400°C (2550°F). Molten salts and liquid metals are being considered for the working fluid. The working fluid would be circulated through a heat exchanger which would boil water to power in a conventional Rankine cycle.

Sec. 10.3   Solar Thermal Power

It is estimated that a 500-MW electric power station would require 100,000 to 125,000 individual heliostats. A tower height of 1500 ft would be typical.

Solar (Photovoltaic) Cells

In a photovoltaic cell, a *p-n* junction in a semiconductor is exposed to electromagnetic radiation. A portion of the incident electromagnetic energy is converted directly into electrical energy. Since there is no intermediate conversion to heat energy, we bypass the Carnot limitation just as we did in fuel cells.

Both short-wavelength radiation produced by radioactive elements and wavelengths present in visible light may be used. Cells for power production are, of course, designed for use with sunlight. Solar cells became a reality in the 1950s with the development of practical selenium cells for generation of power for spacecraft missions. Power requirements have generally been below 1 kW, but units have been designed for up to 50 kW.

The efficiency of present-day solar cells are a few percent at best. Further, while the high costs of such cells are acceptable for space power use, they cannot be borne by central-station power production. Use of solar cells for large-scale power production will require substantial increases in cell efficiency coupled with fabrication costs being reduced by more than an order of magnitude.

Energy Storage or Backup Power Requirements

Solar energy is a very dilute, intermittent form of energy. We observed that it must be concentrated to be useful for electric power production. Due to the intermittent nature of solar radiation, any solar system must also provide for energy storage (always required for nighttime system operation unless solar power is used only to help meet daytime peaks) and possibly for backup power production during prolonged periods of cloudy weather.

The amount of energy that must be stored for cloudy weather use depends on the specific location of the power station. A statistical analyses of the U.S. Weather Bureau daily records can be used to obtain the probability of a certain number of cloud-cover days in succession. In many sections of the country, the possibility of a large number of cloudy days in succession during the winter makes energy storage impractical and backup power generation would be needed. The power plant engineer must include the cost of the energy, storage, and/or backup power, to assure an adequate supply of electrical power in estimating solar power costs. In those sections of the country where backup power production is required, central station solar power is likely to remain uncompetitive.

Cost of Solar Thermal Power Stations

Since solar thermal power is presently in the experimental stage, cost data are not reliable. Most experts agree that a solar-thermal electric power station is highly

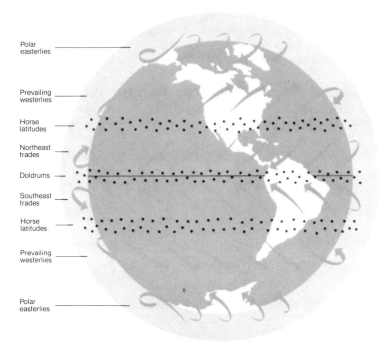

Figure 10.18 Origin of the winds. (From the *EPRI Journal*, March 1980.)

capital-cost intensive. Advocates of solar power argue that the cost can be reduced as the plants are commercialized.

10.4 ELECTRIC POWER FROM WIND ENERGY

Origin of Wind Power

Wind energy is derived from solar radiation absorbed by the earth's atmosphere. Winds result from differences in heating at various locations on the earth's surface. For example, air that is located over a land surface is heated more rapidly than air located over oceans or large bodies of water. This heated air rises and the cooler air over the waters surface flows in to take the place of the heated air. The process is reversed at night when the air over land cools faster than the air over the water.

By a similar process, local wind areas are created in hilly and mountainous regions. The air above mountains or slopes heats and cools more rapidly than air in the plains or plateaus.

In addition to local wind patterns, there is an overall or global wind pattern. The "prevailing westerlies"[†] in the temperate latitudes are attributed to (1) the

---
[†] Winds are named for the direction in which they originate.

Sec. 10.4 Electric Power from Wind Energy

differential heating between the earth's surface in the temperate region and the poles, and (2) the rotation of the earth itself. Figure 10.18 illustrates the origin of the major wind patterns in the world. The cold air near the surface moves toward the equator in a westerly direction, while the warm air moving in the higher atmosphere tends to move toward the poles in an easterly direction. The result is a large counterclockwise circulation of the air around low-pressure areas in the northern hemisphere and clockwise circulation in the southern hemisphere.

Due to the factors described above, the wind energy pattern in the United States varies sharply in different regions. Figure 10.19 shows this variation indicating how the prevailing westerlies are influenced by major geographical features (e.g., mountain chains). The effect of local site features must be superimposed on this general variation.

At any specific site being considered for a wind turbine, wind speed data must be collected and analyzed. Since wind speed varies with height above the ground (up to about 300 m), data should be collected at the height anticipated for the wind turbine.

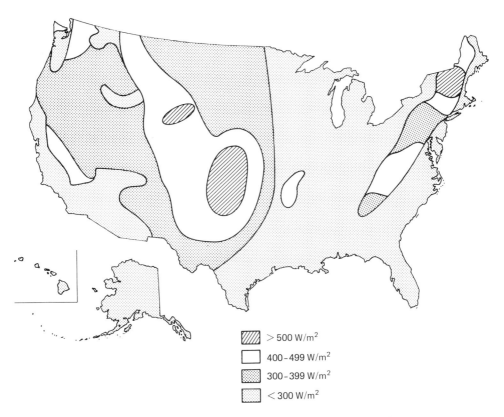

Figure 10.19 Wind power density. (From the *EPRI Journal*, March 1980.)

Calculating the Energy Available from the Wind

The total power available from a wind is

$$P = \frac{KE}{mass} \times \dot{m} = \tfrac{1}{2} u_w^2 \rho u_w A = \tfrac{1}{2} \rho A u_w^3 \tag{10.10}$$

where KE = kinetic energy
$\dot{m}$ = mass flow, kg/s
ρ = density of the air, kg/m^3
A = area normal to the wind, m^2
u_w = wind velocity, m/s
P = power, W

The density of air varies with altitude and atmospheric conditions. At STP conditions, 1000 mbar (29.53 in. Hg) and a temperature of 293°K (68°F), the density is 1.201 kg/m^3. Although the density does decrease slightly with increasing humidity, the decrease is small and is usually neglected in calculations. The density of dry air can be calculated for any temperature and barometric pressure by means of the ideal gas law. By the use of air properties, this simplifies to

$$\rho = \text{density } (\text{kg/m}^3) = \frac{0.349[\text{atmospheric pressure (mbar)}]}{\text{temperature (°K)}} \tag{10.11}$$

Since some proposed sites are at altitudes of 1000 ft or more, both pressure and temperature variations occur and the density correction is necessary. Changes of 10 to 15% in air density are usual at any given site.

In 1927, A. Betz of Göttingen showed that the maximum fraction of the wind energy that could be converted to useful energy by an ideal aeromotor (such as a wind turbine) was 16/27 or 59.3%. This is analogous to the Carnot cycle efficiency for steam power cycles. Thus the maximum, theoretically obtainable power output of a wind device is

$$P = 0.2965 \rho A u_w^3 \tag{10.12}$$

Table 10.5 shows the maximum theoretical power that could be obtained from a wind machine of 50 ft in diameter for various wind speeds. Actual power output

TABLE 10.5 MAXIMUM WIND POWER FROM A 25-METER-DIAMETER WIND TURBINE

Wind speed		Power output
m/s	mi/h	(kW)
5	11.2	21.8
10	22.4	174.3
15	33.6	588.1
20	44.7	1394.0

of real machines of the same size would be less than shown in Table 10.5, due to frictional and electrical losses.

Wind Turbine Theory

Consider a control volume around the wind turbine as shown in Fig. 10.20. The variables shown in this figure are:

$$u_w \equiv \text{wind speed, m/s}$$
$$a_I \equiv \text{interference factor}$$
$$R_{tb} \equiv \text{turbine blade radius, m}$$

The axial thrust on the turbine, F_{th}, derived from a momentum balance is

$$F_{th} = 2\pi R_{tb}^2 \rho u_w^2 a_I (1 - a_I) \tag{10.13}$$

From a momentum and energy balance on the control volume shown in Fig. 10.20, the turbine power would be

$$P = 2\pi R_{tb}^2 u_w^3 a_I (1 - a_I)^2 \tag{10.14}$$

The power coefficient for a wind turbine is defined as the ratio of electric power output to the power available in the wind. The power coefficient, P_c, is then obtained by division of Eq. (10.14) by (10.10). Hence

$$P_c = \frac{2\pi R_{tb}^2 a_I (1 - a_I)^2 u_w^3}{\frac{1}{2} \rho A u_w^3} \tag{10.15}$$

By replacing A by πR_{tb}^2, we have

$$P_c = 4 a_I (1 - a_I)^2 \tag{10.16}$$

Note that P_c will be a maximum at $a_I = \frac{1}{3}$. This maximum value of ($P_c = 0.593$) leads to a maximum value for P of $0.2965\, \rho A u_w^3$, thus confirming Eq. (10.12).

Blade element theory is used to provide a relationship between the blade properties, the power produced by the turbine, the axial thrust, and the interference factor, a_I. No closed-form solutions have been developed that relate all these factors and the usual procedure is to solve the theoretical equations by numerical

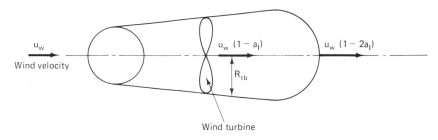

Figure 10.20 Wind turbine schematic analysis.

(a)

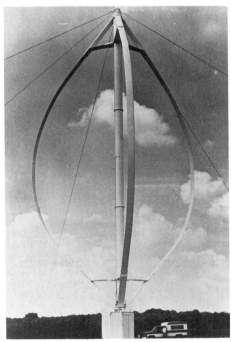

(b)

Figure 10.21 Wind turbine devices: (a) Darrieus and (b) NASA MOD wind machine. (From the *EPRI Journal*, Dec. 1981.)

Sec. 10.4 Electric Power from Wind Energy

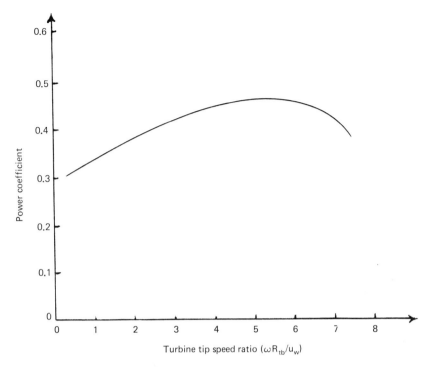

Figure 10.22 Wind turbine power coefficient curve.

analysis techniques. The result of the computer solutions is a series of performance curves for each particular type of wind turbine. For electric power production, the main interest is in a high-speed, two-bladed, horizontal-axis wind turbine such as that shown in the right-hand section of Fig. 10.21(a). The performance curve for this type of machine is illustrated in Fig. 10.22.

Data such as those shown in Fig. 10.22 can be used to solve for the interference factor, a_I. This factor can then be used to predict turbine power from Eq. (10.14). The actual turbine rated power output may be up to 10% less than predicted due to frictional (mechanical) losses and electrical losses. The output that can be obtained from a well-designed wind turbine is approximately 70% of the Betz limit of $0.2965\rho A u_w^3$ or

$$P_{\text{act}} \cong 0.21\rho A u_w^3 \tag{10.17}$$

Wind Speed and Power

One of the fundamental design aspects of a wind turbine is its "rated" or "full-load" wind speed. This is the lowest wind speed at which full-power output is achieved. At higher wind speeds, the output is still limited by the control mechanism to that attainable at this speed. It is presently not practical from an economic basis to design for rated wind speeds greater than 35 mi/h.

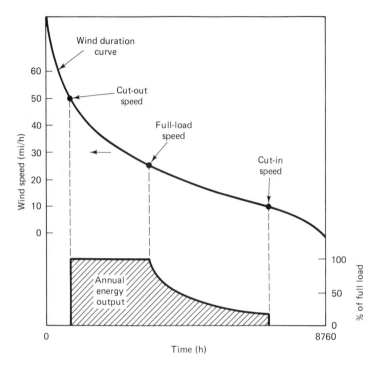

Figure 10.23 Annual power output from a wind turbine.

Information on wind speed is collected at a prospective site and cast into the form of a wind-duration curve (Fig. 10.23). The ordinate of this curve shows the range of wind speed, and the abscissa shows the number of hours in the year for which the wind speed equals or exceeds this value. The average annual energy output is given by the shaded area under the curve.

From Fig. 10.23 we see that the "cut-in" point is the minimum wind speed at which the turbine begins to produce electric power. For large wind turbines, this speed is approximately 10 mi/h. At the full-load speed point, the turbine reaches rated capacity. The turbine then maintains this power output up to the "cut-out" wind speed. At "cut out" the turbine is shut down to avoid damage in high winds (shown here as 60 mi/h).

The capacity factor, C_F, for a wind turbine can be obtained by dividing the annual energy output by the theoretical energy output. In terms of the quantities shown in Fig. 10.23, we would write

$$C_F = \frac{\text{shaded area}}{8760 \times 100} \tag{10.18}$$

A capacity factor of 50% would be considered very good for a wind turbine system.

Sec. 10.4 Electric Power from Wind Energy

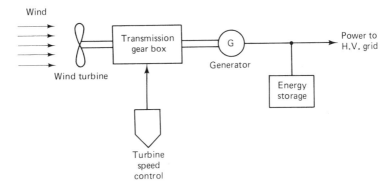

Figure 10.24 Components of a wind power system.

Wind Turbine Systems

The basic components of a wind turbine electric power system are: turbine/generator, tower, energy storage system, electrical conditioning equipment and transmission equipment. The simplified schematic diagram of Fig. 10.24 illustrates this system arrangement. The wind turbines being built and developed in the United States are shown in Fig. 10.25. Note that the largest unit is limited to a 4-MW power output. Hence a large number of such turbines would be required to supply an appreciable fraction of power needs of a given area.

The operation of a wind turbine is basically simple. At "cut-in" speed, the turbine begins to produce power. At low wind speeds, the power generated is a function of the wind speed up to the rated wind velocity where the power output reaches rated power (see Fig. 10.23, the right ordinate). As the wind speed increases, the rotor blade pitch is charged to maintain a nearly constant rpm. When the wind speed exceeds the cutout point, the blades are feathered with the pitch control to shut the turbine down and prevent damage. Pitch control can also be used under emergency conditions to produce a reverse thrust on the rotor. The negative torque developed can shut the system down in as little as 15 s.

In the horizontal-axis machines, a two-bladed turbine drives a synchronous alternator. The blades turn a rotor that is connected to a gearbox. The gearbox increases the rotor speed (from about 40 rpm) to 1800 rpm, which provides 60-Hz three-phase power.

The hub, shown in Fig. 10.26, houses the mechanical components such as gears, bearings, alternator, blade holders, and pitch control mechanisms. The transmission train assembly transmits the torque from the hub to the alternator through the gearbox. From the gearbox, a high-speed shaft transmits the low-torque, high-speed power to an alternator through a belt and pully drive.

The yaw control assembly is a large gear bearing assembly that is capable of rotating or yawing the entire machine on top of the tower. This assembly is used to ensure that the rotor faces directly into the wind.

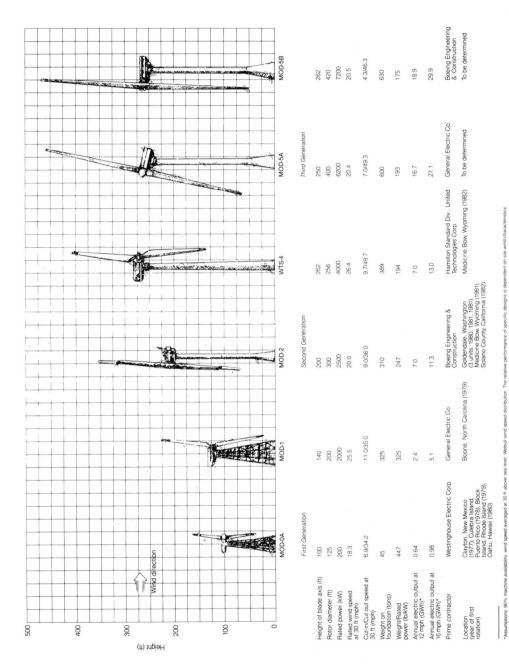

Figure 10.25 Horizontal wind turbine machines under development. (From the *EPRI Journal*, Dec. 1981.)

Sec. 10.4 Electric Power from Wind Energy   485

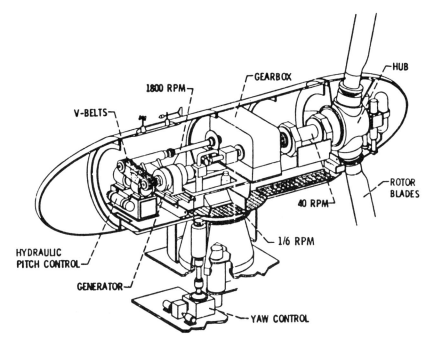

Figure 10.26 100-kW experimental wind turbine drive train assembly. (Courtesy of the National Aeronautics and Space Administration, Lewis Research Center, Cleveland, Ohio.)

Energy Storage and Backup Power

Since the wind is an intermittent source of energy, some provision must be made for energy storage or backup power. As with other intermittent sources, the system designer must weigh storage needs against the increased cost of the system. A wind speed–frequency curve would be needed to determine the probability that a storage system of a given size will be adequate to provide an uninterrupted source of power.

An alternative way of viewing wind power is as a means for saving scarce fuel when wind is available. A primary system, using fossil, nuclear, and hydro, provides the area's power under most circumstances. However, when wind is available, load on the conventional power plants is reduced. Whether the cost of building and maintaining a duplicate power system can be justified by fuel savings has yet to be demonstrated.

Ecological Problems

Wind power is not without its ecological problems. Wind turbines are noisy and many people would consider the large number of turbines required for appreciable power generation to be unsightly. In addition, large numbers of wind turbines could affect bird populations very adversely. However, there may be areas of high wind where the ecological impact is minimal and where significant wind power production is a practical alternative.

10.5 ENERGY SOURCES OF THE FUTURE

It is clear that the establishment and maintenance of a high standard of living for the world's population will require increasing power generation capacity. As power generation facilities become more expensive due to inflation, higher site costs, and greater ecological controls, increased use of energy storage will be needed to reduce the peak generating capacity requirements. Decreased availability of fossil fuels can be expected to lead to increased use of nuclear power and the alternative energy sources discussed in this chapter.

Perhaps the most hopeful avenue to future power abundance is that of nuclear fusion. Although fusion power plants have not been considered in this text since no human-made fusion device has yet produced as much power as it consumed, the technology of this development is well advanced. Devices that can magnetically contain a plasma of deuterium and tritium under conditions which are beginning to approach those needed for a practical device have been built and operated. It is almost certain that large, practical fusion power plants will be operable in the twenty-first century. Since a large amount of deuterium is available in the oceans, and since tritium can be readily produced from lithium in a blanket surrounding the fusion device, an almost unlimited supply of energy will be available for the world's needs once this technology has been developed.

SYMBOLS

a_I	interference factor (turbine property)
A	area, m^2
C_F	turbine capacity factor
D_ϕ	declination angle, degrees
F_{th}	axial thrust on turbine, N
KE	kinetic energy, J
$\dot{m}$	mass flow, kg/s
P	power, W
P_{act}	actual power output, W
P_c	power coefficient (power output/power available from wind)
P_o	net rated power output, W
R_{tb}	turbine blade radius, m

S_0	solar constant, 1376 W/m²
S^1	extraterrestrial normal incident solar radiation, W/m²
S^1_{ave}	average incident solar radiation, W/m²
u	velocity, m/s
u_w	wind velocity, m/s

Greek Symbols

$^0_0\gamma$	gamma-ray photon, MeV
θ_D	declination angle, degrees
θ_p	polar angle, degrees
ρ	density, kg/m³
ϕ_a	azimuthal angle, degrees
ϕ_{ss}	azimuth angle at sunset, degrees
ϕ_{sr}	azimuth angle at sunrise, degrees
ω	angular velocity, rad/s

PROBLEMS

10.1. An energy storage system using underground storage of compressed air gets its power from a coal-fired plant, which has an overall efficiency of 35%. During the energy storage portion of the cycle, air is adiabatically compressed to 80 bar by a 96% efficient compressor driven by a 98% efficient motor. During the power production phase of the cycle, the air is throttled to 35 bar and is fed to a combustor, where it is heated to 600°C. It is then expanded through a 96% efficient turbine which runs a 98% efficient alternator. What is the overall thermodynamic efficiency of the cycle?

10.2. The manager of an industrial power plant, which now has a peak load of 140 MW, finds that due to factory expansion the base load will be increased from 80 MW to 100 MW. It is estimated that the peak load will go to 160 MW. The boiler plant has a capacity of 150 MW, although the turbine and generator can produce 160 MW. Due to the nature of the process being supplied, the manager expects that the peak will consist of a 12-MW spike, which lasts for 10 min each hour, on top of a slowly varying load. It is proposed that, instead of buying outside power for the peak, high-pressure steam be stored to supply the extra demand during peak periods. If the boiler generates 100-bar steam at 200°C superheat, how large a vessel would be required for the steam storage? Do you think the idea is feasible?

10.3. If the voltage output of a lead storage cell is 80% of the ideal voltage output under standard conditions, how many cells in series would be required to produce a 200-V output?

10.4. A geothermal source is to provide water at 185°C to a heat exchanger that boils an organic liquid.
 (a) If the minimum allowable ΔT across the heat exchanger is 12°C, what is the maximum thermodynamic efficiency that could be expected of such a cycle?

Condenser cooling water is available at 30°C and there is a 10°C ΔT across the condenser.

(b) How many liters of geothermal source water are required per megawatt-hour generated at maximum efficiency?

(c) If the turbine generator operates at 80% of the theoretical efficiency and the water from the geothermal reservoir must have its pressure increased by 10 bar to return it to the ground, what is the actual net efficiency of the plant?

10.5. It is believed that a parabolic reflector system can be set up so that the reflectors cover 45% of the land area on which they are placed. Determine the total land area, at about a 40° latitude, required for reflectors sufficient to provide the energy to a system whose 24-h average load is 100 MWe during the winter. You may assume that the power plant operated by the system has 25% efficiency, that the nighttime load is 40% of the total load, and that 10% of the energy that must be stored is lost.

10.6. A utility in a high-fuel-cost area is considering installing wind turbines to reduce its fuel consumption. It presently has a mean load of 1500 MW and the weighted mean thermal efficiency of its plants is 35%. Winds blow at the proposed wind turbine site 48% of the time and at a mean velocity of 37 km/h. Estimate the number of wind turbines with 25-m-diameter blades which would be required to reduce the utility's fuel bill by 10%. Assume that the wind turbine and associated alternator are 90% efficient.

BIBLIOGRAPHY

Solar Power

BENNETT, I., "Monthly Maps of Mean Daily Insolation for the United States," *Solar Energy*, Vol. 9, No. 3, 145–158 (1963).

DUFFY, J., and W. BECKMAN, *Solar Energy Thermal Processes*. New York: Wiley-Interscience, 1974.

"Initial Solar Irradiance Determinations from Numbers 7," *Science*, Vol. 208, No. 18 (Apr. 1980).

LIU, B., and R. JORDAN, "The Long Term Average Performance of Solar Energy Collectors," *Solar Energy*, Vol. 4, No. 3 (1960).

National Aeronautics and Space Administration, *Solar Electromagnetic Radiation*, Report NASA SP-8005, May 1971.

THEKAEKARA, M., and A. DRUMMOND, "Standard Values for the Solar Constant and Its Spectral Components," *Nature (London), Physical Science*, Vol. 229, 6 (1971).

WHILLIER, A., "Solar Radiation Graphs," *Solar Energy*, Vol. 9, 165 (1965).

Geothermal and Wind Energy

CUMMINGS, R., ET AL., "Mining Earth's Heat: Hot Dry Rock Geothermal Energy," *Technology Review*, Vol. 81, No. 4 (Feb. 1979).

FRAAS, A., *Engineering Evaluation of Energy Systems*. New York: McGraw-Hill, 1982.
KRENZ, J., *Energy: Conversion and Utilization*. Boston: Allyn and Bacon, 1976.
SHEAHAN, R., *Alternative Energy Sources*. Aspen Systems Corp., 1981.

Energy Storage

American Association for the Advancement of Science, "Energy Storage I, II," *Science*, Vol. 184, 385 (1974).
ANGRIEST, S., *Direct Energy Conversion*. Boston: Allyn and Bacon, 1971.
BUNDY, F., ET AL., *The Status of Thermal Energy Storage*, Report 76 CRD 041. Schenectady, N.Y.: General Electric Company, Power Systems Laboratory, April 1976.
CULP, A., *Principles, of Energy Conversion*. New York: McGraw-Hill, 1979.
GOLBERSUCH, D., ET AL., *Thermal Energy Storage for Utility Applications*, Report 75 CRD256. Schenectady, N.Y.: General Electric Co., Dec. 1975.
POST, R., and S. POST, "Flywheels," *Scientific American*, Vol. 229, No. 6 (1973).
ROSSINI, F., ET AL., *Selected Values of Chemical Thermodynamic Properties*, U.S. Bureau of Standards Publication 500.
SURFACE, M., "Energy Storage and Exotic Power," *Power Engineering*, Dec. 1977.

Appendix

Conversion Factors and Useful Data

Table A.1   SI Units and Their Conversion
Table A.2   Thermodynamic Properties of Helium
Table A.3   Mollier Diagram for a Steam–Water System
Table A.4   Saturated Steam Properties
Table A.5   Superheated Steam Properties
Table A.6   Properties of Air at Low Pressure

Appendix Conversion Factors and Useful Data

TABLE A.1 SI UNITS AND THEIR CONVERSION

Base Units

Quantity	Name	Symbol	Quantity	Name	Symbol
Length	meter	m	Temperature	degree Kelvin	°K
Mass	kilogram	kg	Luminous intensity	candela	cd
Time	second	s	Plane angle	radian	rad
Electric current	ampere	A	Solid angle	steradian	sr

Derived Units

Quantity	Name	Symbol	Expression in terms of other units
Area	square meter		m²
Volume	cubic meter		m³
Speed, velocity	meter per second		m/s
Acceleration	meter per second squared		m/s²
Viscosity	pascal second		Pa/s
Density	kilogram per cubic meter		kg/m³
Current density	ampere per square meter		A/m²
Magnetic field strength	ampere per meter		A/m
Angular velocity	radian per second		rad/s
Frequency	hertz	Hz	s⁻¹
Force	newton	N	kg · m/s²
Pressure, stress	pascal	Pa	N/m²
Energy, work, quantity of heat	joule	J	N · m
Power, radiant flux	watt	W	J/s
Quantity of electricity, electric charge	coulomb	C	A/s
Electrical potential, voltage, electromotive force	volt	V	W/A
Capacitance	farad	F	C/V
Electric resistance	Ohm	Ω	V/A
Conductance	siemens	S	A/V
Magnetic flux	weber	Wb	V · s
Magnetic flux density	tesla	T	Wb/m²
Inductance	henry	H	Wb/A
Surface tension	newton per meter		N/m
Thermal conductivity	watt per meter		W/m · °K
Specific heat capacity	joule per °K per kg		J/°K · kg

Conversion Factors

To convert from:	To:	Multiply by:	
atmosphere	N/m² (pascal)	1.01325	×10⁵
bar	N/m² (pascal)	1.00000	×10⁵
British thermal unit (Btu)	joule (J)	1.055056	×10³
Btu/h	watt (W)	2.930711	×10⁻¹

TABLE A.1 *Continued*

Conversion Factors

To convert from:	To:	Multiply by:
Btu/ft²-hr-°F (heat-transfer coefficient)	watt/meter²-second-degree Kelvin (W/m² · s · °K)	5.678263 × 10⁰
Btu/ft²-hr (heat flux)	watt/meter²-second (W/m² · s)	3.154591 × 10⁰
Btu/ft-hr°F (thermal conductivity)	watt/meter-second-degree Kelvin (W/m · s · °K)	1.730735 × 10⁰
calorie (International Table)	joule (J)	4.186800 × 10⁰
centimeter of mercury (0°C)	pascal (Pa)	1.333224 × 10³
degree Fahrenheit (°F)	degree Kelvin (°K)	$t_K = (°F + 459.67)/1.8$
degree Rankine (°R)	degree Kelvin (°K)	$t_K = °R/1.8$
dyne	newton (N)	1.0000000 × 10⁻⁵
erg	joule (J)	1.0000000 × 10⁻⁷
foot	meter (m)	3.0480000 × 10⁻¹
foot of water (39.2°F)	pascal (Pa)	2.98898 × 10³
foot pound-force	joule (J)	1.355818 × 10⁰
gallon (U.S. liquid)	meter³ (m³)	3.785412 × 10⁻³
horsepower (550 ft · lb$_f$/s)	watt (W)	7.456998 × 10²
inch	meter (m)	2.540000 × 10⁻²
kilocalorie	joule (J)	4.186800 × 10³
kilogram-force (kg$_f$)	newton (N)	9.806650 × 10⁰
micron	meter (m)	1.000000 × 10⁻⁶
mile (U.S. statute)	meter (m)	1.609344 × 10³
ounce-mass (avoirdupois)	kilogram (kg)	2.834952 × 10⁻²
pint (U.S. liquid)	meter³ (m³)	4.731765 × 10⁻⁴
poise (absolute viscosity)	pascal-second (Pa · s)	1.000000 × 10⁻¹
poundal	newton (N)	1.382549 × 10⁻¹
pound-force (lb$_f$)	newton (N)	4.448222 × 10⁰
pound-force-second/ft²	pascal-second (Pa · s)	4.788026 × 10¹
pound-mass (lb$_m$)	kilogram (kg)	4.535924 × 10⁻¹
pound-mass/foot³ (density)	kilogram/meter³ (kg/m³)	1.601846 × 10¹
pound-mass/foot-second (viscosity)	pascal-second (Pa · s)	1.488164 × 10⁰
psi (pressure)	pascal (Pa)	6.894757 × 10³
slug	kilogram (kg)	1.459390 × 10¹
stoke (kinematic viscosity)	meter²/second (m²/s)	1.000000 × 10⁻⁴
ton (short, 2000 lb$_m$)	tonne (1000 kg)	9.071847 × 10⁻¹
torr (mm Hg, 0°C)	pascal (Pa)	1.333224 × 10²
yard	meter (m)	9.144000 × 10⁻¹

TABLE A.2 THERMODYNAMIC PROPERTIES OF HELIUM[a]

°K	1 atm	10 atm	40 atm	70 atm	100 atm
250	h 1313.0 s 30.46 ρ 0.01950	h 1316.0 s 25.68 ρ 0.1940	h 1326.0 s 23.75 ρ 0.7625	h 1336.0 s 21.65 ρ 1.312	h 1346.0 s 20.91 ρ 1.844
300	h 1573.0 s 31.41 ρ 0.1625	h 1576.0 s 26.63 ρ 0.1618	h 1586.0 s 23.75 ρ 0.6381	h 1596.0 s 22.59 ρ 1.101	h 1606.0 s 21.86 ρ 1.552
400	h 2092.0 s 32.90 ρ 0.01219	h 2095.0 s 28.12 ρ 0.1215	h 2105.0 s 25.25 ρ 0.4811	h 2115.0 s 24.09 ρ 0.8334	h 2125.0 s 23.35 ρ 1.179
500	h 2612.0 s 34.06 ρ 0.009753	h 2615.0 s 29.28 ρ 0.09730	h 2624.0 s 26.40 ρ 0.3861	h 2634.0 s 25.25 ρ 0.6704	h 2644.0 s 24.51 ρ 0.9503
600	h 3131.0 s 35.01 ρ 0.008128	h 3134.0 s 30.23 ρ 0.08112	h 3144.0 s 27.35 ρ 0.3224	h 3153.0 s 26.19 ρ 0.5606	h 3163.0 s 25.45 ρ 0.7959
700	h 3650.0 s 35.81 ρ 0.006967	h 3653.0 s 31.03 ρ 0.06956	h 3663.0 s 28.15 ρ 0.2768	h 3672.0 s 26.99 ρ 0.4818	h 3682.0 s 26.25 ρ 0.6847
800	h 4169.0 s 36.50 ρ 0.006896	h 4172.0 s 31.72 ρ 0.06088	h 4182.0 s 28.84 ρ 0.2424	h 4191.0 s 27.68 ρ 0.4224	h 4201.0 s 26.95 ρ 0.6007
900	h 4689.0 s 37.12 ρ 0.005419	h 4692.0 s 32.33 ρ 0.05413	h 4701.0 s 29.46 ρ 0.2157	h 4710.0 s 28.30 ρ 0.3760	h 4720.0 s 27.56 ρ 0.5351
1000	h 5208.0 s 37.66 ρ 0.004877	h 5211.0 s 32.88 ρ 0.04872	h 5220.0 s 30.00 ρ 0.1942	h 5229.0 s 28.84 ρ 0.3388	h 5239.0 s 28.10 ρ 0.4823
1100	h 5727.0 s 38.16 ρ 0.004434	h 5730.0 s 33.38 ρ 0.04430	h 5739.0 s 30.50 ρ 0.1767	h 5748.0 s 29.34 ρ 0.3083	h 5758.0 s 28.60 ρ 0.4391
1200	h 6247.0 s 38.61 ρ 0.004065	h 6249.0 s 33.83 ρ 0.04061	h 6258.0 s 30.95 ρ 0.1620	h 6267.0 s 29.79 ρ 0.2828	h 6276.0 s 29.05 ρ 0.4029
1300	h 6766.0 s 39.03 ρ 0.003752	h 6769.0 s 34.24 ρ 0.03749	h 6778.0 s 31.37 ρ 0.1496	h 6787.0 s 30.20 ρ 0.2612	h 6795.0 s 29.47 ρ 0.3723
1400	h 7285.0 s 39.41 ρ 0.003484	h 7288.0 s 34.63 ρ 0.03482	h 7297.0 s 31.75 ρ 0.1390	h 7306.0 s 30.59 ρ 0.2427	h 7314.0 s 29.85 ρ 0.4359
1500	h 7805.0 s 39.77 ρ 0.003252	h 7807.0 s 34.99 ρ 0.03250	h 7816.0 s 32.11 ρ 0.1297	h 7825.0 s 30.95 ρ 0.2266	h 7833.0 s 30.21 ρ 0.3231

[a] h = enthalpy, J/g; ρ = density, g/cm^3 × 100; s = entropy, J/g · °K.
Source: Adapted from *U.S. Bureau of Standards Technical Note 631*.

TABLE A.3 MOLLIER DIAGRAM FOR THE STEAM–WATER SYSTEM

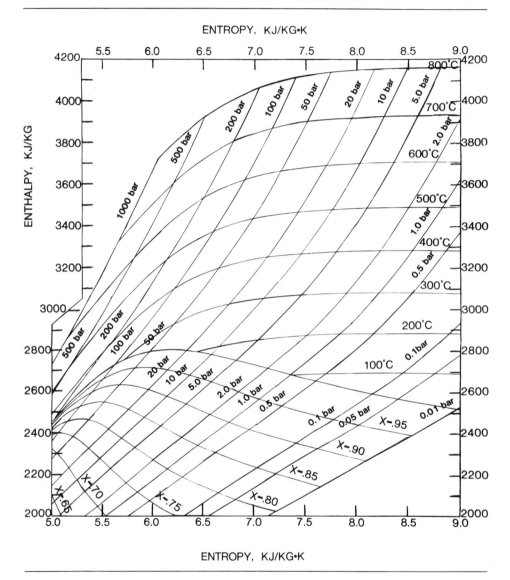

TABLE A.4 SATURATED STEAM PROPERTIES

Pressure, P (bar)	Temperature (°C)	Specific volume		Specific enthalpy		Heat of vaporization, h_{fg} (kJ/kg)	Specific entropy	
		Liquid, v_f (m³/kg × 10⁻³)	Vapor, v_g (m³/kg)	Liquid, h_f (kJ/kg)	Vapor, h_g (kJ/kg)		Liquid, s_f (kJ/kg·°K)	Vapor, s_g (kJ/kg·°K)
0.01	6.9	1.000	129.9	29.32	2514	2485	0.1058	8.975
0.02	17.5	1.001	66.97	73.52	2533	2459	0.2609	8.722
0.04	29.0	1.004	34.81	121.42	2554	2433	0.4225	8.473
0.06	36.2	1.006	23.74	151.50	2567	2415	0.5207	8.328
0.08	41.5	1.008	18.10	173.9	2576	2402	0.5927	8.227
0.10	45.8	1.010	14.68	191.9	2584	2392	0.6292	8.149
0.20	60.1	1.017	7.647	251.4	2609	2358	0.8321	7.907
0.30	69.1	1.022	5.226	289.3	2625	2336	0.9441	7.769
0.40	75.9	1.026	3.994	317.7	2636	2318	1.026	7.670
0.50	81.4	1.030	3.239	340.6	2645	2404	1.091	7.593
0.60	86.0	1.033	2.732	360.0	2653	2293	1.145	7.531
0.70	90.0	1.036	2.364	376.8	2660	2283	1.192	7.479
0.80	93.5	1.039	2.087	391.8	2665	2273	1.233	7.434
0.90	96.7	1.041	1.869	405.3	2670	2265	1.270	7.394
1.0	99.6	1.043	1.696	417.4	2675	2258	1.303	7.360
2.0	120.2	1.060	0.887	504.8	2707	2202	1.530	7.127
4.0	143.6	1.084	0.4624	604.7	2738	2133	1.777	6.897
6.0	158.8	1.110	0.3156	670.5	2757	2086	1.931	6.761
8.0	170.4	1.115	0.2403	720.9	2769	2048	2.046	6.663
10.0	179.9	1.127	0.1946	762.7	2778	2015	2.138	6.587
12.0	188.0	1.138	0.1633	798.3	2785	1987	2.216	6.523
14.0	195.0	1.149	0.1408	830.0	2790	1960	2.284	6.469
16.0	201.4	1.159	0.1238	858.3	2793	1935	2.344	6.422
18.0	207.1	1.168	0.1104	884.4	2796	1912	2.397	6.379
20.0	212.4	1.177	0.09958	908.5	2799	1891	2.447	6.340
25.0	223.9	1.197	0.07993	961.8	2802	1840	2.554	6.256
30.0	233.8	1.216	0.06665	1008.3	2804	1796	2.646	6.186
35.0	242.5	1.235	0.05704	1049.8	2803	1753	2.725	6.125

TABLE A.4 Continued

Pressure, P (bar)	Temperature (°C)	Specific volume		Specific enthalpy		Heat of vaporization, h_{fg} (kJ/kg)	Specific entropy	
		Liquid, v_f (m³/kg × 10³)	Vapor, v_g (m³/kg)	Liquid, h_f (kJ/kg)	Vapor, h_g (kJ/kg)		Liquid, s_f (kJ/kg·°K)	Vapor, s_g (kJ/kg·°K)
40.0	250.3	1.252	0.04977	1087.5	2801	1713	2.796	6.070
45.0	257.4	1.269	0.04404	1122.1	2798	1676	2.862	6.020
50.0	263.8	1.286	0.03944	1154.4	2794	1640	2.921	5.973
55.0	269.9	1.302	0.03563	1184.9	2789	1605	2.976	5.930
60.0	275.6	1.319	0.03243	1213.9	2785	1571	3.027	5.890
65.0	280.8	1.335	0.02973	1241.3	2779	1538	3.076	5.851
70.0	285.8	1.351	0.02737	1267.4	2772	1505	3.122	5.814
75.0	290.5	1.367	0.02532	1292.7	2766	1473	3.160	5.779
80.0	295.0	1.384	0.02352	1317.0	2758	1441	3.208	5.745
85.0	299.2	1.400	0.02192	1340.8	2751	1410	3.248	5.711
90.0	303.3	1.417	0.02048	1363.7	2743	1379	3.287	5.678
95.0	307.2	1.435	0.01919	1385.9	2734	1348	3.324	5.646
100.0	311.0	1.452	0.01803	1407.7	2725	1317	3.360	5.515
110.0	318.0	1.489	0.01598	1450.2	2705	1255	3.430	5.533
120.0	324.6	1.527	0.01426	1491.1	2685	1194	3.496	5.492
130.0	330.8	1.567	0.01277	1531.5	2662	1131	3.561	5.432
140.0	336.6	1.611	0.01149	1570.8	2638	1067	3.623	5.372
150.0	342.1	1.658	0.01035	1610	2611	1001	3.684	5.310
160.0	347.3	1.710	0.00932	1650	2582	932	3.746	5.247
170.0	352.3	1.768	0.00838	1690	2548	848	3.807	5.177
180.0	357.0	1.837	0.00750	1732	2510	778	3.871	5.107
190.0	361.4	1.921	0.00668	1776	2466	690	3.938	5.027
200.0	365.7	2.04	0.00585	1827	2410	583	4.015	4.928
210.0	369.8	2.21	0.00498	1888	2336	448	4.107	4.819
220.0	373.7	2.73	0.00367	2016	2168	152	4.303	4.550
221.2	374.2	3.17	0.00317	2095	2095	0	4.429	4.429

Source: Data derived from *Thermodynamic and Transport Properties of Steam*, ASME, New York, 1967. Metrication and adaptation by the authors.

TABLE A.5 SUPERHEATED STEAM PROPERTIES[a]

Pressure [bar (sat. temp. °C)]		Temperature (°C)														
		50	100	150	200	250	300	350	400	450	500	550	600	650	700	750	800
0.1 (45.8)	v	15.00	17.20	19.52	21.83	24.14	26.46	28.76	31.08	33.39	35.70	38.01	40.32	62.63	44.94		
	h	2592	2688	2783	2879	2977	3077	3177	3280	3384	3490	3598	3707	3818	3931		
	s	8.170	8.442	8.682	8.897	9.094	9.274	9.443	9.601	9.751	9.895	10.031	10.162	10.286	10.405		
0.5 (81.3)	v		3.420	3.889	4.355	4.819	5.284	5.749	6.212	6.674	7.136	7.598	8.058	8.520	8.982		
	h		2683	2780	2877	2975	3076	3176	3279	3383	3489	3597	3707	3818	3930		
	s		7.693	7.936	8.152	8.349	8.531	8.699	8.858	9.008	9.152	9.290	9.419	9.543	9.662		
1.0 (99.6)	v		1.695	1.937	2.172	2.406	2.638	2.871	3.102	3.334	3.565	3.797	4.028	4.259	4.490	4.721	4.952
	h		2676	2777	2876	2975	3074	3175	3278	3383	3488	3596	3705	3816	3928	4042	4159
	s		7.361	7.614	7.828	8.034	8.216	8.386	8.540	8.694	8.835	8.969	9.098	9.218	9.341	9.455	9.565
2.0 (120.2)	v			0.9603	1.080	1.198	1.316	1.433	1.549	1.664	1.781	1.897	2.031	2.130	2.245	2.361	2.476
	h			2769	2870	2970	3071	3173	3276	3381	3487	3595	3705	3816	3928	4041	4157
	s			7.276	7.501	7.702	7.887	8.059	8.219	8.369	8.512	8.648	8.776	8.898	9.018	9.133	9.242
3.0 (133.5)	v			0.6344	0.7161	0.7961	0.8750	0.9534	1.032	1.110	1.187	1.264	1.341	1.418	1.496	1.573	1.650
	h			2762	2864	2966	3068	3171	3275	3380	3486	3594	3704	3815	3927	4041	4157
	s			7.077	7.306	7.509	7.695	7.869	8.030	8.181	8.324	8.460	8.588	8.711	8.830	8.945	9.056
4.0 (143.6)	v			0.4709	0.5341	0.5948	0.6545	0.7137	0.7723	0.8307	0.8890	0.9473	1.0054	1.064	1.121	1.179	1.237
	h			2754	2859	2962	3065	3169	3273	3379	3485	3593	3703	3814	3926	4040	4156
	s			6.928	7.166	7.371	7.560	7.734	7.895	8.047	8.190	8.326	8.455	8.578	8.697	8.812	8.928
5.0 (151.8)	v				0.4249	0.4742	0.5226	0.5700	0.6173	0.6641	0.7108	0.7574	0.8040	0.8404	0.8969	0.9432	0.989
	h				2855	2959	3062	3167	3272	3377	3484	3592	3702	3813	3925	4040	4156
	s				7.057	7.272	7.461	7.634	7.795	7.945	8.088	8.223	8.353	8.477	8.596	8.710	8.821
6.0 (158.8)	v				0.3521	0.3937	0.4342	0.4741	0.5136	0.5528	0.5919	0.6308	0.6697	0.7085	0.7472	0.7859	0.824
	h				2849	2954	3059	3164	3270	3376	3483	3592	3701	3812	3925	4039	4155
	s				6.963	7.175	7.366	7.451	7.704	7.857	8.001	8.137	8.266	8.389	8.508	8.624	8.734
8.0 (170.4)	v				0.2609	0.2930	0.3240	0.3542	0.3842	0.4137	0.4432	0.4725	0.5018	0.5311	0.5601	0.5983	0.618
	h				2839	2947	3054	3160	3267	3373	3481	3590	3699	3810	3924	4038	4155
	s				6.814	7.032	7.226	7.404	7.568	7.722	7.866	8.002	8.132	8.256	8.375	8.490	8.601

TABLE A.5 Continued

Pressure [bar] (sat. temp. °C)		50	100	150	200	250	300	350	400	450	500	550	600	650	700	750	800
10.0 (188.0)	v				0.2060	0.2327	0.2579	0.2824	0.3065	0.3303	0.3539	0.3775	0.4010	0.4244	0.4477	0.4110	0.494
	h				2827	2940	3050	3158	3264	3370	3478	3588	3697	3809	3923	4037	4154
	s				6.692	6.926	7.125	7.303	7.464	7.619	7.763	7.897	8.029	8.154	8.274	8.388	8.498
12.0 (179.9)	v				0.1693	0.1924	0.2139	0.2343	0.2547	0.2747	0.2944	0.3143	0.3359	0.3534	0.3728	0.3923	0.41
	h				2816	2933	3042	3151	3260	3368	3477	3586	3696	3808	3921	4036	4153
	s				6.588	6.824	7.025	7.206	7.373	7.529	6.674	7.811	7.942	8.067	8.187	8.320	8.41
14.0 (201.4)	v					0.1635	0.1823	0.2001	0.2176	0.2349	0.2520	0.2690	0.2858	0.3026	0.3194	0.3361	0.3527
	h					2925	3036	3147	3256	3365	3474	3584	3695	3806	3920	4035	4152
	s					6.741	6.945	7.130	7.299	7.455	7.601	7.739	7.870	7.994	8.114	8.230	8.341
16.0 (195.0)	v					0.1417	0.1585	0.1743	0.1899	0.2051	0.2201	0.2350	0.2490	0.2646	0.2783	0.2940	0.3086
	h					2917	3030	3142	3253	3363	3472	3582	3693	3805	3919	4034	4151
	s					6.667	6.877	7.063	7.233	7.390	7.537	7.675	7.806	7.932	8.052	8.168	8.279
18.0 (207.1)	v					0.1248	0.1401	0.1545	0.1683	0.1819	0.1953	0.2088	0.2219	0.2351	0.2481	0.2613	0.2742
	h					2908	3025	3138	3249	3360	3470	3580	3691	3803	3917	4033	4150
	s					6.601	6.814	7.003	7.175	7.333	7.480	7.619	7.750	7.876	7.996	8.112	8.224
20.0 (212.4)	v					0.1114	0.1255	0.1384	0.1511	0.1634	0.1755	0.1875	0.1995	0.2114	0.2232	0.2350	0.2467
	h					2900	3019	3134	3246	3357	3468	3578	3690	3802	3917	4032	4150
	s					6.539	6.757	6.949	7.122	7.282	7.429	7.569	7.701	7.827	7.947	8.063	8.174
25.0 (223.9)	v					0.08700	0.9893	0.1097	0.1200	0.1300	0.1399	0.1496	0.1592	0.1688	0.1783	0.1877	0.1971
	h					2879	3010	3123	3240	3351	3462	3573	3685	3799	3914	4030	4147
	s					6.408	6.647	6.844	7.017	7.172	7.323	7.463	7.596	7.722	7.843	7.959	8.072
30.0 (233.8)	v					0.07067	0.08119	0.09051	0.09929	0.1078	0.1161	0.1243	0.1325	0.1405	0.1484	0.1564	0.164
	h					2853	2988	3111	3229	3343	3456	3569	3682	3796	3911	4027	4145
	s					6.283	6.530	6.735	6.916	7.080	7.231	7.373	7.506	7.633	7.755	7.872	7.984
35.0 (242.5)	v					0.05877	0.06847	0.07674	0.08448	0.09190	0.09900	0.1062	0.1132	0.1202	0.1269	0.1339	0.1403
	h					2828	2972	3100	3220	3336	3451	3564	3678	3793	3908	4024	4143
	s					6.173	6.437	6.649	6.835	7.002	7.155	7.298	7.432	7.559	7.681	7.799	7.912

40.0 (250.3)											
v	0.05888	0.06639	0.07337	0.08001	0.08642	0.09270	0.09885	0.1049	0.1109	0.1169	0.122
h	2955	3087	3211	3329	3445	3560	3674	3789	3905	4022	4141
s	6.352	6.573	6.762	6.933	7.087	7.231	7.367	7.495	7.618	7.735	7.848
45.0 (257.4)											
v	0.05142	0.05837	0.06474	0.07073	0.07649	0.08215	0.08766	0.09305	0.09841	0.1038	0.109
h	2938	3075	3202	3322	3439	3555	3670	3785	3902	4019	4138
s	6.273	6.504	6.699	6.871	7.028	7.173	7.309	7.438	7.561	7.679	7.792
50.0 (263.9)											
v	0.04530	0.05194	0.05791	0.06325	0.06850	0.07360	0.07860	0.08356	0.08845	0.09329	0.0981
h	2925	3063	3194	3317	3433	3549	3665	3781	3898	4016	4136
s	6.210	6.454	6.650	6.822	6.977	7.121	7.258	7.387	7.511	7.629	7.743
60.0 (275.6)											
v	0.03620	0.04227	0.04742	0.05217	0.05667	0.06103	0.06525	0.06937	0.07347	0.07751	0.0815
h	2880	3039	3174	3299	3421	3540	3658	3775	3893	4012	4132
s	6.060	6.326	6.535	6.716	6.878	7.028	7.165	7.297	7.422	7.542	7.655
70.0 (285.8)											
v	0.02948	0.03532	0.03997	0.04420	0.04817	0.05197	0.05565	0.05923	0.06277	0.06627	0.0697
h	2835	3012	3155	3285	3409	3530	3649	3768	3887	4007	4127
s	5.925	6.222	6.442	6.628	6.795	6.947	7.087	7.220	7.346	7.467	7.580
80.0 (294.9)											
v	0.02429	0.03003	0.03438	0.03821	0.04177	0.04516	0.04844	0.05161	0.05475	0.05785	0.0609
h	2784	2985	3135	3270	3397	3520	3640	3760	3881	4002	4122
s	5.788	6.126	6.358	6.552	6.722	6.876	7.019	7.152	7.280	7.401	7.515
90.0 (303.3)											
v		0.02586	0.03001	0.03354	0.03680	0.03988	0.04285	0.04570	0.04851	0.05130	0.0540
h		2954	3114	3254	3386	3510	3631	3752	3874	3996	4117
s		6.033	6.280	6.481	6.656	6.813	6.957	7.092	7.220	7.343	7.459
100.0 (318.0)											
v		0.02241	0.02641	0.02974	0.03276	0.03560	0.03832	0.04096	0.04355	0.04608	0.0486
h		2925	3097	3243	3374	3499	3622	3744	3867	3989	4112
s		5.943	6.218	6.424	6.599	6.756	6.901	7.037	7.166	7.289	7.406
110.0 (310.0)											
v		0.01967	0.02356	0.02672	0.02954	0.03218	0.03469	0.03709	0.03945	0.04176	0.0440
h		2884	3071	3222	3360	3488	3612	3736	3860	3983	4106
s		5.849	6.138	6.355	6.540	6.702	6.850	6.988	7.118	7.240	7.357
120.0 (324.6)											
v		0.01816	0.02113	0.02414	0.02681	0.02928	0.03163	0.03387	0.3605	0.03820	0.04031
h		2892	3049	3206	3347	3478	3663	3728	3853	3977	4102
s		5.832	6.071	6.298	6.487	6.653	6.083	6.942	7.073	7.196	7.314

TABLE A.5 Continued

Pressure [bar] (sat. temp. °C)		50	100	150	200	250	300	350	400	450	500	550	600	650	700	750	800
130.0 (330.8)	v							0.01514	0.01905	0.02197	0.02450	0.02683	0.02903	0.03113	0.03317	0.03517	0.0371
	h							2799	3026	3189	3334	3467	3594	3721	3847	3972	4097
	s							5.657	6.006	6.243	6.438	6.606	6.758	6.899	7.032	7.156	7.274
140.0 (336.63)	v							0.01325	0.01726	0.02010	0.02252	0.02474	0.02683	0.02881	0.03071	0.03258	0.0344
	h							2750	3000	3172	3321	3456	3585	3713	3841	3966	4092
	s							5.556	5.942	6.190	6.390	6.562	6.716	6.859	6.992	7.118	7.238
160.0 (347)	v								0.00978	0.01429	0.01704	0.01930	0.02132	0.02500	0.02671	0.02834	0.0300
	h								2612	2945	3137	3294	3434	3698	3829	3955	4082
	s								5.302	5.816	6.090	6.303	6.482	6.786	6.922	7.049	7.171
180.0 (357.0)	v								0.01194	0.01467	0.01678	0.01867	0.02043	0.02204	0.02357	0.02511	0.0265
	h								2884	3100	3267	3412	3549	3683	3815	3944	4072
	s								5.688	5.995	6.221	6.407	6.572	6.720	6.858	6.987	7.110
200.0 (365.0)	v								0.00995	0.01271	0.01477	0.01655	0.01816	0.01967	0.02111	0.02249	0.0238
	h								2819	3064	3247	3394	3535	3669	3804	3935	4065
	s								5.558	5.909	6.145	6.337	6.504	6.655	6.795	6.927	7.051
220.0 (373.7)	v								0.00828	0.01112	0.01312	0.01481	0.01631	0.01773	0.01907	0.02034	0.0216
	h								2736	3017	3207	3367	3512	3653	3789	3922	4053
	s								5.406	5.813	6.070	6.276	6.449	6.606	6.747	6.880	7.007
240.0	v								0.00676	0.00977	0.01174	0.01334	0.01478	0.01612	0.01735	0.01856	0.0197
	h								2638	2971	3174	3343	3493	3637	3776	3910	4043
	s								5.236	5.723	5.999	6.216	6.394	6.554	6.697	6.832	6.960
260.0	v								0.00529	0.00861	0.01056	0.01213	0.01350	0.01477	0.01593	0.01706	0.0181
	h								2511	2922	3141	3319	3473	3622	3763	3899	4033
	s								5.028	5.631	5.932	6.157	6.340	6.505	6.651	6.788	6.916
280.0	v								0.00387	0.00760	0.00956	0.01108	0.01239	0.01360	0.01471	0.01577	0.0167
	h								2348	2870	3107	3293	3453	3606	3749	3888	4023
	s								4.769	5.539	5.865	6.101	6.389	6.457	6.607	6.746	6.875

300.0	v	0.00283					0.01017	0.01144	0.01258	0.01365	0.01465	0.0156
	h	2157					3277	3440	3595	3739	3879	4018
	s	4.489					6.038	6.240	6.413	6.556	6.700	6.839
350.0	v	0.00211	0.00673	0.00868			0.00834	0.00952	0.01056	0.01152	0.01242	0.0133
	h	1992	2815	3083			3205	3395	3556	3707	3852	3995
	s	4.222	5.449	5.797			5.917	6.119	6.300	6.458	6.604	6.740
400.0	v	0.001910	0.00496	0.00693			0.006982	0.008088	0.009053	0.009930	0.01075	0.01152
	h	1934	2676	2986			3152	3346	3507	3674	3825	3972
	s	4.119	5.203	5.635			5.783	6.014	6.214	6.370	6.421	6.601
450.0	v	0.001801	0.003675	0.005616			0.005934	0.006984	0.007886	0.008700	0.009452	0.01016
	h	1901	2815	2906			3086	3297	3478	3642	3798	3948
	s	4.055	4.981	5.476			5.665	5.914	6.115	6.290	6.445	6.588
500.0	v	0.001729	0.002913	0.004625			0.005113	0.006113	0.006960	0.007720	0.008420	0.00907
	h	1878	2384	2813			3021	3248	3438	3610	3770	3925
	s	4.008	4.747	5.323			5.552	5.821	6.033	6.214	6.375	6.522
	v	0.001729	0.002492	0.003882								
	h	1878	2293	2723								
	s	4.008	4.602	5.178								

[a]Units: $v = \text{m}^3/\text{kg}$; $h = \text{kJ/kg}$; $s = \text{kJ/kg} \cdot °\text{K}$.

Source: Data derived from *Thermodynamic and Transport Properties of Steam*, ASME, New York, 1967. Metrication and adaptation by the authors.

TABLE A.6 PROPERTIES OF AIR AT LOW PRESSURE

Temperature				
°C	°K	h (kJ/kg)	p_{re}[a]	$\bar{\phi}$ (kJ/kg)
10	283	283.46	1.13685	1.3555
50	323	323.62	1.8025	1.43910
100	373	374.02	2.9870	1.51954
150	423	424.63	4.6514	1.59029
200	473	475.63	6.9186	1.65350
250	523	527.03	9.9107	1.71091
300	573	578.88	13.7952	1.76358
350	623	631.56	18.7294	1.81240
400	673	684.66	24.922	1.85791
450	723	738.36	32.582	1.90066
500	773	786.60	41.964	1.94100
550	823	847.58	53.322	1.97922
600	873	909.24	68.643	2.01948
650	923	959.08	83.229	2.05018
700	973	1015.60	102.456	2.08330
750	1023	1072.63	125.018	2.11500
800	1073	1130.15	151.358	2.14500
850	1123	1188.08	181.898	2.17480
900	1173	1239.93	217.14	2.2030
950	1223	1305.14	257.57	2.23030
1000	1273	1364.24	303.74	2.25660
1050	1323	1423.64	356.26	2.28202
1100	1373	1483.35	415.70	2.306617
1150	1423	1543.36	482.70	2.33045
1200	1473	1603.62	558.12	2.35360

[a] p_{re} = relative pressure [see Eq. (8.44)].

Index

Ac power: dispatching of, 430; generation of, 410–421; reaction power, 415–416, 427; real power, 415–417; transmission of, 426–427
Active coal storage, 160
Actual flame temperature, 141
Air circulation in power plants, 217: dynamic pressure head, 218; gas velocity, 218; pressure measurement, 218
Air heaters, 223: heat transfer, 228; overall coefficient of heat transfer, 226; overall design, 223; pressure losses, 222; recuperative, 223–227; regenerative, 225, 227–228
Air pollution: emission standards, 247; nitrogen oxides, 249, 250–252; solid particles, 253–255; sulfur dioxide, 249, 251
Air properties, 502
Air storage, 453–455
Air-to-fuel ratio: actual, 106; theoretical, 104
Alkane hydrocarbons, 65
Alternating current: frequency, 412, 414, 428; generation of, 410–421; phase relationships, 413–414
Alternator: basic principles of, 410–417; configuration, 420–422; control, 419; cooling, 421; current frequency, 412, 414; design, 412–419; power output, 415–419
Anthracite coal, 81
As-burned mass fraction, 84
Ash: disposal, 255; pit, 113; softening temperature, 83; specific heat, 120
Attemperation, 209
Atomic mass unit, 265
Atomic number, 265
Atomic particles, 264–265
Availability factor, 14
Azimuthal angle, 469

Barrier base, 64
Base load, 15, 17
Batteries, 457–459
Benson boilers, 173–174
Beta-alumina, 458
Betz limit, 478–499, 481
Binary cycle, 463
Binding energy, 266–268
Bituminous coal, 79
Blowdown (cooling tower), 241
Boiler efficiency, 174
Boilers (*see* Steam generators)
Boiling heat transfer, 317–318
Boiling point, 43
Boiling water reactor, 288, 293–296
Brayton cycle: closed, 352–358; ideal, 50–52; nonideal, 360–363; open, 358–363; part-load operation, 386–388; temperature–entropy diagram, 51; thermodynamic efficiency, 52, 354–357
BTHR, 348
Buckling (in nuclear reactor core), 282
Burners: coal, 108–110; oil, 111
Burnup, 31
BWR, 288, 293–296

Calorifie value of coal, 85
CANDU reactor, 303–305
Cap rock, 64

Capacity factor, 14, 26
Capital cost (of electric power stations), 26
Carnot cycle: basic assumptions, 45; efficiency, 46; temperature-entropy diagram, 46
Carryover velocity, 188, 189
Catalytic methanation, 92
Cation exchange, 231
Cellular fill, 235
Chimney (*see* Stack design)
Clausius statement of the second law, 40
Clean Air Act, 246–247
Coal, 76: classification, 80, 82; cleaning, 87; composition, 85; cost, 30; gasification, 90; handling equipment, 160; heating valve, 85; liquefaction, 96; properties, 81; proximate analysis, 83; pulverizing, 108; rank, 81; resources, 77, 79; solvent refining, 796; types of coal, 82; ultimate analysis, 84
Coke production, 89
Combined cycle, 363–365
Combustion: actual air required, 106; chemical equilibrium, 132–241; excess air required, 104; fluidized bed, 148; heat of combustion, 103; reactions, 102, 132; temperatures of combustion, 135–144
Combustion products: determination of, 132; emissivity of, 146; specific heats of, 119
Complex number, 414
Compressed air storage, 453–455
Compressed liquid, 43
Compressor: characteristics, 387; pressure ratio, 52; work done by, 354, 358–359
CONCEPT, 27
Condensers: condensate treatment, 230–231; design and functions, 195; heat-transfer coefficients, 200
Control: alternator, 419; gas turbine, 393; line voltage, 427; load frequency, 427–430; nuclear reactor, 286, 289; steam turbine, 389–393
Control rods, 286, 289
Cooling efficiency, 240
Cooling range, 239, 242
Cooling tower: counterflow, 235, 239; cross-flow, 235; draft, 243; dry tower design, 234; mechanical draft tower, 234; natural draft tower, 234; packing height, 243; types of, 236; wet tower design, 234, 237
Cooling-water losses, 240
Core losses, 421
Costs: capital, 26; fuel, 29–32; overall, 24; solar power, 475–476
Critical heat flux (CHF), 319–320
Cross sections, nuclear, 274–279
Crud, 231
Crude oil, 65–66
Cut-in and cut-out speeds, 482

Deaerating heater, 195
Declination angle, 469
Demineralized water, 231
Desulfurization: of flue gases, 250; of synthetic gas, 94; of synthetic liquid fuel, 97
Distribution ratio, 189
District heating, 463
Dittus–Boeltner equation, 185
Diurnal variation in power demand, 12
DNBR, 320

503

Doubling time, 6
Draft: balanced draft, 217; forced draft, 217; induced draft, 217; stack draft, 219
Dry ash-free quantities, 84
Dry cooling tower, 234
Dry steam, 94
Dulong–Berthelot formula, 86
Dust control systems, 162

Economics: capital costs, 26; fuel costs, 29–32; of power production, 24; of solar power, 475–476; power dispatch, 430
Economizer: basic design and function, 192; heat-transfer analysis, 199
Efficiency: Brayton cycle, 52, 354–357; Carnot cycle, 46; Rankine cycle, 49, 347; regenerative cycle, 55
Electric power demand: characterization of, 14; historical behavior, 6; load–duration curve, 15; variation, 12–14
Electric reliability regions, 13
Electrical system, power plant, 422–425
Electrostatic precipitators, 253
Emission control systems, 246
Emissivity: combustion products, 146; of flame, 145; of fuel particles, cloud, 147, 148; of soot, 147
Endothermic reaction, 127
Energy costs, 24–32, 475–476
Energy demand: demand for electricity, 4; energy resource estimates, 3; exponential growth of electric power, 6; future projections, 10
Energy sources, 11
Energy storage: battery, 457–459; compressed air, 453–455; latent heat, 455–457; necessity for, 451–453; pumped water, 401–403; thermal energy, 455–457
Enthalpy, 39–41
Entropy, 38–41
Equilibrium composition (in combustion), 133–135
Equilibrium constant: in combustion process, 132–133; in terms of partial pressures, 130; variation with temperature, 131
Equivalent diameter, 226
Excess air (in combustion), 105, 251
Exciter, 412
Exothermic reaction, 127
Extensive system properties, 41

Fabric filters, 255
Fans, 221
Faraday effect, 410
Faraday's constant, 442
Fast fission factor, 281
Feedwater heaters: basic design and function, 192, 193; thermal analysis, 199
Feedwater treatment, 233
Fermi age, 283
Fill pack, 235
Film coefficients, 186, 187
Firing methods for coal, 110
First law of thermodynamics, 37, 38
Fission, 264–268, 270–274
Fission gas, 312–313
Fission products, 271–273
Flame: actual temperatures, 141; emissivity, 145; mean beam length, 146; net heated radiated, 143; theoretical temperature, 136
Flue gas: composition, 114, 249, 251; particulate removal from, 252–255; recirculation, 251; scrubbing, 249–250; waste disposal caused by treatment of, 257–259
Fluidized-bed combustion, 148
Fly ash, 252
Forced-draft cooling tower, 234
Fouling, 199–227
Four-factor formula, 281
Free-burning coal, 83
Free energy: change in combustion reaction, 129; change in fuel cell, 440–441; definition of, 126; formation, 127, 440; relation to equilibrium constant, 128–130

Friction factor: in-line banks, 222; stacks or chimneys, 219; staggered tube banks, 222
Froth flotation method, 89
Fuel, fossil: composition, 64–66, 74, 83–84; formation, 76–77, 63–64; heating value, 71–72, 73, 85–87; properties, 69–71, 73, 81–83; resources, 3, 66–67, 74, 77–79; synthetic, 90–99
Fuel, nuclear: costs, 31–32; design and analysis, 305–316; temperatures, 317–318
Fuel cells: analysis, 440–441; basic concept, 439–440; free-energy change in, 441–442; voltage, 442
Fuel costs: fossil-fuel stations, 29–30; nuclear, 31–32
Fuel elements, nuclear: behavior, 308–316; cladding, 306, 309–310, 313–315, 317–318; configuration, 295, 299, 302, 307; design, 305–308; fabrication, 333; fission gas release, 312–313; reprocessing, 333–337; thermal analysis, 317–320
Fuel-handling systems, 158
Fugacity, 128, 131
Full-load speed, 482
Furnace: energy balance, 118, 124; heat loads, 178; heat transfer in, 174–177, 179–180; mass balance, 111; wall-area effectiveness factor, 180
Fusion: carbon cycle, 466; deuterium–tritium, 486; in sun, 466–467; magnetic containment, 486; proton–proton cycle, 466–467

Gas centrifuge, 331–332
Gas-cooled reactors, 288, 300–304
Gas diffusion, 325–331
Gas tempering, 208
Gas turbine plants (see Turbines, gas)
Gaseous fuels from coal, 90
Generator, electric (see Alternator)
Geothermal gradient, 464
Geothermal power: binary cycle, 463; district heating, 463; geopressurized water, 460, 466; geothermal gradient, 464; hot dry rock, 460, 463–465; hydrothermal, 460–463
Gibbs free energy, 440–441
Gibbs phase rule, 42
Gravity separation process, 89
Grindability of coal, 83

Half-life, 269
Hall effect, 434–436
Heat balance for a steam cycle, 55
Heat exchanger analysis, 199; economizer, 202; friction loss, 204; natural convection systems, 202; Nusselt's equation, 201; reheater analysis, 201; shell-and-tube type, 199; water-cooled surface condensers, 200
Heat of combustion, 71
Heat rate, 347–348
Heat transfer: boiler screen tubes, 181–182; convective, 178–179; economizer, 199; furnace boiler region, 179–180; radiative, 143–148; superheat region, 183; total to the boiler, 183
Heating value: from heats of combustion, 103; higher heating value (HHV), 71, 85; lower heating value (LHV), 71, 85
Heavy water, 288, 303–305
Heliostat, 474–475
Helium: in gas-cooled reactors, 300; thermodynamic properties, 483
HTGR, 288, 300–304
Hydroelectric power: emergency storage, 401–403; generators, 421–422; power plant, 24; turbines, 393–401
Hydrogasification, 92
Hydrogenation process, 97
Hydropyrolysis of coal, 97
Hydrothermal power, 460–463

Impedance, 415
Incremental power cost, 33
Inequality of Clausius, 38

Index

In-situ retorting, 67
Intensive system properties, 41–44
Intermediate load, 15, 17
Internal energy, 39–41
Irreversible processes, 42

Joule, James, 33
Joule's constant, 39

Kelvin–Planck statement of second law, 39
Kerogen, 67
Kerosene, 72

Latent heat storage, 456, 457
Levelized fixed-charge rate, 27, 28
Light-water nuclear power plants, 22, 288–296, 303–305
Lignite, 86
Lime/limestone SO_2 removal, 250
Liquefaction of coal, 96–97
Liquefied natural gas, 75
Load–duration curve, 15
Load factor, 14
Load-following plants, 18
Log mean temperature difference, 185
Log normal frequency function, 16
Loss-of-coolant accident (LOCA), 320
Lurgi process, 92, 93
LWR, 288

Magnetohydrodynamics (MHD): basic concept, 430–432; cycles, 365, 432; generators, 432–437; Hall effect, 434–436; system, 437–438
Makeup water, 230, 241
Mass defect, 266
Mean beam length, 182
Mean driving force, 244
Methane, 65
MFBR, 297–300
Mollier diagram (steam), 44, 494
Multiplication factor (in nuclear reactor), 281–285

Natural-draft cooling tower, 234
Natural gas: composition, 74; heating value, 75; properties, 73; resources, 74
Nernst equation, 442
Net rated power output, 26
Neutron: cross sections, 274–279; flux, 275–276, 286; in atom, 264–265; velocity, 278
Nitrogen oxides: control, 251; emission standards, 247
Nonleakage probability, 282
Nonsalient rotor, 412
NO_x emissions: in fossil plant, 250; in MHD plant, 438; standards for, 247
Nuclear fission, 264–268, 270–274
Nuclear power plants, 264–341: boiling-water reactor, 22, 293–296; gas-cooled reactor, 23, 288, 300–304; heavy water, 303–305; liquid-metal breeder, 297–300; postulated accident, 320; pressurized water reactor, 23, 289–293
Nuclear reactors: breeder, 297–300; core analysis, 274–286; dynamics, 286–287; fuel, 295, 299, 302, 305–317; multiplication factor, 281–285; reactor systems, 286–305; thermal analysis, 317–320; thermal reactor, 286–287
Number density, 276–284
Nussett's equation, 201

Oil trap, 64
Once-through boilers, 173–174
Operating and maintenance costs, 28–29
OPPHR, 348
Opposed firing, 109
Orsat gas analysis, 106

Packing height, 243
Parabolic solar collector, 473–474
Parr formulas, 84
Particulate emissions, 252

Peaking load, 15, 17
Peat, 76–80
Pelton wheel, 394
Perimeter ice, 237
Petroleum: burning equipment, 111; chemistry, 64; molecular compounds, 66; physical properties, 69, 70; resources, 67
Petroleum gas, 75
Pipeline gas, 92
Plasma: in fusion device, 486; in MHD generator, 433; in sun, 466–467; in theromionic generator, 443
Plutonium, 273, 279, 288, 300
Polar angle, 409
Polishing (of condensate), 230
Pollution (see Air pollution)
Power angle, 415
Power cost: components of, 24; effect of load duration, 18; fixed charges, 25, 27; fuel costs (general), 25; nuclear fuel costs, 32; operating and maintenance costs, 25, 28; solar power, 475–476; total production cost, 32–33
Power plants, coal: auxiliary systems, 217–258; burners, 108–110; configuration, 19–20; costs, 24–27; major components of, 157; primary system, 156–210; steam cycles for, 345–352
Power plants, gas turbine: configuration, 21; control, 343; part-load operation, 386–388; thermodynamic cycle, 354–362
Power plants, nuclear (see Nuclear power plants)
Pressure losses: furnace, 222; recuperative air heater, 222; regenerative air heater, 222
Pressure ratio, 52
Pressurized water reactor, 288, 289–293
Primary air, 109
Primary regulation, 429
Proportional controller, 207
Proximate analysis, 83
Pulverized coal, 108, 163
Pumped water storage, 401–403
PWR, 288, 289–293

Quad, 464
Quality, 43

Radiation conductance, 183, 184
Radiative heat transfer, 142–148, 179, 183–184
Radioactive decay, 268–269
Random variation in power demand, 12
Rankine cycle: actual, 198–199, 345–352; basic, 47–49; binary, 463; efficiency, 49, 347; temperature–entropy diagram, 48; work output, 49
Reactance, 415
Reactive power, 415
Reactivity, 287
Recuperative air heaters, 223–227
Refuse: carbon in refuse, 114; disposal of, 255–258; enthalpy, 125; heating valve, 125; mass formed in burning, 114
Regenerative air heaters, 225, 227–228
Regenerative feedwater heating, 54
Reheaters: design and function, 199; heat-transfer coefficient, 201
Relative pressure, 361, 502
Reservoir rock, 63
Resonance escape probability, 281
Resonance, nuclear, 278–279
Retorting process, 67
Richardson–Dushman equation, 444

Salient poles, 411, 412
Saturated steam, 44
Saturation pressure, 43
Saturation temperature, 43
Screen tubes, boiler, 181
Scrubbing processes (SO_2), 249

Seasonal variation in power demand, 12
Second law of thermodynamics, 38–39
Secondary air, 109
Secondary regulation, 429
Separative work, 330–331
Shale oil, 67–68
Shape factors, 143–144
Shift reaction, 92
SI units, 491–492
Site power consumption, 26
Slag-screen tubes, 181
Sludge, 257
Solar constant, 467
Solar power: backup power, 475–476; collectors, 472–475; economics, 475–476; energy production in sun, 466–467; heliostat, 474–475; incident radiation, 469–472; photovolataic cells (solar cell), 475; solar constant, 467; spectral distribution, 467–469
Solvent refining of coal, 96–97
SO_2 removal: from fossil-fuel plant, 249–250; in MHD plant, 437
Source rock, 63
Space charge, 444
Stack design, 219
Standard free energy of formation, 127
Steam: corrosion by, 383–384; cycle, 47–49, 57, 198–199, 345–382; drum, 190; power plants, 20, 26, 40; properties, 394–501; rate, 351–352; scrubbing (drying), 188; separation and purification, 187
Steam generator control, 205: burner control, 208; closed-loop, 205; error signal, 206; feedwater control, 207; gas recirculation, 209; integral control, 206; proportional control, 206; reset control, 206; steam demand, 206; superheat control, 208
Steam generators (fossil-fired boiler): efficiency, 174; forced circulation, 173; heat losses in the boiler, 175; history of design, 166; modern type, 170; natural circulation, 169; pump-assisted circulation, 172; radiant type, 169, 179; screen tubes, 181; water-cooled, 168
Steam generators, nuclear: once-through, 292–293; recirculating, 189–292
Steam properties: saturated, 495–496; superheated, 497–501
Stress corrosion, 383–384
Sulfur: effect of sulfur on furnace mass balance, 117; in flue gas, 114, 249–250
Sulfur dioxide, 249: lime/limestone scrubbing, 250; recovery processes, 249; removal processes, 249, 437; solium/calcium process, 250; throwaway processes, 249
Superconductivity, 422, 438
Supercritical fluid, 44
Superheated steam, 44
Switchgear, 422–424
Synchronization, 414–428
Synchronizing power, 428
Synchronous generator (see Alternator)
Synchronous motor, 402, 419
Synthane coal gasification process, 94
Synthetic fuels: chemistry, 90; gasification processes, 91; Lurgi process, 92; producer gas, 90; Synthane process, 94–95; synthesis gas, 91; underground gasification, 95; water gas, 90

Tangential firing, 109
TCHR, 348
Tertiary regulator, 429
Theoretical flame temperatures, 135
Thermal utilization factor, 281
Thermionic generators: analysis of, 444–445; basic concept, 443–444; collector, 443; efficiency, 446; emitter, 443; work function 444
Thermodynamic efficiency, 41
Thermodynamic laws, 38–40
Thermodynamic properties: air, 502; helium, 483; steam, 495–501
Topping cycle: Brayton, 363–364; MHD, 432, 437–438; thermionic generator, 445
Transmission lines, 422, 426–427
Trough collectors, 473
Turbine pressure ratio, 52
Turbines, gas: configuration, 371; control, 393; ideal, 50–52; part-load operation, 386–388; plant configuration, 21; pressure ratio, 52; thermodynamic efficiency, 354–357; use of air in, 361–362; work, 354–359, 362
Turbines, hydraulic: analysis, 393–399; configurations, 399–401; Francis, 399–401; impulse, 394–396; Kaplan, 400–401; Pelton, 394; reaction, 396–399
Turbines, steam, 384–386: analysis, 371–379; configuration, 365–370; control, 388–393; design, 374–383; end losses, 376; heat rate, 347–349; impulse, 366–369; materials problems, 371–379; mechanical efficiency, 371–376; part-load operation, 384–386; reaction, 367–369; reheat in, 198, 199, 377–378; thermodynamic efficiency, 347; thermodynamics of, 347–351; vibration, 384; work, 347
Turbines, wind: analysis, 479–482; configurations, 480, 483–485; cut-in and cut-out speeds, 482; full-load speed, 482

Ultimate analysis, 84
Underground gasification (in-situ gasification), 95
Unit conversion, 491–492
Uranium, 301: carbide, 276–279, 284; cross sections, 276–274; enrichment, 325–332; fission, 323–337; fuel cycle, 323–324; isotopes, 271, 325; mining and milling, 306, 308–309; oxide, 317; resources, 322, 323

Void fraction in steam/water, 203

Waste disposal: from fossil-fuel plants, 255–258; from nuclear plants, 336–337
Water gas, 90
Water treatment, 229–231
Weatherability of coal, 83
Wellman–Lord process, 249, 250
Wet-bulb approach temperature, 238, 243
Wet cooling tower, 234, 237
Wet ponding, 258
Wet recovery slurry process, 250
Wind power: available energy, 478; Betz limit, 478–479, 481; ecological problems, 485–486; turbines, 479–485; wind directions, 476; wind power density, 477; wind speed, 481–482; work function, 444

Zircaloy, 296, 306, 309–310, 317, 320